AIR POLLUTION CONTROL

HOW TO ORDER THIS BOOK

BY PHONE: 800-233-9936 or 717-291-5609, 8AM–5PM Eastern Time

BY FAX: 717-295-4538

BY MAIL: Order Department
Technomic Publishing Company, Inc.
851 New Holland Avenue, Box 3535
Lancaster, PA 17604, U.S.A.

BY CREDIT CARD: American Express, VISA, MasterCard

BY WWW SITE: http://www.techpub.com

REVISED EDITION

AIR POLLUTION CONTROL

Traditional and Hazardous Pollutants

HOWARD E. HESKETH, Ph.D., PE, DEE

Professor Emeritas / Consultant
College of Engineering
Southern Illinois University
Carbondale, IL 62901

CRC Press
Taylor & Francis Group
Boca Raton London New York

CRC Press is an imprint of the
Taylor & Francis Group, an **informa** business

Contents

Preface   xiii

Preface to the First Edition   xv

Acknowledgements   xvii

1. **INTRODUCTION** 1
   1.1   Air Pollution   1
   1.2   Particulates   2
         1.2.1   General Relationships 2
         1.2.2   Size Distribution 4
         1.2.3   Special Properties 10
   1.3   Gases   13
         1.3.1   General Relationships 13
         1.3.2   Relevant Factors 16
   1.4   Particulate-Gas Mixtures   19
   1.5   Standards   20
         1.5.1   Ambient Air Quality Standards 21
   1.6   Toxic Air Pollutants   24
         1.6.1   Toxic Releases and Reporting 24
         1.6.2   Sources of Toxic Air Pollutants 26
         1.6.3   Control of Air Toxics 28
         1.6.4   Polychlorinated Biphenyls (PCBs) 28
         1.6.5   Toxic Waste Sites 33
   1.7   Emission Standards   34
   1.8   Risk Assessment   38

1.9   Indoor Air Quality   41

1.10 Atmospheric Dispersion   44
      1.10.1 Stacks 44
      1.10.2 Atmospheric Diffusion 51
      1.10.3 Modeling 58
      1.10.4 Evacuation 60

1.11 Environmental Auditing   61

1.12 Chapter Problems   69
      1.12.1 Particulate Matter 69
      1.12.2 Flue Gas 69
      1.12.3 Stacks 69
      1.12.4 Dispersion 70

2. **BASIC ENERGY SOURCES, CONSERVATION PROCESSES,
    AND EMISSION PRODUCTS** 73

2.1   Energy and Conservation   73

2.2   Combustion   75
      2.2.1 Flue Gas Composition 76
      2.2.2 Excess Air 78
      2.2.3 Air-Fuel Ratio 81
      2.2.4 Flue Gas Quantity 81
      2.2.5 Energy Losses 82

2.3   Coal Combustion   84
      2.3.1 Coal 84
      2.3.2 Coal Burners 87
      2.3.3 Combustion Products 91

2.4   Oil Combustion   99
      2.4.1 Oil 99
      2.4.2 Oil Burners 99
      2.4.3 Combustion Products 101

2.5   Gas Combustion   104
      2.5.1 Gases 104
      2.5.2 Gas Burners 105
      2.5.3 Combustion Products 107

2.6   Transportation   107
      2.6.1 Motor Fuels 109
      2.6.2 Motor Fuel Combustion 111
      2.6.3 Combustion Products 112

2.7   Waste Combustion   114
      2.7.1 Waste as Fuel 117
      2.7.2 Solid Waste Combustion 123

2.7.3 Combustion Products 124

2.8 Incineration   128
    2.8.1 Incineration Terminology 129
    2.8.2 Types of Incinerators 131
    2.8.3 Design Factors 138
    2.8.4 Combustion Products 138

2.9 Other Processes   144
    2.9.1 Gasification 145
    2.9.2 Liquefaction 145

2.10 Heating Value   145

2.11 Combustion Temperature   148

2.12 Correction Factors   152

2.13 Emission Minimization from Combustion Processes   154
    2.13.1 Primary Combustion Chambers 155
    2.13.2 Secondary Combustion Chambers 155
    2.13.3 Waste Recycling 156
    2.13.4 Minimizing NO_x 157
    2.13.5 Operation and Instrumentation 157

2.14 Combustion Flashback   158

2.15 Unconfined Sources   158

2.16 Chapter Summary   159

2.17 Chapter Problems   160
    2.17.1 Flue Gases 160
    2.17.2 Coal Combustion 160
    2.17.3 Oil Combustion 161
    2.17.4 Incineration 161
    2.17.5 Incineration of Wet Waste 161
    2.17.6 Incinerator Emissions 162
    2.17.7 Combustion Temperature 162
    2.17.8 FBC Mass-Heat Balance 162

3. **PARTICULATE CONTROL MECHANISMS** 165

3.1 Introduction   165

3.2 Dimensionless Numbers   167
    3.2.1 Flow Reynolds Number (Re_f) 167
    3.2.2 Particle Reynolds Number (Re_p) 167
    3.2.3 Knudsen Number (Kn) 168
    3.2.4 Cunningham Correction Factor (C) 168
    3.2.5 Stokes' Number (St)—*Impaction Parameter* (K_I) 169

3.3 Linear–Steady-State Motion   170

    3.3.1 *Under Influence of Gravity* 170
    3.3.2 *Nonspherical Particles* 173
    3.3.3 *Aerodynamic Diameter* 175
    3.3.4 *Particle Diffusion* 177
3.4 Centrifugal–Steady-State Motion   178
3.5 Nonsteady-State Motion   179
    3.5.1 *Relaxation Time* 179
    3.5.2 *Accelerating Particles* 180
    3.5.3 *Decelerating Particles* 181
    3.5.4 *Distance Traveled* 182
3.6 Forces and Their Effects   183
    3.6.1 *Gravitational Settling* 183
    3.6.2 *Diffusive Deposition* 184
    3.6.3 *Inertial Force* 185
    3.6.4 *Electrostatic Force* 189
    3.6.5 *Phoretic Forces* 195
    3.6.6 *Boundary Layers* 197
3.7 Particle Collection   199
    3.7.1 *Gravitational Settling* 199
    3.7.2 *Centrifugation* 200
    3.7.3 *Impactor Systems* 200
    3.7.4 *Spray Systems* 204
    3.7.5 *Filtration* 205
    3.7.6 *Electrostatic Precipitation* 206
    3.7.7 *Coalescence* 208
    3.7.8 *Cut Diameter* 209
    3.7.9 *High Temperature and Pressure Effects* 210
3.8 Chapter Problems   214
    3.8.1 *Terminal Settling* 214
    3.8.2 *Particle Behavior* 215
    3.8.3 *Charges on Particles* 215
    3.8.4 *Gravity Settling Chambers* 215
    3.8.5 *Filtration* 216
    3.8.6 *Coalescence and Elevated Temperature* 216

4. **GAS CONTROL MECHANISMS** 219
4.1 Diffusivities   220
    4.1.1 *Gas Diffusivity in Gases* 220
    4.1.2 *Gas Diffusivity in Liquids* 223
4.2 Mass Transfer   224

4.3 Gas Absorption   225
   4.3.1 Solution Laws 226
   4.3.2 Mass Transfer Coefficients and Henry's Law Constant 231
   4.3.3 Interfacial Area 231
   4.3.4 Log Mean Pressure Difference 233
   4.3.5 Vapor Liquid Equilibrium 233
   4.3.6 Absorption Driving Force 238
   4.3.7 Absorber Operating Lines 240
   4.3.8 Contact Stages and Stage Efficiency 245
   4.3.9 Mass Transfer Coefficients and Stage Efficiency 246

4.4 Moisture Content of Gases   248

4.5 Gas Adsorption   255
   4.5.1 Adsorption and Desorption 256
   4.5.2 Heat of Adsorption 257
   4.5.3 Adsorbents 258
   4.5.4 Capacity 264
   4.5.5 Application 264

4.6 Chemical Removal Processes   265
   4.6.1 Gas to Particulate Conversion 265
   4.6.2 Other Chemical Processes 265

4.7 High Temperature and Pressure Effects   266
   4.7.1 Gas Diffusivity 266
   4.7.2 Thermal Conductivity 267

4.8 Chapter Problems   268
   4.8.1 Diffusivity of Gases in Air 268
   4.8.2 Flue Gas Humidity 268
   4.8.3 Ammonia Absorption 268
   4.8.4 Butane Adsorption 269

5. **CONTROL DEVICES** **271**

5.1 Mechanical Collectors   272
   5.1.1 Cyclone Separators 272
   5.1.2 Cyclone Dimensions 273
   5.1.3 Pressure Drop 276
   5.1.4 Cyclone Efficiency 277

5.2 Electrostatic Precipitators   281
   5.2.1 Unit Specifications 282
   5.2.2 Collection Efficiency 288
   5.2.3 Design Parameters 292

5.3 Filters   294

     5.3.1   *Baghouses* 294
     5.3.2   *Pressure Drop* 298
     5.3.3   *Collection Efficiency* 300
     5.3.4   *Design and Operating Parameters* 302
5.4 Scrubbers   308
5.5 Wet Scrubbers   309
     5.5.1   *Atomization* 310
     5.5.2   *Gas Atomized Spray—Venturi* 318
     5.5.3   *Cut Diameter* 321
     5.5.4   *Contacting Power* 322
     5.5.5   *Pressure Drop* 326
     5.5.6   *Collection Efficiency* 327
     5.5.7   *Venturi Absorption* 329
     5.5.8   *Venturi Design Parameters* 331
     5.5.9   *Mist Elimination* 333
     5.5.10 *Chemical Additives* 341
     5.5.11 *Recent Innovations in Wet Scrubbers* 341
     5.5.12 *Other Conventional Wet Scrubbers* 343
5.6 Dry Scrubbers   347
     5.6.1   *Dry/Dry Scrubbers* 348
     5.6.2   *Wet/Dry Scrubber* 348
     5.6.3   *Particle Removal* 349
     5.6.4   *Chemicals* 350
5.7 Absorbers   350
     5.7.1   *Absorption Towers* 351
     5.7.2   *Absorption Tower Capacity* 356
     5.7.3   *Absorption Tower Efficiency* 360
     5.7.4   *Relative Pressure Drop of Packing* 364
5.8 Adsorbers   366
     5.8.1   *Adsorber Arrangements* 366
     5.8.2   *Adsorber Capacity* 368
     5.8.3   *Adsorber Pressure Drop* 373
     5.8.4   *Adsorber Efficiencies* 374
     5.8.5   *Adsorber Design* 374
5.9 Hybrids   376
     5.9.1   *Electrostatic Filters* 376
     5.9.2   *Moving Bed Filters* 376
     5.9.3   *Wet Filters* 378
     5.9.4   *Charged Scrubbers* 379
     5.9.5   *Ejector Scrubbers* 380

5.10 Gas Conditioning   380
      5.10.1 Quenching 380
      5.10.2 Using Condensation Force 385

5.11 Special Case—Particle Collection in Spray Towers   386
      5.11.1 Particle Collection Forces 386
      5.11.2 Applicability 388

5.12 Device Sizing   391

5.13 Chapter Problems   392
      5.13.1 High-Efficiency Cyclone Optimization 392
      5.13.2 ESP and Filter 392
      5.13.3 Venturi Scrubber 393
      5.13.4 Venturi Scrubber 393
      5.13.5 Packed Tower Absorber 393
      5.13.6 Wet Scrubbing 394

6. **CONTROL SYSTEMS** **399**

6.1 Upstream of the Control System   401
      6.1.1 Process Modifications 401
      6.1.2 Mechanical Collectors 402
      6.1.3 Gas Conditioning 403

6.2 Downstream of the Control System   405
      6.2.1 Stack Gas Reheat 406
      6.2.2 Chimneys 406

6.3 Electrostatic Precipitators   408
      6.3.1 Model Testing 408
      6.3.2 System Sizing 409
      6.3.3 Operation Parameters 410
      6.3.4 Installation 410
      6.3.5 Maintenance 413
      6.3.6 Costs 413

6.4 Filters   418
      6.4.1 System Sizing 418
      6.4.2 Operation Parameters 420
      6.4.3 Installation 420
      6.4.4 Maintenance 421
      6.4.5 Costs 421
      6.4.6 Gas Removal 423

6.5 Scrubbers   425
      6.5.1 Wet Particulate Scrubbers 425

      6.5.2 *Wet FGD and Acid Gas Control Systems* 432
      6.5.3 *Lime Spray Dryer* 441
      6.5.4 *Dry Injection* 441
      6.5.5 *Absorbers* 443
      6.5.6 *Mist Eliminators* 444

  6.6 Adsorbers   444
      6.6.1 *Sizing and Operating* 444
      6.6.2 *Installation and Maintenance* 445
      6.6.3 *Costs* 445

  6.7 Incinerators   448
      6.7.1 *Capital Cost* 449
      6.7.2 *Operating Cost* 450

  6.8 Operation and Maintenance   451
      6.8.1 *Operation* 451
      6.8.2 *Maintenance* 453

  6.9 Costs   453
      6.9.1 *Comparative Costs* 453
      6.9.2 *Generalized Costs* 454
      6.9.3 *Cost Extrapolation* 456

  6.10 Chapter Problems   461
      6.10.1 *Cyclone Costs and Gas Cooling* 461
      6.10.2 *Gas Cooling* 461
      6.10.3 *ESP Costs* 461
      6.10.4 *Filter Costs* 462
      6.10.5 *Scrubber Costs* 462
      6.10.6 *Adsorber Costs* 463
      6.10.7 *Cost Extrapolation* 463
      6.10.8 *Large Pollution-Control System* 463

Appendix A: Table of Most Commonly Used Symbols   467

Appendix B: Conversion Factors   472

Appendix C: Toxic Chemicals Subject to Section
            313 (Title III of SARA) Reporting   479

Index   489

Preface

SINCE the first edition was printed in 1991, there have only been minor changes in air regulations. The opposing "trenches" used by environmental regulation proponents and opponents have deepened as each side increases their database. Agencies and environmental groups have backed off a little in issues such as bubble policies and enforcement time tables. This has made it extremely difficult for equipment vendors to anticipate industry requirements. Overall, the current market projections are not very favorable for the new equipment suppliers.

In contrast, the service organizations are seeing increased need for their help in areas such as dispersion modeling, troubleshooting and testing. Existing systems are being improved upon to keep them in operation. There remains a continuous need for up-to-date references and training materials to serve these needs, and it is for this purpose this revised edition is dedicated.

The heart of the book is the first edition. In this revised edition, the introductory chapter adds puff dispersion material, which is especially important when dealing with toxic chemical accidents and explosions. The Gulf Desert Storm conflict and the recent Japanese terrorist activities accentuate this importance.

In order to gain improved equipment performance and keep operating costs low, guidelines for design and use of cyclone collectors, packed towers, mist eliminators and quench tanks are added to the "Control Devices" chapter. A special section is also included giving particle collection efficiency equations and examples for co-current and counter-current spray towers.

The cost information in the last chapter, Chapter 6, is expanded to include quencher costs and typical data on costs of fuel, utilities, raw materials and waste treatment. Data are also presented to reflect basic pollution control system one-time costs and some typical annual regulatory expenses that must be borne by industrial operations.

There is also a selfish reason for this revised edition—to try to correct errors that escaped detection before printing (even with *numerous* proof readings). It is my hope that this book will be even more useful and valuable to you, the users.

HOWARD E. HESKETH, PH.D., PE, DEE
Carbondale, IL

Preface to the First Edition

ENVIRONMENTAL issues are one large melting pot where the subjects of air, water, and land pollution interrelate and interact. Historic air pollution control cannot be discussed without reference to air toxics, hazardous air emissions, indoor air pollution, industrial hygiene, heating ventilation, and air conditioning, as well as hazardous waste management, atmospheric meteorology, ground water decontamination, soil cleanup, and other related activities. Fortunately, we find that many of the same basic control technologies are applicable to most of these problems. Recognition of the basic principles underlying these technologies better enables one to choose the optimum control device for a particular application.

This book has been prepared in an effort to present both theory and application data. A good background is provided relevant to behavior theories and control techniques for capturing gaseous and particulate air pollutants. The significant application data available have been digested and combined with the theories to provide a needed tie between the two. Numerous, detailed example problems are worked throughout the book to serve as guides in the use of both the theoretical relationships and the data. These guides help to make this volume a valuable resource tool and handbook. The book also contains much practical-experience data, to expose the reader to some of the more important internal requirements which are part of actual air pollution control facilities. Problems are provided at the end of each chapter.

The first two chapters supply general information related to air pollution control. Chapter 1 presents reasons for controlling emissions, including a general discussion of environmental regulations from the U.S. EPA, RCRA, TSCA, and other countries. Problems such as how pollutants are formed, how much must be controlled, how they are dissipated, and what can be done to minimize pollution emissions are addressed. Chapter 2 is a detailed presentation of both the fossil fuel combustion and incineration. Techniques for estimating combustion temperatures and for fuel consumption are developed and presented here along with a summary of factors to adjust emission concentrations.

Theoretical foundations of particulate and gaseous pollutant control comprise Chapters 3 and 4, respectively. A basic understanding of chemistry and physics is useful in some places but not necessary. These are important chapters. Familiarity with this material provides knowledge of how the pollutants behave and what mechanisms could be considered for their control.

Chapter 5 is the application of theory to actual control devices. It could well be a complete book by itself. All major types of control devices are discussed and many of the operating features are related to the control theories New equations are developed and presented to provide ways for estimating device quench water needs, electrostatic precipitator efficiency, venturi scrubber design, and waste fuel combustion. Some practical design and operating issues are presented at this point to help explain the deviations from the theoretically expected results.

Chapter 6 examines the practical application of control techniques to gaseous and particulate control systems. Information related to cost estimating, installing, operating, and maintaining these systems is included. This should help the reader to develop a hands-on familiarity with these systems, although it cannot substitute for actual experience. Chapter 6 is important, as these practical applications are what make all the research, calculations, and design work—or fail.

The reader should find this book valuable in the transition from English to SI Metric Units. Quantities and equations are presented for the most part, in both English and cgs Metric System (centimeter, gram, second) units. Appendix B contains conversion factors for all systems, as well as other useful information.

It has been said that to read effectively, one must be alert; to speak effectively, one must think; and to write effectively, one must be exact. Hopefully both you, the reader, and I, the writer, are alert, thinking, and exact in our interaction of communicating this material. Best wishes to you for your effective utilization of it.

HOWARD E. HESKETH, PH.D., PE, DEE
Carbondale, IL

Acknowledgements

FRANK W. Sherman (Environmental Engineering Consultant from Springfield, IL) helped review the galleys and provided valuable technical and editorial assistance. Thank you, Frank. I also sincerely appreciate the help of Kyle Russell (Missouri Department of Natural Resources), Shane Webber and all of my class students who reviewed the manuscript. Special thanks go to Amy, Judi and the other secretaries of the Department of Mechanical Engineering and Energy Processes, Southern Illinois University, for helping prepare the manuscript.

CHAPTER 1

Introduction

1.1 AIR POLLUTION

LIGHT is the most important entity in our physical world. Next to it and nearly equal in significance is air. Air surrounds us both night and day, unless we purposefully exclude it. The quality of the air affects the amount and type of light we receive and affects the quality of life, as we know it, both directly and indirectly.

The extent of civilization and the lifestyles of the people dictate how much and what type of atmospheric degradation will be present. In the past we have been most concerned about the ambient levels of the *traditional* pollutants: sulfur dioxide (SO_2), total suspended particulates (TSP), nitrogen oxides (NO_x), carbon monoxide (CO), and photochemical oxidants. Of the first three (SO_2, TSP, NO_x), predictions made in 1977 [1] reported that by the year 2000 only TSP emission would decrease. Currently we have added many new chemicals to our list of concerns. These include lead, asbestos, mercury, arsenic, acids (such as H_2SO_4, HCl, HF), halogens, dioxins, furans, and PCBs. The list of toxic chemicals numbers 308 individual substances (see Appendix C). These can be present in urban, rural, and wilderness areas.

Atmospheric problems are compounded when local winds cannot blow away and disperse the pollutants, and when reactions take place in the atmosphere. California, for example, has a local problem of maintaining good air quality because of the high emissions generation rate and the inability of the atmosphere to cleanse itself. Another example, Antarctica, has a problem of global significance because emissions of fluorocarbons and other halogens are reacting with and destroying the upper atmosphere ozone layer. Rain forests generate oxygen while simultaneously emitting

1

hydrocarbons which can react with nitrogen oxide in the presence of sunlight to form destructive lower level atmospheric ozone.

Concern over emissions of substances not normally considered as pollutants has increased during the last quarter of the 20th century. The CO_2 buildup is of major importance. Rotty [2] notes that as of 1980, the amount of CO_2 production in the atmosphere had increased by 4.3% over the levels two decades prior. This is mainly due to fossil fuel combustion. Rate of CO_2 production from this source is predicted to increase until about the year 2040; however, atmospheric concentrations could continue to increase after that time due to accumulation. Rotty predicts that the current atmospheric CO_2 concentration of mid-300 ppm will increase to 450. Concerns such as the 1988 year moisture (excessive drought and rains), climate heating and cooling trends, and other issues (such as ozone depletion) make it important for mankind to monitor all atmospheric changes closely while striving to control and regulate emissions.

It is possible to reduce man-made emissions to the atmosphere by good control procedures and by process modifications. This chapter shows us what air pollutants are, current levels of regulations, ways to evaluate emissions, indoor vs. outdoor concerns, and ways emissions disperse after they enter the atmosphere. Later chapters deal with theories, devices, systems, and economics of controlling both traditional and hazardous air pollutants.

1.2 PARTICULATES

1.2.1 GENERAL RELATIONSHIPS

Particulate air pollution consists of solid and/or liquid matter in air or gas. Size of particulate matter ranges upward from near molecular size. Particulate size is expressed in micrometers, μm. The lower practical limit for control is about 0.01 μm. Fine particles are defined as those particles 3 μm or smaller because of the increased difficulty in controlling emission of these particles. Concentrations of particulate matter, unless otherwise specified, are by mass. For example, particulates in a gas would be g/m^3 and in liquid or solid phases would be g/g total mass.

Very small particles, liquid particulate matter and particulates formed from liquids are likely to be spherical in shape. In order to express the size of irregular (nonspherical) particles as a diameter, several important relationships are discussed.

1.2.1.1 Aerodynamic Diameter

Aerodynamic diameter, d_a, is the diameter of a sphere of unit density

(ϱ_p = 1 g/cm³) that attains the same terminal settling velocity, v_s, at low particle Reynolds number in still air as the actual particle under consideration. Particle Reynolds number is defined by Equation (3.2).

1.2.1.2 Equivalent Diameter

Equivalent diameter, d_e, is the diameter of a sphere having the same volume as the irregular particle and is given by

$$d_e = \left(\frac{6V}{\pi}\right)^{1/3}$$ (1.1)

where V is volume of the particle.

1.2.1.3 Sedimentation Diameter

Sedimentation diameter, d_s, is the diameter of a sphere having the same terminal settling velocity and density as the particle. In the case of unit density particles, it is called the reduced sedimentation diameter, making it the same as aerodynamic diameter. Sedimentation diameter is also known as Stokes diameter.

1.2.1.4 Cut Diameter

Cut diameter, d_c, is the particle diameter for which half the particles are collected, i.e., individual efficiency ϵ_i = 0.5, and half penetrate through the collector, i.e., penetration Pt_i = 0.5.

$$Pt_i = 1 - \epsilon_i$$ (1.2)

1.2.1.5 Dynamic Shape Factor

Dynamic Shape Factor, χ, is a dimensionless proportionality constant relating the equivalent and sedimentation diameters

$$\chi = \left(\frac{d_e}{d_s}\right)^2$$ (1.3a)

The d_e equals d_s for spherical particles, so χ for spheres is 1.0. Approximate values for various-shaped particles calculated by the Pettyjohn [3] procedure are noted on Figure 1.1 with a line representing the average of Heiss' [4] data for various cylindrical and parallelapiped-shaped particles in free-fall in still air. Free-falling particles usually have particle Reynolds numbers $Re_p < 0.1$. For large particles or aggregates of particles with $Re_p > 0.1$, values of χ as large as 2.0 have been reported [5]. As a first approximation in determining χ for any shaped particle, the linear portion of Figure 1.1 can be used knowing the ratio of horizontal to vertical axis length, r. For $r > 0.5$ this can be expressed as

$$\chi = 8.82 \times 10^{-2}r + 0.967 \qquad (1.3b)$$

1.2.2 SIZE DISTRIBUTION

Normal particulate matter emissions and atmospheric particle samples have a geometric (log normal) size distribution. As such, the size distribution of this material can be described by plotting on log-probability coordinates as, e.g., log of particle diameter versus cumulative percent by mass, area or number less than stated diameter on the probability ordinate. This is shown in Figure 1.2 by mass distribution.

The data in Figure 1.2 represent a typical log-probability size distribution of coal combustion particulate matter leaving an "efficient" wet scrubber. The solid line in Figure 1.2 is a cumulative percent by weight less than stated diameter and can be obtained by cascade impactors. The data for

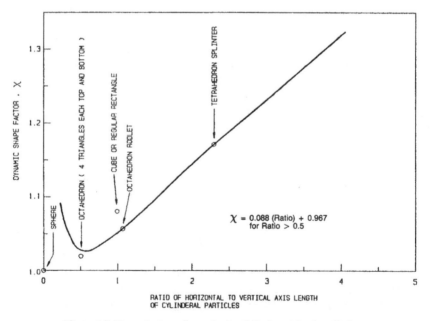

Figure 1.1 Dynamic shape factor for free-fall of particles in still air.

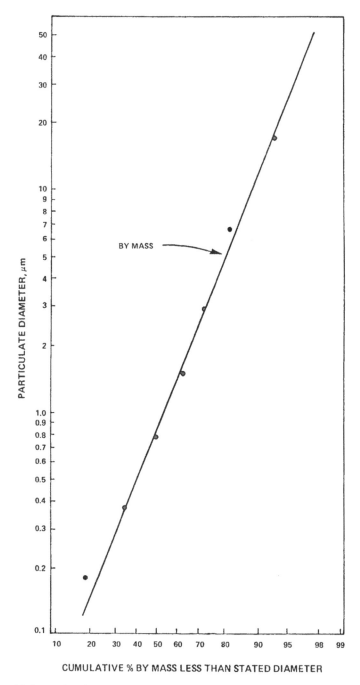

Figure 1.2 Log probability cumulative distribution of particulate matter in a coal combustion flue gas leaving a wet scrubber.

this are presented in Table 1.1. The aerodynamic cut diameter, d_{50}, size depends on construction of the device and test operating conditions. Values of d_{50} are usually obtained from the equipment suppliers' calibration curves. Cut diameters for the filter can only be obtained by special techniques such as by use of porous metal filters and electron microscopy. Mass of sample collected per stage is determined by careful analytical weighing. Percent on any stage is stage mass divided by total sample mass including the filter catch. Cumulative percent undersize is obtained by simply adding up the stage mass percents starting with the smallest. Remember that d_{50} represents the size particle which is captured with a 50% efficiency. Therefore, the cumulative percent undersize column cannot be directly plotted against d_{50}. Rigorous procedures have been developed to obtain the true value of cumulative percent to plot, however, a simple technique that provides good results is to assume that half the mass on any stage plus all mass on smaller stages is less than d_{50}. This gives the last column (cumulative percent by mass less than d_{50}) by

$$11 + \frac{15}{2} = 18.5 \quad \text{for } d_{50} = 0.18$$

$$26 + \frac{17.5}{2} = 34.8 \quad \text{for } d_{50} = 0.37$$

These data are plotted in Figure 1.2 and a single straight best-fit line is drawn.

Shown in Figure 1.3 are size distributions by mass, by area, and by num-

TABLE 1.1. Typical Source Test Size Data Using a Cascade Impactor in a Wet Scrubber Effluent Gas.

Stage No.	Stage d_{50}, μm	Stage Plate Mass Gain, g	Mass % on Stage	Cumulative % Undersize on Stage	Cumulative % by Mass Less than d_{50}
1	17	0.01171	13.5	100.0	93.3
2	6.4	0.00734	8.5	86.5	82.3
3	2.9	0.00821	9.5	78.0	73.3
4	1.5	0.00967	11.0	68.5	63.0
5	0.77	0.01245	14.0	57.5	50.5
6	0.37	0.01506	17.5	43.5	34.8
7	0.18	0.01328	15.0	26.0	18.5
Filter		0.00937	11.0	11.0	5.5
Total		0.08709	100.0		

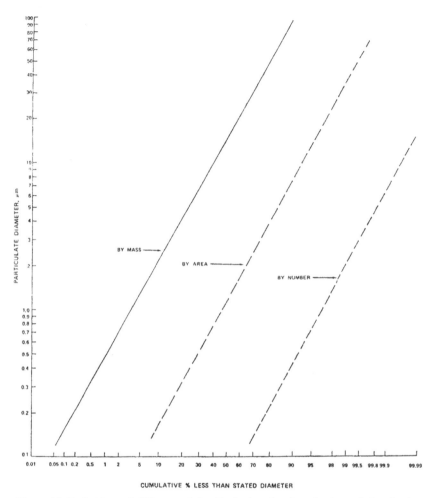

Figure 1.3 Typical log probability cumulative distribution of coal combustion emissions leaving a low efficiency electrostatic precipitator.

ber. These data can be obtained from the original data by simple mathematical procedures. However, these procedures are subject to error for smaller-sized particles. The procedures assume that density of all size particles in the sample is a constant. This may not be true. For example, large particles of fly ash formed by solidification of molten slag can contain hollow centers. These are called cenospheres and they would have a lower density. Smaller-size particles have proportionately larger area-to-mass ratios. This encourages chemical oxidation, reduction, and other reactions which could result in a surface of different density material. For particles no smaller than those in the Figure 1.3 example, the problem is usually not significant.

To convert size-distribution data to other bases, assume spherical particles and constant density. Then for a given diameter particle

$$\text{mass} = (V)(\varrho_p) = n(1/6\pi d^3)\varrho_p \qquad (1.4)$$

where

$V =$ volume
$\varrho_p =$ particle density
$n =$ number of particles

As each data point ultimately becomes divided by the total, it is only necessary to use the stated assumptions and

$$\text{mass} \; \alpha \; nd^3 \qquad (1.5)$$

or for determining number of particles rearrange for

$$n \; \alpha \; \frac{\text{mass}}{d^3} \qquad (1.6)$$

Similarly, area can be determined using area of a sphere equals πd^2 as

$$\text{area} \; \alpha \; nd^2 \qquad (1.7)$$

or directly from the above data

$$\text{area} \; \alpha \; \frac{\text{mass}}{d} \qquad (1.8)$$

Table 1.2 summarizes this method using the mass size distribution data given as the upper line in Figure 1.3. These data represent a typical sample of particulate matter leaving a low-efficiency (97%) electrostatic precipitator where the flue gases come from a large utility boiler burning pulverized sub-bituminous western coal. To convert these data, the procedure must start by using some arbitrarily chosen particle size range. If too large a size range is used then the average diameter of the group will not be representative. If too small a range is chosen, the iterative calculations become excessive unless a computer program is used. For a starting point,

TABLE 1.2. Procedures for Converting Mass Size Distribution Data from Figure 1.3 to Area Size Distribution.

Cumulative Mass % Undersize, Range Equals	Mass % in Range	Range Av. d by Mass, μm	Area $\left(\alpha \dfrac{mass}{d} \right)$	Area, % in Range	Cumulative % by Area Less than Av. d
98–100	2	—	—	—	100.00
94–98	4	220	0.02	0.10	100.00
98–94	8	110	0.07	0.36	99.90
70–86	16	50	0.32	1.66	99.53
50–70	20	22	0.91	4.73	97.87
30–50	20	10	2.00	10.41	93.13
14–30	16	4.4	3.64	18.94	82.73
6–14	8	2.1	3.81	19.82	63.79
2–6	4	1.0	4.00	20.81	43.96
0–2	2	0.45	4.45	23.15	23.15
		Total	19.22	100.00	

a range size of 2% is suggested for the large and small ends. Double this successively and consistently as shown in the second column of Table 1.2. This produces column 1. The average diameter by mass in column 3 for each range is obtained from the Figure 1.3 log-probability plot of the mass size distribution data. Some points here are approximated by extrapolating the line. Calculate area using the relation given as (1.8) for column 4, then divide each by the total and multiply by 100 to obtain the area percent data in column 5. Note that consistent dimensions cancel out. The last column is the summation of individual range areas. These values are plotted against the range average diameter in Figure 1.3. Size distribution by number is obtained by a similar procedure.

Particle size and size distribution can be specified using a mean diameter, $\bar{d}$, and a standard deviation. The standard geometric deviation is found by

$$\sigma_g = \frac{d_{84.13}}{\bar{d}} = \frac{\bar{d}}{d_{15.87}} \tag{1.9}$$

where $d_{84.13}$ is the value of cumulative percent undersize at 84.13%, etc. For Figure 1.3, the mean diameter by *mass*, $\bar{d}_M$, is 14.7 μm and σ_g is 14.7/3.3 or 4.45.

Note that the slope of all curves in Figure 1.3 should be essentially equal. This makes it possible to relate the mean diameter by mass to the mean diameter by number, $\bar{d}_N$, etc. Orr [6] suggests a general procedure of

$$\frac{\bar{d}_M}{\bar{d}_N} = \exp[c_1 \ln^2 \sigma_g] \qquad (1.10)$$

where c_1 is a constant with a value of about 2.66 (this value would predict a $\bar{d}_N$ equal to 0.04 for Figure 1.3 data which is close to the actual value).

Most particulate matter consists of more than one type of material. There could be two or more materials, or the same material with two physical size distributions as a result of partial thermal or mechanical fracturing. If the amount of the second (or more) type of material is significant, then the size distribution actually becomes bimodal and the plots on log-probability coordinates result in two (or more) straight intersecting lines. For example, careful collection and evaluation of data can result in the curves shown in Figure 1.4. The left set of curves is before and the right set is after a "low-efficiency" wet scrubber. Figure 1.5 shows a similar bimodal type of distribution for particulate matter in ambient air [7].

1.2.3 SPECIAL PROPERTIES

Some properties such as surface irregularities and density have been noted. Therefore, the apparent density of airborne particulate matter may differ from true density because of inclusions of air pockets or surface reactions to form layers of a different material. In addition, aggregates of particles have lower apparent densities than single particles. Apparent densities of aggregates may be up to 10 times less than true densities. Bulk density is the mass per unit volume of loosely packed solid particulate matter and depends on particle shape and gas space included between particles. This is less than true density and usually differs from both true and apparent densities.

Particles may contain positive, negative or no charge. Natural charges usually do not exist on particles less than 0.1 μm in diameter. Larger particles average about one elementary charge (1 electron) as a result of natural diffusion charging. Probability of more than one natural charge increases with particle size.

The presence of small suspended particulate matter results in light scattering. Amount of scattering varies directly with particle size. Scattered short wavelength light (blue end of the spectrum) results in only red colors penetrating. This is why sunrises and sunsets are more red at times of high particulate pollution levels.

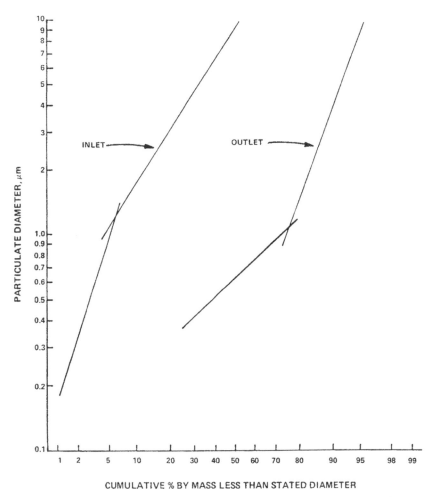

Figure 1.4 A coal combustion particulate matter size distribution at inlet and outlet of a low-efficiency wet scrubber.

Particles can absorb, reflect, and refract light. Part of the light striking an opaque particle is absorbed. This warms the particle on the side toward the light which, in turn, warms gas molecules and causes them to push the particle away from the light. Transparent particles transmit light to the far side resulting in a warming of that side and this ultimately results in movement toward the light. Semitransparent particles may exhibit either of these phenomena.

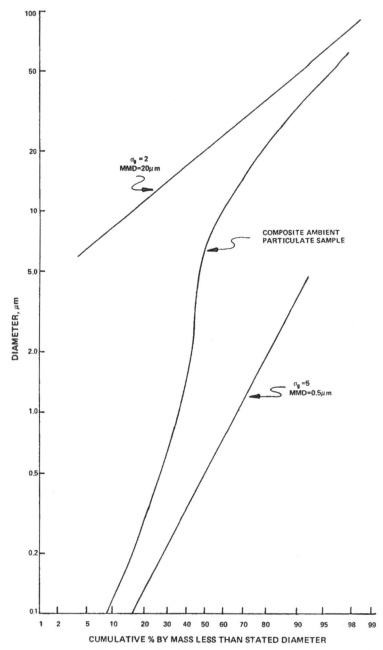

Figure 1.5 Ambient particulate matter sample.

12

As particle size decreases below about 0.1 μm, extreme variations in physical properties may occur. As an example, Hesketh [7] summarizes data to show that for water droplets < 0.1 μm, the surface tension and heat of vaporization may show a drastic decrease and vapor pressure increases exponentially. This is shown in Figure 1.6.

The great increase in surface area results in both desirable and undesirable characteristics. The large area produces valuable adsorption sites, but it also increases area for oxidation reactions to occur in air. This makes small particulate dusts subject to explosions. Controlled explosions of a finely pulverized coal dust in a boiler are used to produce heat energy, but explosions as in a grain storage facility are destructive to life and property. The amount of explosiveness varies with particle size. As an example for caution, the explosiveness of atomized aluminum dust can be greater than that of the dangerous grain dusts.

Ultrafine particles such as those formed by gaseous impurities subjected to radiation have properties that are more dependent on particle diameter than on chemical composition. Particle geometry and crystal structure can even be dependent on type and temperature of the radiant energy, and color is likely to be characteristic of the gaseous material from which it is produced.

1.3 GASES

1.3.1 GENERAL RELATIONSHIPS

Gases are important fluids not only from the standpoint that a gas can be a pollutant, but also because gases *convey* the particulate and gaseous pollutants. Often, the main gas stream is considered to be air which is mainly nitrogen. Two terminologies are commonly used to describe gas conditions: standard temperature and pressure, STP, and standard conditions, SC. STP represents 0°C (32°F) and 1 atm. SC is more commonly used and represents typical room conditions of 20°C (70°F) and 1 atm. SC units of volume are commonly given as normal cubic meters, Nm³, or standard cubic feet, scf.

The majority of air pollution calculations can be made assuming ideal gas behavior. Air is an ideal gas at normal ranges of temperature and pressure and usually is considered to be a binary system of nonreacting gases containing about 79% nitrogen and 21% oxygen. The molecular weight of dry air is 28.96 g/g mole or about 29. Even high molecular weight gaseous air pollutants behave as ideal gases because they are so dilute in concentration.

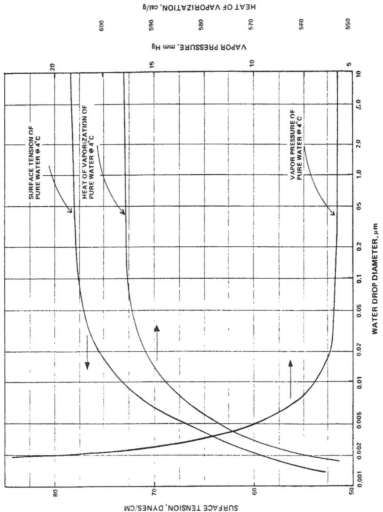

Figure 1.6 Surface effects as related to particle size of water droplets.

14

1.3.1.1 Gas Laws

The familiar form of the ideal gas law derived from Charles' and Boyle's laws shows that for *one mole* of an ideal gas

$$P_T V = RT \qquad (1.11)$$

where

P_T = total pressure
V = volume
R = universal gas constant
T = absolute temperature

In cgs units, the value for R is 83.14×10^6 g cm^2/(sec^2 g mole K). At STP this equates to 22.4×10^{-3} m^3/g mole (or 359 ft^3/lb mole). Using the ideal gas law, this value can be converted from STP to SC by the absolute temperature ratio noting that gas volume increases directly with temperature and absolute temperature in Kelvin is 273.16 + °C:

$$\frac{22.4 \times 10^{-3} \text{ m}^3}{\text{g mole}} \frac{273 + 20}{273} = 24.0 \times 10^{-3} \text{ Nm}^3/\text{g mole}$$

Pressure corrections can be made in a similar manner, but note from Equation (1.11) that pressure and volume are inversely related. For example, the actual volumetric flowrate of 1,000 m^3/min at 150°C and 100-cm water gauge pressure could be converted to flowrate at standard conditions remembering that standard pressure is about 1,034 cm water:

$$\left(\frac{1,000 \text{ m}^3}{\text{min}}\right)\left(\frac{273 + 20}{273 + 150}\right)\left(\frac{1,034 + 100}{1,034}\right) = 760 \text{ Nm}^3/\text{min}$$

The total pressure of a mixture of ideal gases can be determined by Dalton's law which states that the sum of the individual partial pressures equals the total pressure, P_T. For a binary mixture composed of gases A and B this would be

$$p_{AG} + p_{BG} = P_T \qquad (1.12)$$

where p_{AG} and p_{BG} are partial pressures of A and B respectively in the gas mixture. Combining the ideal gas law with Dalton's law and Amaget's law of additive volumes, one can obtain for ideal nonreacting gases the extremely useful gas relationships of

volume ratio = mole ratio = pressure ratio

In symbols for component A this is

$$\frac{V_{AG}}{V_T} = \frac{N_{AG}}{N_T} = \frac{p_{AG}}{p_T} \tag{1.13}$$

where N is the number of moles of gas.

It is possible to obtain the partial pressure of a gaseous component using the ideal gas law to express Dalton's law in the form

$$p_{AG} = y_A P_T \tag{1.14}$$

where y_A is mole fraction A in the gas (moles of A divided by moles of total gas). For example, assume air at SC contains 1% water vapor. The partial pressure of the water vapor then becomes

$$p_{AG} = (0.01)(0.101325 \text{ MPa}) = 1.01 \times 10^{-3} \text{ MPa}$$

(Note: 1 atmosphere expressed in SI units of megapascals equals 0.101325 MPa. There are 10^6 pascal in a MPa and a pascal is 1 Newton/m^2.)

Gas concentrations, unless otherwise noted, are expressed by *volume*. From the previous discussion it is obvious that this is the same as by mole or by pressure. A concentration of 0.05 moles SO_2 in 100 moles of flue gas (0.05%) could then be expressed as

$$\left(\frac{0.05}{100}\right)(10^6) = 500 \text{ ppm}$$

where ppm stands for parts per million by volume.

1.3.2 RELEVANT FACTORS

It has been stated that most of the time gases in these systems can be considered ideal gases. However, at high pressures and temperatures this is not true and methods to account for deviation from ideal behavior must be used. This includes use of compressibility factors and/or departure tables or some appropriate equations of state as given in numerous thermodynamics references [8,9].

1.3.2.1 Incompressible Fluids

Gases in these systems are considered incompressible fluids when the Mach number, Ma, is $\ll 1$. Mach number is the ratio of gas velocity, v_g, to acoustic velocity, v_a:

$$Ma = \frac{v_g}{v_a} \qquad (1.15)$$

Acoustic velocity is the speed of sound in the fluid at local conditions and in an ideal gas this is

$$v_a = \sqrt{\frac{\gamma RT}{M}} \qquad (1.16)$$

where γ is the ratio of gas specific heats (C_p/C_v) which equals 1.40 for ideal gases.

1.3.2.2 Molecular Weight

The gas molecular weight, M, of a pure compound can be established by the chemical formula. If the gas is a mixture of ideal noninteracting gases, an average molecular weight can be found by the chain rule

$$M = \frac{N_A M_A + N_B M_B + \ldots N_X M_X}{N_T} \qquad (1.17)$$

where N_X and M_X are number of moles and molecular weight respectively of species X and N_T is total number of moles. For example, a wet flue gas containing 16% CO_2, 3% O_2, 75% N_2, and 6% H_2O has an average molecular weight of

$$M = \frac{(16)(44) + (3)(32) + (75)(28) + (6)(18)}{100} = 30.08 \text{ g/g mole}$$

Note that this flue gas would have a greater molecular weight dry. In this case it would be 30.85.

1.3.2.3 Density

Density of an ideal gas, ϱ_g, can be determined using the molecular weight and the ideal gas law by

$$Q_g = \frac{MP}{RT} = \frac{NM}{V} \qquad (1.18)$$

A convenient value for the universal gas constant, R, is 82.05 atm cm³/ (g mole K). It is necessary to know densities of gases, especially when using Reynolds numbers.

1.3.2.4 Mean Free Path

It was noted that the gases were important fluids because they also conveyed the particulate matter. The extent to which a particle slips through a gaseous medium is a function of the gas mean free path, λ_g. Gas molecules are in constant motion at a speed dependent on the type of gas and the temperature and pressure. The kinetic molecular theory predicts that gas mean free path can be estimated by

$$\lambda_g = \frac{\mu_g}{0.499P} \sqrt{\frac{\pi RT}{8M}} \qquad (1.19)$$

where P is pressure. At SC, this estimates the value for air to be about 6.60×10^{-6} cm.

Combining the kinetic theory equation to the Glasstone [10] viscosity relationship gives a simplified expression for air near SC:

$$\lambda_{air} = 2.26 \times 10^{-9}(T/P_1) \qquad (1.20)$$

where P_1 is pressure in megapascals, MPa. At SC (0.1013 MPa) this gives the mean free path of air as 6.54×10^{-6} cm or 6.54×10^{-2} μm.

1.3.2.5 Viscosity

Gas viscosity values are required in many useful relationships. Viscosity is needed in Equation (1.19) and will be needed in other extremely important equations (e.g., Reynolds numbers). Viscosity of a gas increases with temperature. In contrast, viscosity of a liquid decreases with temperature. Values of gas viscosities are sometimes difficult to obtain, especially at elevated temperatures. Values for air at standard pressure are given in Figure 1.7. An alternate procedure is to obtain a gas viscosity at SC from ref-

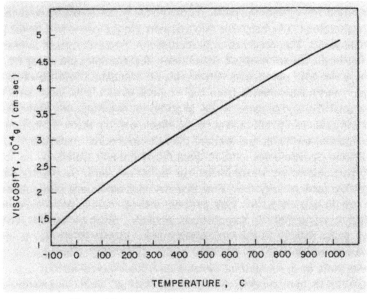

Figure 1.7 Viscosity of air at 1 atmosphere pressure.

erence literature and calculate elevated temperature viscosity values using

$$\mu_g = \frac{T^{1.5}}{mT + b} \times 10^{-6} \qquad (1.21)$$

where μ_g is in poise [i.e., g/(cm sec)]. T is in K and m and b are empirical constants for various gases. If the constants are unknown for a particular gas, obtain two values of μ_g and solve for the constants.

Calculated values for several gases are:

	m	b
air	0.070	7.9
CO_2	0.065	14.7
N_2	0.072	7.5
CH_4	0.102	16.2

1.4 PARTICULATE-GAS MIXTURES

Particulate matter, as related to air pollution, is significant because it is transported by a gaseous media. Therefore it is necessary to understand the properties of both the particulates and the gases as discussed in the previous sections. The behavior of particulates in the gases will be discussed in Chapter 3. The presence of particulates in gases and gases in gases can result in the creation of new conditions, for example: chemical reactions of the various components can produce particulates from gases or a different particulate matter from the original particulates; and particulates can volatilize and become gases. In addition, combinations of particulates and/or gases can result in synergistic effects making the combination more detrimental to health and welfare than the individual materials.

Primary particulates exist in the atmosphere as emitted, while secondary particulates are those formed in the atmosphere. Secondary particulates are usually very fine. Fine primary and secondary particulates tend to remain suspended for long periods of time. Sulfate particulate matter, which exists mainly in the northeastern U.S., results when SO_2 oxidizes and reacts with cations in the atmosphere. Ultimately, limits on ambient sulfates such as $8-10$ μ_g/m^3 could in turn limit SO_2 emissions.

Aerosols or aerocolloidal systems have been given various definitions. They can include condensation aerosols (formed when supersaturated vapors condense) and dispersion aerosols (formed directly by mechanical action). One definition is those particles with a Reynolds number of less than 1 and a surface-to-volume ratio of greater than 10^3 cm^{-1}. This includes materials as concentrated as fluidized beds or as dilute as interplanatary dust.

Various terms are used to describe specific types of natural and man-made mixtures of particles and gases. Smokes are both liquid and solid particles from oxidation processes while fumes are specifically smoke-condensed from metallic vapors. Dust is a dispersion aerosol of particles naturally generated by wind–ground interactions. Mists are liquid droplet supsensions and fog is a concentrated mist. Hazes are mists and dusts while smog originally was a contraction indicating the presence of both *smoke* and *fog*.

1.5 STANDARDS

Standards relative to air pollutants may be set by federal, state or local regulatory bodies. Regulations set by any regulatory body must be as strict as or stricter than those established at the next higher level of government.

Federal regulations are set by Congress and are published daily in the *Federal Register*. Regulations properly issued and published in the *Federal*

Register have the force and effect the law. *The Code of Federal Regulations* is a codification of current regulations of the various federal agencies.

Each year the *Federal Register* is codified into the *Code of Federal Regulations*, *CFR*. The *CFR* contains only final regulations promulgated as of each July. Title 40 is "Protection of the Environment." Currently, it is contained in ten volumes and has all of the EPA regulations. Parts and subject matter are

1– 29	General EPA information
30– 46	Grants and other federal assistance
50– 99	Air programs
100–117	Water programs
121–124	EPA permits
125–149	Water programs
150–180	Pesticide programs
190–192	Radiation protection programs
201–211	Noise abatement programs
220–237	Ocean dumping
240–257	Solid waste
260–272	Hazardous waste
280	Undergound storage tanks
300–355	Superfund
400–471	Effluent guidelines
600–610	Car mileage requirements
702–730	TSCA
761	PCBs
762–799	TSCA

The U.S. Environmental Protection Agency (EPA) is the federal agency responsible for interpreting the laws and setting standards. EPA documents that appear in the *Federal Register* are: (1) environmental rulemaking such as Advance Notice of Proposed Regulations, Final Regulations, and Standards; (2) policy such as Environmental Decision Statements signed by the EPA Administrator. The EPA is organized into ten regional offices, each with a regional administrator who reports to the agency administrator in Washington, DC.

1.5.1 AMBIENT AIR QUALITY STANDARDS

Federal standards to protect the quality of the air were established by the Clean Air Act (CAA) Amendments of 1970. These were changed by the

TABLE 1.3. National Ambient Air Quality Standards (NAAQS) as of 1987.

| Pollutant | Symbol | 40 CFR Section | | Primary Standards | Secondary Standards |
		Primary	Secondary		
Sulfur Dioxide	SO_2	50.4	50.5	80 $\mu g/m^3$ (0.03 ppm) annual arith. mean 365 $\mu g/m^3$ (0.14 ppm) 24-hr max.—once/yr	1300 $\mu g/m^3$ (0.5 ppm) 3-hr max.
Particulates	PM-10	50.6	50.6	50 $\mu g/m^3$ annual arith. mean 150 $\mu g/m^3$ 24-hr max.	(same)
Carbon Monoxide	CO	50.8		9 ppm 8-hr max.—once/yr 35 ppm 1-hr max.—once/yr	(same)
Oxidants	O_3	50.9	50.9	0.12 ppm (235 $\mu g/m^3$) as O_3 1-hr max.—once/yr	(same)
Nitrogen Oxides	NO_x	50.11	50.11	0.053 ppm (100 $\mu g/m^3$) as NO_2 annual arith. mean	(same)
Lead	Pb	50.12	50.12	1.5 $\mu g/m^3$ arith. mean per calendar quarter	(same)

1987 Amendments for the Criteria Pollutants to the values in Table 1.3. PM-10 represents respirable particles and the suspended particulate matter 10 μm or smaller in size. Ambient criteria pollutants represent concentrations of pollutants regardless of their origins. The primary standards are intended to protect health, and the secondary standards protect welfare.

Non criteria pollutants (NCPs) are those for which no ambient air quality standards exist. However, a number of NCPs have either emission limits or testing procedures. These include mercury, beryllium, asbestos, sulfuric acid, vinyl chloride, fluorine, hydrogen sulfide and reduced sulfur compounds [11]. In contrast, air toxics are materials normally regulated by federal laws, the control of which is implemented at the state level. These are discussed in the next several sections.

The Clean Air Act Amendments of 1977 (PL 95-95) developed provisions for the prevention of significant deterioration (PSD) of air quality, in order to protect areas with already clean air. The regulation required that every part of the country be assigned to one of three classes. All areas are initially designated Class II, except for a few Class I areas. Class I is the most restrictive with respect to the PSD and includes national wilderness areas and memorial parks over 5,000 acres and national parks over 6,000 acres. Class III allows most deterioration. Table 1.4 lists allowable increments (which may not exceed ambient air quality standards) for the various classes. Special agreements exist whereby increments can exceed those listed by 15% maximum in high terrain for up to 18 days and by 8% maximum in low terrain if the governor and federal land manager agree or if the President so decides. It could be noted that PL 95-95 also includes an opacity restriction for Class I areas which could be more restrictive than the other limits.

TABLE 1.4. Maximum Allowable Increments Permitted to U.S. Ambient Air Quality.[a]

Pollutant	Increments in μg/m³ to Class		
	I	II	III
Particulate annual mean	5	19	37
Particulate 24-hr max.	10	37	75
SO_2 annual mean	2	20	40
SO_2 24-hr max.	5	91	182
SO_2 3-hr max. (secondary std.)	25	512	700

[a]1977 Clean Air Amendments to prevent significant degradation to ambient air quality. Incremental increases noted here are permitted if ambient air quality standards are not exceeded.

1.6 TOXIC AIR POLLUTANTS

Over 5 billion pounds of chemicals are currently produced annually in the U.S. There are about 1,400 chemical accidents per year in which, on the average, 30 persons die, 300 are injured, and 44,000 persons are evacuated from their homes. Chlorine chemicals caused the most accidents and evacuations between 1981 and 1986.

The U.S. has the highest cancer rate per capita. It is believed that heavy industry accounts for about one quarter of these, and motor vehicles plus combustion sources account for over half of them. Vehicle emissions contribute a significant amount of air toxics in urban areas, especially lead, diesel particulates, benzene, and polycyclic organic materials (POM). Schroeder et al. [12] prepared a review of some toxic elements and compared values from the U.S., Canada, and Europe. These are summarized in Table 1.5. Note that in the high ranges the U.S. has higher concentrations for most categories.

Both "toxic" air pollutants and "hazardous" air pollutants (HAPs) are noted in various regulations. For all practical purposes, these terms are synonymous and the expression "toxic air pollutants" is used to relate to both. Section 112 of the Clean Air Act calls for the listing of HAPs and the enactment of National Emission Standards for Hazardous Air Pollutants (NESHAPs). As of the end of 1988, only eight chemicals or chemical categories were listed (see Section 1.5) and NESHAPs for only seven of these had been promulgated. Because of the disappointing effectiveness of this regulation in establishing health-based emission standards, new incentives were created to help control the routine releases of toxic chemicals. This is discussed in the following section.

1.6.1 TOXIC RELEASES AND REPORTING

Toxic air pollutants come from both routine emissions and from periodic accidental releases. The federal Emergency Planning and Community Right-to-Know Act of 1986 established a procedure at state and local levels to assist in the planning for chemical emergencies. This act is also known as Title III of SARA (Superfund Amendments and Reauthorization Act). One of the requirements outlined in Section 313, as published in the *Federal Register*, February 1988, is that certain businesses must report annually releases of toxic chemicals to the air, water, and land. Obviously, this could become a very significant air pollution control concern. This information is then made available to the public and hence Section 313 is known as the Community Right-to-Know Act. The first of these reports

TABLE 1.5. Concentration Ranges of Elements in Atmospheric Particulate Matter of Urban Areas (in ng/m³) (Data from Reference [12]).

Country	As	Cd	Ni	Pb	V	Zn	Co
U.S.	2-2320	0.2-7000	1-328	30-96,270	0.4-1460	15-8328	0.2-83
Canada	7.7-626	2-103	4-371	353-3416	10-130	55-1390	1-7.9
Europe	5-330	0.4-260	0.3-1400	10-9000	11-73	130-8340	0.4-18.3

Country	Cr	Cu	Fe	Hg	Mn	Se	Sb
U.S.	2.2-124	3-5140	130-13,800	0.58	4-488	0.2-30	0.5-171
Canada	4-26	17-500	700-5,400	<5	20-270	—	13-125
Europe	3.7-227	13-2760	294-13,000	0.1-5	23-850	0.01-127	2-470

TABLE 1.6. Standard Industrial Classification Industry
Groups Subject to Section 313 Reporting.

SIC No.	Industry Group	SIC No.	Industry Group
20	Food	30	Rubber and Plastics
21	Tobacco	31	Leather
22	Textiles	32	Stone, Clay, and Glass
23	Apparel	33	Primary Metals
24	Lumber and Wood	34	Fabricated Metals
25	Furniture	35	Machinery (Excluding Electrical)
26	Paper	36	Electrical and Electronic Equipment
27	Printing and Publishing	37	Transportation Equipment
28	Chemicals	38	Instruments
29	Petroleum and Coal	39	Miscellaneous Manufacturing

was due July 1, 1988 for the 1987 calendar year and subsequent annual
reports are due each July 1.

A facility must report if it conducts manufacturing operations included
as SIC (Standard Industrial Classification) codes 20 through 39 as listed in
Table 1.6, has ten or more full-time employees, and manufactures or processes over 25,000 lb/year of any of the listed toxic chemicals. Alternately,
if a facility uses any of the listed chemicals without incorporating them
into a product and if the use rate is 10,000 lb/yr or more, the facility must
report it. Currently, 308 individual toxic chemicals and 20 categories of
chemicals are listed under Section 313. These are given in Appendix C of
this book.

The major exemptions to Section 313 reporting are:

- if toxic chemical concentration is less than 1% of the mixture or,
  for OSHA defined carcinogens, if it is less than 0.1% (this is
  called the "de minimus" concentration, and above de minimus it is
  all considered as that toxic chemical)
- if the chemical is used as a structural component; for routine jani-
  torial or maintenance; for food, drugs, or other personal use; in
  vehicles; or in the laboratory
- the chemical is in process water as drawn from the environment
  or from municipal sources or in air used for compressed air or
  combustion

In each of these cases, no report need be filed.

1.6.2 SOURCES OF TOXIC AIR POLLUTANTS

Toxic (or hazardous) air pollution emissions can come from essentially all sources. It is considered that 50% comes from transportation-combustion sources, 25% from heavy industry, and 25% "other." The following is a partial listing of potential sources other than transportation sources.

Material Storage and Handling

Raw material delivery and storage
Final product and waste transport
Raw material, intermediate product, and final product storage tanks
Temporary storage (e.g., railroad cars)
Drum storage and handling/spillage
Tank car and truck cleaning
Weight scales and balances
Material storage piles

Process Sources

Chemical process vents: reactors, separation devices, product recovery
Product sterilizers (chambers, offgassing), product dryers
Cooling towers
Metal processing: mining, ore processing, primary and secondary smelters, foundries, coke ovens, machining, metal plating
Lime, cement kilns
Solvent-recovery stills, condensors

Equipment Leaks

Valves, flanges, pumps, compressors
Open-ended valves or lines
Sampling connections
Pressure-relief devices

Solvent Evaporation

Surface coating
Surface cleaning/degreasing
Printing
Graphic Arts
Adhesive application
Dry cleaning

Combustion Sources (not incl. home furnaces/fireplaces)

Boilers (coal, gas, oil)
Process heaters
Waste oil combustion
Incineration of organic solvent vapors
Municipal solid waste combustion

Waste Treatment and Disposal

Waste solvent reclamation
Landfills
Waste piles
Surface impoundments
Soil aeration/landfarming
Wastewater treatment
Groundwater cleanup (e.g., air stripping)

Table 1.7 shows typical sources of some gaseous toxic air pollutants and Table 1.8 is a partial breakdown of some expected sizes of toxic particulate pollutants with possible sources [13]. The distinction between fine and coarse particles occurs at a size of about 3 μm.

1.6.3 CONTROL OF AIR TOXICS

All conventional control devices are applicable to the control of toxic air pollutants. Some of the common conditions for control of organic vapor stack emissions are listed in Table 1.9. Table 1.10 shows some ways to control fugitive organic emissions with the suggested control effectiveness one could expect.

1.6.4 POLYCHLORINATED BIPHENYLS (PCBs)

Material as important and toxic as PCB must at least be mentioned. PCB is a dielectric, nonflammable, synthetic cooling oil used in electrical transformers especially in areas where fire safety is a main concern. These transformers are owned primarily by electric utilities, but are also found in industrial, governmental, and commercial facilities such as public buildings, hospitals, and schools. Over one billion pounds of PCBs were produced from 1929 to 1979 when the U.S. EPA banned their production.

Askarel is a generic name for an oil which contains from 40% to 80% PCBs. Askarel is sold under trade names such as Inerteen, Pyranol, Chlorextel, and Arochlor. In addition to being alleged carcinogens, PCBs

TABLE 1.7. Major Sources for Some Gaseous Toxic Air Pollutants (Reference [13]).

	Manufacturing	Transportation	Combustion
Halogenated HC's			
Chloroform	Fluorocarbon production		
Carbon Tetrachloride	Fluorocarbon production		
Methylene chloride	Solvent		
Ethylene di-bromide		Gasoline Pb scavenging	
Ethylene di-chloride	Vinyl chloride monomer production	Gasoline	
Perchloroethylene	Dry cleaning, solvent	Pb scavenging	
Trichloroethylene	Metal cleaning		
Aromatics			
Benzene		Auto exhaust— gasoline additive	
Toluene		Auto exhaust— gasoline additive	
Xylene		Auto exhaust— gasoline additive	
Ethyl benzene	Styrene production	Auto exhaust— gasoline additive	
Miscellaneous			
Formaldehyde*		Auto exhaust	Refuse incineration/ coal combustion
HCl			
Hg	Mercury processing		Sludge incineration/ coal combustion
Acrylonitrile	Acrylic fiber/resin production		

*Formaldehyde is also generated in photochemical smog.

29

TABLE 1.8. Classifications of Major Toxic Particulate Species According to Size and Source (Reference [13]).

Species Condition			Species Source	
Normally Fine	Normally Coarse	Variable	Fine Source	Coarse Source
NO_3^-			Aerosol, oil combustion	
$SO_4^=$			Fossil fuel, aerosol	
Elemental carbon			Fossil fuel	
Organic carbon			Aerosol	
Se			Coal combustion	
Pb			Auto incineration	Road dust
PAH (polycyclic aromatic HC)				
Aliphatic HC			Fossil fuel	
Carboxylic acids			Natural	
			Natural	

TABLE 1.8. (continued).

Species Condition			Species Source	
Normally Fine	Normally Coarse	Variable	Fine Source	Coarse Source
	Fe		Ferrometal production	Road dust, soil
	Al			Soil, road dust
	Mg			Road dust, soil
		Mn	Coal combustion	Road dust, soil
		Ni	Oil combustion	Soil
		Cu	Incineration	Non-ferrous metal production
		As	Fossil fuel / Pesticides	Soil
		Cd	Incineration, auto	Non-ferrous metal production
		Zn	Incineration	Metal production
		Cr	Fossil fuel / Incineration	Metallurgy / Soil
		V	Oil combustion	Soil

TABLE 1.9. Emission Stream Characteristics and Requirements for Organic Vapor Control.

Control	Organic Content, ppm	Heat Content, Btu/scf	Moisture Content, %	Flow Rate, scfm	Temp. Limits, °F
Thermal incinerator	>20 or <25% LEL			<100,000	
Catalytic incinerator	50–10,000 or <25% LEL			<100,000	
Flare		>300		<2,000,000	
Boiler/process heater		>150		steady flow	
Carbon adsorber*	1000–10,000 or <25% LEL			300–100,000	100–200
Absorber	250–10,000		50%	1000–100,000	
Condenser	<5,000			<2000	

*Molecular weight range limited from 45 to 130.
LEL stands for Lower Explosive Limit.

TABLE 1.10. Control of Organic Fugitive Emission Sources.

Emission Source	Control Technique	Control Effectiveness, %
Pumps	Monthly leak detection and repair	60
	Sealless pumps	100
	Dual mechanical seals	100
	Closed vent system	100
Valves		
Gas	Monthly leak detection and repair	75
	Diaphragm valves	100
Light liquid	Monthly leak detection and repair	50
	Diaphragm valves	100
Pressure relief	Rupture disk	100
valves	Closed vent system	100
Open-ended lines	Caps, plugs, blinds	100
Compressors	Mechanical seals with vented degassing reservoirs	100
Sampling connections	Closed purge sampling	100

are toxic and are regulated by the Toxic Substance Control Act (TSCA). They can be poisonous by the subcutaneous route and toxic by ingestion, leading to liver and pulmonary complications.

The EPA requires containers (including transformers) to be labeled if they contain over 50 ppm PCBs. At over 500 ppm the label states it is PCB-containing and at 50–500 ppm it is PCB-contaminated. In 1982, the EPA prohibited use of PCB-containing equipment in areas where there was an exposure risk to food or feed. In addition, weekly inspections are required to prevent exposure due to leaking. In 1985, PCB regulations were promulgated which prohibited use of > 500 ppm PCB oils in or near commercial buildings by October 1990. Procedures to accomplish this could be to remove the PCB oil from the container and replace it with another dielectric liquid. However, the PCBs must still be disposed of safely (e.g., by incinerating to 99.9999% destruction and with <0.001 g/kg of PCBs in the feed). Penalties for noncompliance with PCB regulations to date have ranged from fines of $25,000/day to incarceration.

1.6.5 TOXIC WASTE SITES

The number of hazardous waste sites on the EPA Superfund National

Priority List (NPL) in the U.S., including territories, is 1,173. As of early 1989, only 43 had been cleaned up. They range from two areas having none (Nevada and Washington, DC) to one with over 100 (New Jersey) as follows:

No. Toxic Waste Sites/ State	No. Areas	States (or Location)
0	2	Nevada, Washington, DC
1-10	18	Alaska, Hawaii, Oregon, Idaho, Montana, Wyoming, North Dakota, South Dakota, Nebraska, Arizona, New Mexico, Arkansas, Mississippi, West Virgnina, Maryland, Vermont, Maine, Rhode Island
11-20	11	Utah, Colorado, Kansas, Oklahoma, Louisiana, Alabama, Georgia, Tennessee, Kentucky, Connecticut, New Hampshire
21-30	8	Texas, Iowa, Missouri, North Carolina, South Carolina, Virginia, Delaware, Massachusetts
31-40	5	Minnesota, Wisconsin, Illinois, Indiana, Ohio
41-50	2	Washington, Florida
71-80	1	New York
81-90	2	California, Michigan
91-100	1	Pennsylvania
>100	1	New Jersey

1.7 EMISSION STANDARDS

Emission standards for some new plants have been established by the EPA in an effort to achieve the ambient air quality standards. These are called the New Source Performance Standards (NSPS) and are listed in Table 1.11. State and local regulations exist for new sources which are more stringent than the federal standards. Emissions from existing plants are limited by state regulations. They are required by the EPA as part of the State Implementation Plans (SIPs) to show how they will achieve the ambient air quality standards.

TABLE 1.11. Partial Summary of U.S. New Source Performance Standards.

Source	Particulate	Opacity	SO₂		NOₓ
Fossil fuel-fired steam generators >63 × 10⁶ kcal/hr (>250 × 10⁶ Btu/hr) heat input	0.054[a] (0.03b)	20% max. on 6 min av.			
			Max. of 1.4[a] (0.80[b]) & 90% control for >0.35[a] (>0.20[b])	←Gas→	0.36[a] (0.20[b])
			Max. of 1.4[a] (0.80[b]) & 90% control for >0.35[a] (>0.20[b])	−Oil→	0.54[a] (0.30[b])
			Coal cleaning credit w/ max. of 2.1[a] (1.2[b]) up to 90% control for ≥1.08[a] (≥0.60[b]); min. of 70% control for <1.08[a] (<0.60[b])	←Solid→	1.08[a] (0.60[b])
Coal preparation plants					
Thermal dryer	0.070[c] (0.031[d])	<20%			
Pneumatic cleaning	0.040[c] (0.018[d])	<10%			
Processing and conveying		<20%			
Nitric acid plants		<10%			1.5[e] of NO₂ (3f)
Sulfuric acid plants	0.075[e] of acid mist (0.15f)	<10%	2e (4)		
Petroleum refineries					
Some processes and including 0.05% CO	1 kg/1,000 kg coke (1 lb/1,000 lb)	<30%	0.230[c] (0.10[o])		
Asphalt concrete plants					
Some processes	0.090[c] (0.04[d])	<20%			
Portland cement plants					
Kiln	0.150[g] (0.30[h])	≤20%			
Clinker cooler	0.050[g] (0.10[h])	≤10%			
Other		<10%			

(continued)

35

TABLE 1.11. (continued).

Source	Particulate	Opacity	SO_2	NO_x
Basic oxygen furnace	0.050ᶜ (0.022ᵈ)			
Steel plants ore furnace	0.012ᶜ (0.0052ᵈ)	<3%		
Primary copper smelters				
Dryer	0.050ᶜ (0.022ᵈ)			
Roaster, others			0.065%	
Primary lead smelters				
Some processes	0.050ᶜ (0.022ᵈ)	20%	0.065%	
Primary zinc smelter	0.050ᶜ (0.022ᵈ)	20%		
Secondary brass and bronze				
Reverberatory furnace	0.050ᶜ (0.022ᵈ)	<20%		
Blast and electric furnace		<10%		
Secondary lead smelters				
Blast and reverberatory furnace	0.050ᶜ (0.022ᵈ)	<20%		
Pot furnace		<10%		
Incinerators	0.18ᶜ (0.08ᵈ) at 12% CO_2			

TABLE 1.11. (continued).

Source	Particulate	Opacity	SO$_2$	NO$_x$
Sewage treatment plants				
Incinerator	0.65 g/kg dry sludge (1.30 lb/ton)	<20%		
Phosphate fertilizer industry				
Wet phosphoric acid— 10[i]				
Superphosphoric acid—5[i]				
Diammonium phos- phate—30[i]				
Primary aluminum				
Soderberg plants— 100[j]				
Potroom prebake— 0.95[j]		<10%		
Anode bake—0.05[j]		<20%		

Key:
( ) = English units
a = g/10^6 cal
b = lb/10^6 Btu
c = g/dscm = grams per dry standard cubic meter
d = gr/dscf = grains per dry standard cubic foot
e = kg/metric ton 100% product
f = lb/ton 100% product
g = kg/metric ton feed
h = lb/ton feed
i = g fluoride/ton P$_2$O$_5$ feed
j = kg fluoride/ton product

Total emissions for the U.S. as estimated by the EPA [14] are given in Table 1.12. The EPA emission trends report [15] finds that from 1977 to 1986 ambient, total suspended particulates (TSP) decreased 23%; SO_2 decreased 32%; CO decreased 32%; NO_2 decreased 14% overall (there were rises and falls in levels during this period); O_3 decreased 21% (however, a calibration procedure change could make the decrease closer to 13%); and Pb decreased 87%. The large lead reduction is due to the decrease in lead antiknock compounds in gasoline. In 1986, it was reduced to 1.1 g Pb/gal and since January 1987, the limit has been 0.1 g/gal. During the same period, this report notes emissions changed: SO_2 decreased 21%; CO decreased 26%; NO_2 decreased 8%; and Pb decreased 84%. Controlled and uncontrolled emission factors are used in emission estimates. Values for these can be found in publications such as AP-42 as supplemented [16].

1.8 RISK ASSESSMENT

Standards are set based on atmospheric concentrations of air pollutants and their adverse effects on human life. Table 1.13 defines several indices for this as used by ACGIH (American Conference of Governmental Industrial Hygienists) and NIOSH (National Institute of Occupational Safety and Health). These ACGIH and NIOSH values were developed for workplace (8 hr) exposures. The TLV (Threshold Limit Value) approach is used in most states, as shown in Table 1.14. This approach is based on obtaining a TLV concentration for each air toxic pollutant. These concentrations are usually reduced to provide additional degrees of safety. The reduced TLV is then used as an ambient concentration which can then be related, using a dilution factor, to a stack emission concentration depending on the source, local meteorology, and terrain.

The procedure of assessing risks is extremely complex in view of the variety of persons in any country and the interacting related factors. Society assigns different levels of risk acceptability depending on whether individuals face the risks voluntarily or involuntarily. Voluntary risks are less tightly controlled. Benefits also influence risk assessment—for example, benefits are derived from living in an industrial society. Table 1.15 shows deaths resulting from exposure to different activities [17,18]. These risk values can differ significantly for smokers and non-smokers. The particulate and SO_2 criteria documents [19] show that smoking impairs long-term dust clearance from the lungs. Smokers had about 50% of the exposed dust remaining in their lungs twelve months after inhalation, whereas non-smokers had about 10%. The peanut butter risk in Table 1.15 is in regard to the aflatoxins from fungus growth on aged, improperly stored peanuts.

TABLE 1.12. Estimated Air Pollution Emissions in the United States in 1975 [14].

Source Category	Particulates	Sulfur Oxides	Nitrogen Oxides	Hydrocarbons	Carbon Monoxide	Total
			in 10⁶ ton/yr and (%)			
Transportation	1.3	0.8	10.7	11.7	77.4	101.9
	(0.6)	(0.4)	(5.3)	(5.8)	(38.3)	(50.4)
Fuel combustion, stationary	6.6	26.3	12.4	1.4	1.2	47.9
	(3.3)	(13.0)	(6.1)	(0.7)	(0.6)	(23.7)
Industrial	8.7	5.7	0.7	3.5	9.4	28.0
	(4.3)	(2.8)	(0.3)	(1.7)	(4.6)	(13.8)
Solid waste disposal	0.6	<0.1	0.2	0.9	3.3	5.0
	(0.3)	(~0)	(0.1)	(0.4)	(1.6)	(2.5)
Miscellaneous	0.8	0.1	0.2	13.4	4.9	19.4
	(0.4)	(~0)	(0.1)	(6.6)	(2.4)	(9.6)
Total	18.0	32.9	24.2	30.9	96.2	202.2
	(8.9)	(16.3)	(12.0)	(15.3)	(47.6)	(100.0)

TABLE 1.13. Guidelines for Assessing Adverse Effects.

PEL	Permissible Exposure Limit	Time-weighted averages and ceiling concentrations similar to (and in cases taken from) the threshold limit values published in 1968.	ACGIH
REL	Recommended Exposure Limit	Time-weighted averages and ceiling concentrations based on NIOSH evaluations.	NIOSH
IDLH	Immediately Dangerous to Life or Health	The maximum level from which a worker could escape without any escape-impairing symptoms or any irreversible health effects.	NIOSH
TLV	Threshold Limit Value	Any of the following:	
TLV-TWA	Threshold Limit Value— Time-Weighted Average	The time-weighted average concentration for a normal eight-hour workday and a forty-hour workweek, to which nearly all workers may be repeatedly exposed without adverse effect.	ACGIH
TLV-STEL	Threshold Limit Value— Short-Term Exposure Limit	A fifteen-minute time-weighted average exposure that should not be exceeded at any time during the work day.	ACGIH
TLV-C	Threshold Limit Value— Ceiling	The concentration that should not be exceeded even instantaneously.	AGCIH

TABLE 1.14. Determination of Acceptable Ambient
Concentration by State Agencies.

EPA Region	State	Factor Applied to TLV	Averaging Time
I	New Hampshire	1/420	annual
	Vermont	1/300	24-hour
II	New York	1/50, 1/300	annual
III	Virginia	1/20	hourly
	Maryland	1/300	24-hour
IV	Kentucky	1/42	hourly
	Alabama	1/40	hourly
V	Michigan	1/100	annual, 8-hour, hourly
	Illinois	1/30 to 1/300	24-hour
	Indiana	1/33 to 1/200	
	Wisconsin		24-hour
VI	Texas	1/100	
	Arkansas		24-hour, annual
	Louisiana	1/20	24-hour
VIII	Wyoming	1/42, 1/50, 1/30	annual, 24-hour, hourly
	Montana	1/42	annual
IX	Nevada	1/100	24-hour,-hourly
X	Oregon		annual, 24 hour, hourly

One can see from Table 1.14 that different states assess different levels of risks to their citizens. The state of Michigan uses a procedure that tries to assess a risk of one death in one million persons resulting from exposure to various atmospheric pollutants. Table 1.16 shows how this is resolved to ground-level concentrations (glc) of several materials based on available data and current interpretations of these data. Emissions of 361 acute toxic compounds are back-calculated by ambient dispersion modeling so that the glc are 1% of TLV. For the 24 known human carcinogens and the 47 suspected carcinogens, Michigan requires the best commercially available control technology without consideration of cost. This state further specifies maximum concentration and mass emissions rates for these.

1.9 INDOOR AIR QUALITY

Indoor air quality (IAQ) could be an entire book by itself and several excellent books, such as the one by Godish [20], are available. Attention to this potential problem is rapidly increasing. "Sick building" or "tight build-

TABLE 1.15. Risks Associated with Various Activities [17,18].

Activity	Lifetime (70-year) Risk per Million Population
Mining and quarrying	66,500
Construction	42,700
Mountain climbing	42,000
Agriculture	42,000
Police killed in line of duty	15,400
Motor vehicle accident (traveling)	13,900
Home accidents	7,700
Smoker plus asbestos	6,016
Manufacturing	5,740
Frequent airline traveler	3,500
Pedestrian hit by motor vehicle	2,940
Alcohol, light drinker	1,400
Smoking	1,226
Non-smoker plus asbestos	584
Eating peanut butter, four tablespoons per day	560
Electrocution	371
Tornado	42.0
Lightning	35.0
Living near a waste-to-energy plant	10.0
Earthquake (Southern California residents)	1.00
Drinking water containing trichloroethylene at maximum allowable EPA limit	0.14

TABLE 1.16. Ground Level Concentrations for 1 in a Million Risk for Potential Carcinogenic Pollutants.

Pollutants	Ground Level Concentrations, mg/m^3
Cadmium	54.3×10^{-5}
Chromium	8.5×10^{-5}
Arsenic	24.7×10^{-5}
Dioxins	9.8×10^{-7}
Furans	59.3×10^{-7}

ing" syndrome can cause problems for the occupants such as: eye irritation, sinus congestion, headache, fatigue, dizziness, nausea, sneezing, dermatitis, chest tightness, difficulty wearing contact lenses, aching joints, heartburn, drowsiness, and complaints of being too hot or too cold. Any of these can result because of poor indoor air quality, and are aggravated when people stay confined in the building for longer periods of time.

The major areas of concern are microbial contamination, formaldehyde, cigarette/cigar/pipe smoke, asbestos, and radon. "Legionnaire's Disease" is an infamous example of microbial contamination, where a microbe lived and bred in a hotel water-cooling tower. Asbestos is the most notorious building material contaminant. Radon comes from the natural decay of radium found in soil, rock, and groundwater, which releases cancer-causing alpha particles. In 1991, passive smoke was reported as the third leading cause of death due to indoor pollution and was banned in most work and public places.

The sources of indoor air contamination include both local indoor materials or activities, plus outside air pollution. The five major sources are operation of equipment and cleaning; building furnishings and construction materials; outside air pollutants; microbiological colonies; and smoke and gases from people.

Examples of specific emissions include:

> tetrachloroethylene from dry-cleaned clothes
> benzene from tobacco smoking
> solvents from paints, cleaners, cosmetics and sprays
> formaldehyde from carpets, wood panels, plywood
>   preservatives, dyes, binders, urea insulation,
>   and other materials
> hydrocarbons from adhesives, resins, insulation and
>   furnishings
> odors from cooking, people, outdoor activities, and
>   other sources
> ozone from motors and other electrical discharges
> CO and CO_2 from vehicles in garages or near air in-
>   takes
> fumes from stored chemicals
> asbestos from insulation
> radon from basement gas

The above are just a few of the concerns of IAQ. Many of the procedures used to make indoor air acceptable are direct applications or adaptations of air pollution control systems.

1.10 ATMOSPHERIC DISPERSION

Section 1.7 reports emission data and notes the reported improvements in the reduction of emissions to the atmosphere. Despite this, it is estimated [21] that over 33 million persons in the U.S. live in nonattainment areas, where overall air quality has actually deteriorated since 1981. In the U.S., as of the end of 1988, there were sixty-seven O_3 nonattainment areas and forty CO nonattainment areas.

Pollution control limits of 99.7% for particulates and 90% or more for gaseous pollutants are examples of specific levels of control as will be used for certain processes in the following chapter. However, there are emissions that escape control, as well as uncontrolled emission releases, which pass into the atmosphere. Natural removal mechanisms ultimately clear many of these pollutants from the atmosphere (by means such as producing acid rain). Immediately upon release, these uncontrolled pollutants are dispersed by the wind, but this is a dilution and not a control procedure. Stacks are useful in removing pollutants from the immediate vicinity of both animate and inanimate objects and in aiding the dilution procedure. Dispersion and stacks, although not control mechanisms, are important in describing what happens to emissions from air pollution control systems.

1.10.1 STACKS

The physical arrangement of a stack directs the exhausted pollutants away from potential receptors. In addition, a minimum exit velocity of about 20 m/sec (60 ft/sec) is desired to provide a greater effective height for the plume. It is not possible to obtain this rate of exit gas velocity by natural draft so forced draft fans must be used in all but the smaller commercial and residential stacks. There will obviously be increased friction losses as stack-gas velocities increase. At 20 m/sec, flow is highly turbulent. However, this reduces gas heat losses which are typically less than 2°C/m of stack height.

The concept of an acceptable stack height called "Good Engineering Practice" (GEP) has been developed by the EPA. In dispersion modeling studies, the effect of nearby structures must be considered if a stack height is less than GEP height. GEP height equals the height of the tallest obstruction within $5L$ of the stack plus $1.5L$ where L is the lesser of the height (or the projected width if greater) of the obstruction. However, if the proposed stack is greater than 65 meters high, then only a GEP height of 65 meters may be used in dispersion modeling analyses.

The following example problem shows how stack heat losses and pressure drops can be estimated.

Example Problem 1.1

A hospital stack is constructed for the boiler flue gases. The 2.14-m ID stack has a single flue and is nontapered except for the 15° truncated cone on the top 0.61 m of the 62.8-m high stack. This is so designed as to produce a 22-m/sec exit velocity. This double steel shell stack has an insulant factor (overall heat transfer coefficient based on stack ID) of 1,244 cal/(min m² °C). Gas enters the stack from a breeching 9 m above the base at a temperature of 250°C. Tests on this stack have shown that at 10°C ambient temperature, when the stack hot face is 500°C, the cold face is 90°C. Determine the flue gas temperature drop, the volumetric flowrate at inlet and outlet, and the stack pressure drop when the ambient temperature is 10°C.

Solution

This can be solved by a trial-and-error solution. Assume the gases cool 1.8°C/m. Therefore exit temperature would be

$$250°C - (1.8°C/m)(62.8 - 9 \text{ m}) \cong 153°C$$

Plotting a linear relationship for hot face-cold face temperatures from 500 and 90°C to 0 and 0°C indicates that when the inlet gas is 250°C, the cold face is 45°C and at the exit the cold face is 28°C when the ambient temperature is 10°C. The convective heat transfer equation shows that the heat loss, q, through the stack would be

$$q = UA\Delta T_{\text{ln}} \tag{1.22}$$

where U is overall heat transfer coefficient based on inside area, A is area and log mean temperature difference is

$$\Delta T_{\text{ln}} = \frac{\Delta T_{\text{bottom}} - \Delta T_{\text{top}}}{\ln \dfrac{\Delta T_{\text{bottom}}}{\Delta T_{\text{top}}}}$$

$$= \frac{(250 - 45) - (153 - 28)}{\ln \dfrac{205}{125}} = 162°C \tag{1.23}$$

Then

$$q = \left(\frac{1{,}244 \text{ cal}}{\text{min m}^2 \text{ °C}}\right) [(\pi 214 \text{ m})(62.8 - 9 \text{ m})](162\text{°C})$$

$$= 7.29 \times 10^7 \text{ cal/min}$$

All of this heat comes from the flue gases and equals

$$q = mC_p\Delta T \qquad (1.24)$$

where

m = mass of gas
C_p = specific heat at constant pressure of the gas

Assume C_p is constant and equal to 0.26 cal/(g K). Solving for m gives

$$m = \frac{7.29 \times 10^7 \text{ cal/min}}{\dfrac{(0.26 \text{ cal})}{\text{g K}}(250 - 153)} = 2.89 \times 10^6 \text{ g/min}$$

Under these conditions if the flue gas molecular weight is 29 and noting that the 15° top taper gives an exit ID of 1.83 m, the exit velocity is

$$v_e = \left(\frac{2.89 \times 10^6 \text{ g}}{\text{min}}\right)\left(\frac{\text{min}}{60 \text{ sec}}\right)\left(\frac{22.4 \text{ }\ell}{29 \text{ g}}\right)$$

$$\times \left(\frac{\text{m}^3}{1{,}000 \text{ }\ell}\right)\left(\frac{273 + 153}{273}\right)\left(\frac{4}{\pi 1.83^2 \text{ m}^2}\right)$$

$$= 22.1 \text{ m/sec}$$

The next iteration could assume a slightly lower exit temperature, but this is close enough to the specified exit velocity of 22 m/sec to allow us to proceed.

Volumetric flowrate out, Q_{out}, in actual cubic meters per minute is

$$Q_{out} = \left(\frac{22 \text{ m}}{\text{sec}}\right)\left(\frac{\pi 1.83^2 \text{ m}^2}{4}\right)\left(\frac{60 \text{ sec}}{\text{min}}\right) = 3{,}472 \text{ Am}^3/\text{min}$$

and

$$Q_{in} = (3,472)\left(\frac{273 + 250}{273 + 153}\right) = 4,262 \text{ Am}^3/\text{min}$$

The average flue gas velocity, v, in the stack is

$$v = \left(\frac{3,472 + 4,262 \text{ Am}^3/\text{min}}{2}\right)\left(\frac{\text{min}}{60 \text{ sec}}\right)\left(\frac{4}{\pi 2.14^2 \text{ m}^2}\right)$$

$$= 17.9 \text{ m/sec}$$

From fluid flow, the average stack flow Reynolds number, Re_f, is about

$$Re_f = \frac{Dv\varrho_g}{\mu_g}$$

$$= \frac{(214 \text{ cm})(1,790 \text{ cm/sec})(0.74 \times 10^{-3} \text{ g/cm}^3)}{(2.48 \times 10^{-4} \text{ g/cm sec})}$$

$$= 1.14 \times 10^6 \tag{1.25}$$

Tables of friction factors show that at this Re_f, the Fanning friction factor, f, for smooth new steel pipe is about 0.003 (for rough concrete f would be about 0.006 at this Re_f). Pressure drop due to flowing gas can be found using the Fanning equation

$$\Delta P = 2f\left(\frac{v^2 L\varrho_g}{D}\right) \tag{1.26}$$

where L is the length of system. Resolving the entrance elbow as equivalent length, standard tables give this as (32)(D) or (32)(214 cm) which equals 6,850 cm. Stack wall flow plus elbow entrance losses equal about

$$\Delta P = (2)(0.003)\left(\frac{\begin{array}{c}(1,790 \text{ cm/sec})^2[(6,280 - 900 \text{ cm}) + 6,850 \text{ cm}] \\ \times (0.74 \times 10^{-3} \text{ g/cm}^3)\end{array}}{(214 \text{ cm})}\right)$$

$$= 813 \text{ g/(cm sec}^2)$$

A g/(cm sec^2) is a dyne per cm^2 or a microbar. One microbar equals about 1.02 × 10^{-3} cm of water gauge pressure. [Note that if English units are used, g_c must be included in the denominator of both Equations (1.26) and (1.27)].

Enlargement pressure loss is

$$\Delta P = \frac{v^2 \varrho_g}{2}$$

$$= \frac{(1,790 \text{ cm/sec})^2 (0.74 \times 10^{-3} \text{ g/cm}^3)}{2} = 1,185 \text{ g/(cm sec}^2)$$

$$(1.27)$$

Total pressure loss due to the stack *only* is

$$813 + 1,185 = 1,998 \text{ microbars}$$

or

$$2.04 \text{ cm H}_2\text{O} \ (0.80 \text{ in H}_2\text{O})$$

Natural draft would tend to reduce this pressure, and can be approximated in cm of H$_2$O as $(100)(H')(\Delta \varrho_g)$ where H' is stack height in meters and $\Delta \varrho_g$ is difference in stack gas and ambient air densities. Using stack top conditions this would amount to about 2.5 cm H$_2$O draft.

1.10.1.1 Plume Rise

Plume height above the ground is the sum of stack height plus plume rise minus downwash. The exit velocity helps provide plume rise. This amount of plume rise equals the vertical distance the plume travels after leaving the top of a stack until it becomes horizontal. Meteorological factors of wind speed, wind shear and eddy currents influence amount of plume rise. Stack exit temperature and ambient temperature and pressure affect plume buoyancy. Plume composition and hence density is also significant.

Several techniques are used to predict amount of plume rise, ΔH. Holland's equation for a single stack (where ground-level concentrations to be estimated are 300 m or further away) gives

$$\Delta H = \frac{v_e D_s}{\bar{u}_s} \left[1.5 + 2.68 P_2 \frac{\Delta T D_s}{T_s} \right] K_s \qquad (1.28)$$

where

ΔH = plume rise above stack exit, m
v_e = stack exit velocity, m/sec
D_s = stack i.d., m
$\bar{u}_s$ = mean wind speed at stack height, m/sec [see Equation (1.29)]
P_2 = atmospheric pressure, bar (1 atm = 1.013 bar)
T_s = stack temperature, K
$\Delta T = T_s$ = ambient temperature, K
K_s = function of atmosphere stability (see Table 1.18)

This value predicts plume-rise conditions over a short time period, specifically for a 10-min averaging time.

Atmospheric stability relates to how much movement exists in the atmosphere. Unstable conditions result in a greater plume rise. Pasquill's stability categories [22] for 10-min sampling times are given in Table 1.17 as related to surface wind speed, $\bar{u}$, at 10-m height and either incoming solar radiation for daytime conditions or cloud cover for night. These stability classes are related to K_s and to temperature lapse rates (change in temperature with change in height, $\partial T / \partial z$) as noted in Table 1.18. Adiabatic atmospheric lapse rate is $-2°C/100$ m.

Mean wind speed at the top of the stack can be estimated using the relationship

$$\bar{u}_2 = \bar{u}_1 \left(\frac{z_2}{z_1} \right)^a \qquad (1.29)$$

where z is vertical height and subscripts 1 and 2 represent the lower and higher locations. Touma [23] shows that values of the exponent a vary with stability class. His values for this are given in Table 1.18.

Stack tip downwash can be approximated by $2D_s[(v_e/\bar{u}_s) - 1.5]$. A negative result is the amount the release height is decreased in meters. Any value > zero is ignored.

Example Problem 1.2

Determine the plume centerline height, H, during the daytime when the gas exit velocity, v_e, is 20 m/sec and temperature is 150°C. The 10-m wind speed, $\bar{v}$, is 4 m/sec and air temperature is 0°C. The stack height, H', is 50 m and inside diameter, D_s, is 3.5 m.

TABLE 1.17. Key to Stability Categories.

Surface Wind Speed at 10 m Height (m/sec)	Insolation Stability Classes				
	Day			Night	
	Strong[a]	Moderate[b]	Slight[c]	Thinly Overcast or > ½ Cloud[d]	Clear to < ½ Cloud
< 2 < (4.5 mph)	A	A–B	B	—	—
2–3 (4.5–6.7)	A–B	B	C	E	F
3–5 (6.7–11)	B	B–C	C	D	E
5–6 (11–13.5)	C	C–D	D	D	D
>6 (>13.5 mph)	C	D	D	D	D

[a]Sun > 60° above horizontal; sunny summer afternoon; very convective.
[b]Summer day with few broken clouds.
[c]Sunny fall afternoon; summer day with broken low clouds; or summer day with sun from 15–35°C with clear sky.
[d]Winter day.
Insolation = amount of sunshine.

TABLE 1.18. Atmospheric Stability Factors.

Atmospheric Stability	Stability Class	Stability Function, K_s	Avg. Values of Exponent, a[23]	$\partial T /\partial z$ in C/m[23]
Very unstable	A	1.2	0.141	−1.9
Unstable	B	1.2	0.176	−1.9 to −1.7
Slightly unstable	C	1.1	0.174	−1.7 to −1.5
Neutral	D	1.0	0.209	−1.5 to −0.5
Slightly stable	E	0.9	0.277	−0.5 to 1.5
Stable	F	0.8	0.414	1.5 to 4.0

Solution

Under these conditions it would be a winter day and insolation stability class from Table 1.17 is "D." These data specify neutral conditions so K_s in Table 1.18 equals 1.0.

Wind speed at stack exit is found from Equation (1.29) using the a value of 0.209 from Table 1.18.

$$\bar{u}_s = (4)\left(\frac{50}{10}\right)^{0.209} = 5.6 \text{ m/sec}$$

Plume rise using Equation (1.28) is

$$\Delta H = \frac{(20)(3.5)}{(5.6)}\left[1.5 + 2.68(1.013)\left(\frac{150 - 0}{150 + 273}\right)(3.5)\right](1.0)$$

$$= 61 \text{ m}$$

Answer: plume centerline height is $H = 50 + 61 = 111$ m.

1.10.2 ATMOSPHERIC DIFFUSION

Once a pollutant is emitted from a source, it is dispersed by atmospheric diffusion. Procedures for estimating downwind concentrations have been developed. Several of the more useful equations that have been developed are presented in this section. These techniques are for flat terrain with a single emission source and no local interfering structures. Using the Pasquill stability classes and empirical data by Gifford [24] for flat terrain, one can estimate downwind concentrations using variations of Sutton's [25] diffusion equation. These equations use quantities as defined in the following and as shown by Figure 1.8.

$C_{(x,y,z)}$ = concentration downwind from source at position x, y, z, g/m³; or in odor units (ou)

Q_s = source strength of emission point, g/sec; or in ou/sec

$\bar{u}_s$ = mean wind speed at emission source height, m/sec

H = effective centerline height of plume, m

x = distance downwind from source, m

y = horizontal distance crosswind from plume centerline, m

z = distance above ground, m

σ_y = horizontal standard deviation of plume from centerline, m

σ_z = vertical standard deviation of plume from centerline, m

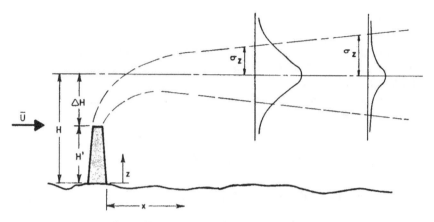

Figure 1.8 Terms defined for diffusion equations.

The values of σ_z and σ_y determined by Gifford are plotted against down-wind distance in Figures 1.9 and 1.10.

An estimation of *maximum* downwind ground-level concentration of gases from an elevated source can be obtained by finding

$$\sigma_z = \frac{H}{\sqrt{2}} \qquad (1.30)$$

Using this σ_z, find x from Figure 1.9 and find σ_y using Figure 1.10 for a specific stability class. Maximum concentration is then

$$C_{(x,0,0)\max} = \frac{0.117 Q_s}{\bar{u}_s \sigma_y \sigma_z} \qquad (1.31)$$

An alternate procedure for finding $C_{(x,0,0)\max}$ by direct calculation was developed by Ranchoux [26]. Using the same parameters, this becomes

$$C_{(x,0,0)\max} \frac{Q_s}{\bar{u}_s} \exp[b + c \ln H + d \ln^2 H + e \ln^3 H] \qquad (1.32)$$

where the constants b, c, d and e are given in Table 1.19 as a function of the stability class.

Downwind ground-level concentration at any location from an elevated

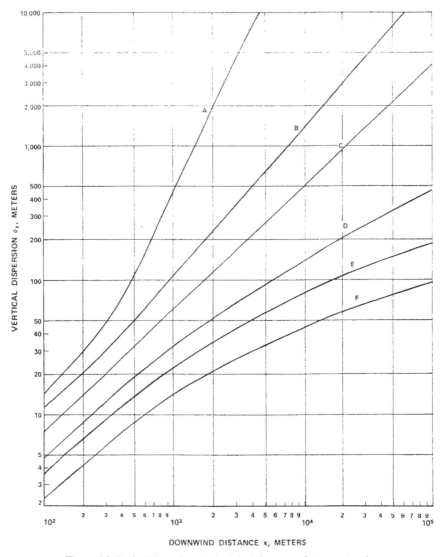

Figure 1.9 Vertical dispersion standard deviations over flat, open terrain.

53

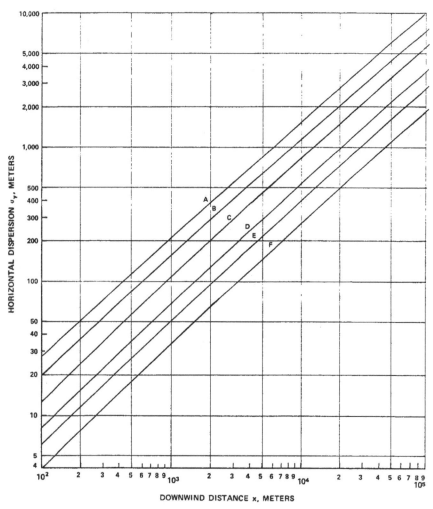

Figure 1.10 Horizontal dispersion standard deviations over flat, open terrain.

54

TABLE 1.19. Values of Constants for Equation (1.32).

Stability Class	Constants			
	b	c	d	e
A	− 1.0563	− 2.7153	0.1261	0
B	− 1.8060	− 2.1912	0.0389	0
C	− 1.9748	− 1.9980	0	0
D	− 2.5302	− 1.5610	− 0.0934	0
E	− 1.4496	− 2.5910	0.2181	− 0.0343
F	− 1.0488	− 3.2252	0.4977	− 0.0765

source of emission is

$$C_{(x,y,o)} = \frac{Q_s}{\pi \bar{u}_s \sigma_y \sigma_z} \exp - \left[\frac{y^2}{2\sigma_y^2} + \frac{H^2}{2\sigma_z^2} \right] \qquad (1.33a)$$

Remember that $H = 0$ for ground level source.

The Pasquill-Gifford sampling times as well as the Holland sampling times used to develop the values given here were for 10-min intervals. If it is necessary to convert from this averaging time to another, that can be accomplished using

$$C_{(x,y,z)2} = C_{(x,y,z)1} \left(\frac{t_1}{t_2} \right)^{0.165} \qquad (1.34)$$

where t is time in minutes. For example, $C_{(x,y,z)1}$ could be data calculated by the above procedures, then $t_1 = 10$ min.

Estimated centerline downwind concentration of *particulate matter* from an elevated source can be found using the tilted plume model

$$C_{(x,0,0)} = \frac{FQ_s}{\bar{u}_s \sigma_y \sigma_z} \exp - \left(\frac{A^2}{2\sigma_z^2} \right) \qquad (1.35)$$

where

F = weight fraction of effluent in a particular mass size range
$A = H - v_s x / (100 \bar{u}_s)$
v_s = terminal settling velocity for the average particle within each specific weight fraction [see Equations (3.10), (3.12) and (3.15)], cm/sec

For example, a weight fraction of 0.2 means five separate calculations are needed at each value of x to obtain the total ground-level concentration. Values of v_s are obtained for the average-size particle representing each of the separate weight fractions (e.g., if $F = 0.2$, there would be one v_s each for the particles >0.8, 0.6–0.8, 0.4–0.6, 0.2–0.4, and <0.2). If F were 0.1, ten calculations would be needed, each with an appropriate v_s. Procedures for obtaining v_s are given in Chapter 3.

Dispersion equations are only estimating techniques and depend on terrain, accuracy of dispersion parameters and techniques and accuracy by which the numerous factors are obtained. Variations of 2 to 5 times "true" values are typical.

Puff Release Model

In the special case of a puff release of pollutants (e.g., an explosion), where there is an "instantaneous" release over a less than 3 minute time period, Equation (1.33a) can be modified. Note that the initial center of gas moves downwind so a time, t, in seconds is needed. The puff dispersion equation is:

$$C_{(x,y,z)} = \frac{2Q_T}{(2\pi)^{3/2}\sigma_x\sigma_y\sigma_z} \exp - \left[\frac{1}{2}\left(\frac{x - \bar{u}t}{\sigma_x}\right)^2\right]$$

$$\times \exp - \left[\frac{1}{2}\left(\frac{H}{\sigma_z}\right)^2\right] \exp - \left[\frac{1}{2}\left(\frac{\gamma}{\sigma_y}\right)^2\right] \qquad (1.33b)$$

where for Equation (1.33b):

Q_T = total mass of release, g
σ_x = standard deviation of downwind dispersion, equals approximately σ_y

For ground level release and downwind centerline concentration on ground elevation, Equation (1.33b) becomes:

$$C_{(x,0,0)} = \frac{2Q_T}{(2\pi)^{3/2}\sigma_x\sigma_y\sigma_z} \exp - \left[\frac{1}{2}\left(\frac{x - \bar{u}t}{\sigma_x}\right)^2\right] \qquad (1.33c)$$

Example Problem 1.3

Assume that emissions are 80 g SO_2/sec and 12 g particulates/sec. Plume H is 134 m and corresponding $\bar{u}$ is 5.6 m/sec. Particulate size distribution is given by Figure 1.3. Find where the maximum downwind centerline ground-level concentrations occur under conditions of Problem 1.1 and estimate what they would give as 24-hr averages.

Solution

The location of maximum SO_2 and maximum particulate concentration will differ, with the particulate maximum being closer to the source. Location of maximum SO_2 concentration is estimated using Equation (1.30) and assuming flat terrain, no other contributing sources and no interferences. The value of σ_z at maximum SO_2 concentration is

$$\sigma_z = \frac{134}{\sqrt{2}} = 95 \text{ m}$$

From Figure 1.9, insolation "D," the maximum SO_2 concentration occurs at about 5,500 m.

Then from Figure 1.10, σ_y equals about 330 m and, using Equation (1.31), the maximum SO_2 concentration is

$$C_{(x,0,0)\,\text{max}} = \frac{(0.117)(80)}{(5.6)(330)(95)} = 5.33 \times 10^{-5} \text{ g/m}^3$$

An alternate solution using Equation (1.32) gives

$$C_{(x,0,0)\,\text{max}} = \frac{80}{5.6} \exp[-2.5302 - 1.5610 \ln 134$$

$$- 0.0934 (\ln 134)^2 + 0]$$

$$= 5.79 \times 10^{-5} \text{ g/m}^3$$

The 24-hr concentration can be predicted using Equation (1.34)

$$C_{(x,0,0)\,24\text{-hr max}} = 53.3 \ \mu\text{g/m}^3 \left(\frac{10}{24 \times 60}\right)^{0.165}$$

$$= 23.5 \ \mu\text{g/m}^3$$

or

$$\left(\frac{2.35 \times 10^{-5} \text{ g } SO_2}{\text{m}^3}\right)\left(\frac{22.4 \text{ l}}{64 \text{ g}}\right)\left(\frac{293}{273}\right)\left(\frac{\text{m}^3}{10^3 \text{ l}}\right)(10^6) = 0.0088 \text{ ppm}$$

The particulate concentration is found using Equation (1.35) and a trial-and-error solution. With F equal to 0.2, the average diameter of the largest group is about 100 μm from Figure 1.3. Terminal settling velocity of a 1 g/cc sphere this size would be about 24 cm/sec. Simple vector analysis assuming a constant wind speed and no wind eddy movement shows that this size particle should reach the ground at about

$$x = (134 \text{ m})\left(\frac{5.6 \text{ m/sec}}{0.24 \text{ m/sec}}\right) = 3,127 \text{ m}$$

Concentrations can be calculated for example at 3,000, 3,500, etc. meters and a plot drawn of concentration versus distance to obtain the location of the maximum. At 3,000 m, σ_y is 200 m and σ_z is 63 m. For the largest 0.2 fraction group

$$A = 134 - \frac{(24)(3,000)}{(100)(5.6)} = 5.43$$

and

$$C_{(x,0,0)} = \frac{(0.2)(12)}{(5.6)(200)(63)} \exp - \left[\frac{5.43^2}{2(63)^2}\right] = 3.4 \times 10^{-5} \text{ g/m}^3$$

These calculations are summarized in Table 1.20 and the total concentration at 3,000 m is about 52.1 μg/m^3.

The 24-hr average at this location is estimated as

$$C_{(x,0,0)24 \text{ hr}} = (52.1)\left(\frac{10}{24 \times 60}\right)^{0.165} = 23.0 \text{ } \mu\text{g/m}^3$$

This calculation can be refined using smaller variable values of F. (If F is varied during a calculation, results must be weighted appropriately before adding to find the total concentrations.) It is left for the reader to determine location and concentration of maximum ground-level particulates.

1.10.3 MODELING

Modeling is important for estimating downwind concentrations and for estimating emissions. The two basic types of modeling are dispersion modeling (as discussed in Section 1.10.2) and receptor modeling. Dispersion modeling with source sampling data helps specify where ambient monitors should be placed. In contrast, ambient monitoring data are used

TABLE 1.20. Example Problem 1.3—Calculations of Particulate Ground-Level Concentration at 3,000 m.

Mass Fraction (F)	Cumulative Mass Represented by Group	Average Diameter of Group (μm)	Terminal Settling Vel. of 1 g/cm^3 Spheres (cm/sec)	Calculated Factor (A)	Downwind Ground-Level $C_{(x,0,0)}$ of Fraction (μg/m^3)
0.2	0.8–1.0	~100	24	5.43	34.0
0.2	0.6–0.8	33	4	112.57	6.9
0.2	0.4–0.6	15	0.8	129.71	4.1
0.2	0.2–0.4	7	0.16	133.14	3.6
0.2	0–0.2	2	0.014	133.92	3.5
					Total $C_{(x,0,0)}$ = $\overline{52.1}$ μg/m^3

for receptor modeling to back-calculate source emissions. Dispersion modeling is frequently adequate, but there are cases (e.g., Connecticut) where receptor modeling for dioxins and furans is required in the permitting of incineration resource recovery installations. Computer dispersion models are commercially available and seventeen of these are listed by the U.S. EPA [27].

In order to conduct accurate receptor modeling, good ambient monitoring data must be available. Good data can only be obtained with proper use of quality control procedures, checks of instrument calibration, correct use of zero and span, checks of controls, and use of calibration and span gases that are traceable to the National Bureau of Standards gases or Standards Reference Material. Ambient monitoring is receiving renewed attention because of the emphasis on toxic/hazardous air pollutants and because of growing health concerns. Basically, ambient monitoring can be divided into three groups: criteria pollutant measurement, air toxics, and indoor air quality (IAQ). The criteria group includes particulates, SO_2, CO, NO_x, ozone, and lead, and is dictated by the Clean Air Act Amendments of 1987 and PSD. All PSD monitoring except TSP and lead require use of continuous emission instrumentation and a minimum of four months of data. A comparison of data requirements for dispersion and receptor models as listed by the EPA is shown in Table 1.21 [27].

With these data, receptor models are used to interpret the chemical fingerprints and the variability of the ambient pollutants, as well as to back-calculate and specify source emissions. Overall, five steps are used in this modeling procedure: development, verification, evaluation, application and validation. Part of this process can consist of a mass balance technique which requires knowledge of both ambient and source compositions. It is important to know the time sequence for both emissions and the monitoring data, and complementary methods should be used to insure reliability of conclusions. Receptor models were used during the development of SIPs to demonstrate the effectiveness of various control strategies in attaining the NAAQS. Table 1.22 is a summary of receptor model information [27].

1.10.4 EVACUATION

When releases become excessive, there can be immediate danger to persons and materials in the vicinity. Examples of this are accidents such as tank ruptures, truck wrecks, fires, and explosions. These activities require immediate isolation and evacuation of the area surrounding and downwind. The U.S. Department of Transportation (DOT) guidebook lists relevant distances for both "small" and "large" releases. Some of these are

TABLE 1.21. Comparisons of Data Requirements
for Dispersion and Receptor Models [27].

Data	Dispersion Models	Receptor Models	
		Qualitative Data	Quantitative Data
Emissions			
Emission rates	X		
Release heights	X		
Source configuration parameters	X		
Source locations	X	X	
Size distributions	X		X
Particle compositions	X		X
Meteorology			
Wind directions	X	X	
Wind speeds	X	X	
Atmospheric stabilities	X		
Mixing heights	X		
Temperatures	X	X	
Precipitation		X	
Synoptic scale pattern		X	
Ambient Concentrations			
Measured concentrations			X
Particle compositions			X
Particle sizes			X
Spatial & temporal variations		X	

given in Table 1.23. This type of listing supersedes the immediate need to perform dispersion calculations and comparisons with allowable toxic levels. Those can be done later, to provide supplemental data which would assure that safe conditions exist further downwind and within the listed distance.

1.11 ENVIRONMENTAL AUDITING

Environmental audit programs can help assure a company that it is in compliance with federal, state, and local regulations. The audit covers all of air, water, and waste emissions and generation. In addition, it includes contracted environmental companies who install or service equipment, or dispose of hazardous waste. These contractors must be responsible and

TABLE 1.22. **Summary of Receptor Model Information** [27].

Model	Features	Input Data	Output Data
1. Chemical mass balance	Quantitative estimates of source contributions	Ambient concentrations of particle properties from one or more samples	Source contributions	Fine and coarse samples with x-ray fluorescence (XRF), ion, and other analyses
2. Factor analysis	Has potential to identify presence of source contributions Requires multiple samples	Ambient concentrations of particle properties from >50 samples	Number of distinguishable source types Source compositions	Fine and coarse samples with x-ray fluorescence (XRF), ion, and other analyses
3. Multiple linear regression	One unique "tracer" property is chosen for each source considered; it should have negligible concentrations in other source emissions. Requires multiple samples	Number of source types Unique tracer property for each source type Ambient concentrations from >50 samples	Quantity of tracer property in source composition Source contributions	Fine and coarse samples with x-ray fluorescence (XRF), ion, and other analyses

TABLE 1.22. (continued).

Model	Features	Input Data	Output Data
4. Optical microscopy	Provides estimates of source contributions Provides size classification	Physical/optical properties of source particles Ambient samples (particles >1–2 μm only)	Source contributions Size classifications	Hi-vol. samples
5. Scanning electron microscopy	Provides estimates of source contributions Provides size classification	Elemental/physical properties of source particles Ambient samples	Source contributions Size classifications	Hi-vol. samples
6. X-ray diffraction	Provides estimates of source contributions of crystalline materials No sample preparation	X-ray diffraction properties of crystalline source particles Ambient samples	Source contributions	Hi-vol. samples

TABLE 1.23. US DOT Table of Isolation & Evacuation Distances.

Name of Material Spilling or Leaking (ID No.)	INITIAL ISOLATION Spill or Leak from (Drum, Smaller Container, or Small Leak from Tank) Isolate in All Directions feet	INITIAL EVACUATION Large Spill from a Tank (or from Many Containers, Drums, etc.) First Isolate in All Directions, feet	Then Evacuate in a Downwind Direction Width, miles	Then Evacuate in a Downwind Direction Length, miles
Acrolein (1092)	550	1140	3.0	4.7
Acrylonitrile (1093)	30	60	0.1	0.2
Ammonia, anhydrous (1005)	100	200	0.4	0.7
Ammonia solution, not less than 44% (2073)	100	200	0.4	0.7
Boron trifluoride (1008)	320	670	1.7	2.6
Bromine (1744)	300	620	1.5	2.4
Carbon bisulfide (1131) Carbon disulfide (1131)	30	70	0.2	0.2
Chlorine (1017)	250	520	1.3	2.0
Dimethylamine, anhydrous (1032)	80	170	0.4	0.6
Dimethyl sulfate (1595)	80	170	1.4	2.2
Epichlorohydrin (2023)	40	80	0.2	0.3

TABLE 1.23. (continued).

Name of Material Spilling or Leaking (ID No.)	INITIAL ISOLATION Spill or Leak from (Drum, Smaller Container, or Small Leak from Tank) Isolate in All Directions feet	INITIAL EVACUATION Large Spill from a Tank (or from Many Containers, Drums, etc.) First Isolate in All Directions, feet	Then Evacuate in a Downwind Direction Width, miles	Then Evacuate in a Downwind Direction Length, miles
Ethylene imine (1185)	270	570	1.4	2.2
Ethylene oxide (1040)	40	70	0.2	0.2
Fluorine, liquid (1045)	460	880	2.5	3.9
Hydrochloric acid, anhydrous (1050)	190	450	1.0	1.4
Hydrogen chloride, anhydrous (1050)				
Hydrogen chloride, refrigerated liquid (2186)				
Hydrocyanic acid (1051)	90	190	0.5	0.7
Hydrogen cyanide, anhydrous (1051)				
Hydrofluoric acid (1790)	240	490	1.2	1.8
Hydrogen fluoride, anhydrous (1052)				
Hydrogen sulfide (1053)	120	240	0.6	0.9

(continued)

TABLE 1.23. (continued).

Name of Material Spilling or Leaking (ID No.)	INITIAL ISOLATION Spill or Leak from (Drum, Smaller Container, or Small Leak from Tank) Isolate in All Directions feet	INITIAL EVACUATION Large Spill from a Tank (or from Many Containers, Drums, etc.) First Isolate in All Directions, feet	Then Evacuate in a Downwind Direction Width, miles	Then Evacuate in a Downwind Direction Length, miles
Methylamine, anhydrous (1061)	110	220	0.5	0.8
Monomethylamine, anhydrous (1061)				
Methyl bromide (1062)	50	90	0.2	0.3
Methyl chloride (1063)	30	60	0.1	0.2
Methyl mercaptan (1064)	370	770	1.9	3.0
Methyl sulfate (1595)	80	170	1.4	2.2
Nitric acid, fuming	100	210	0.5	0.7
Nitric acid, red fuming (2032)				
Nitric oxide (1660)	110	220	0.5	0.8
Nitrogen dioxide (1067)				
Nitrogen tetroxide and mixtures (1067)				

TABLE 1.23. (continued).

Name of Material Spilling or Leaking (ID No.)	INITIAL ISOLATION Spill or Leak from (Drum, Smaller Container, or Small Leak from Tank) Isolate in All Directions feet	INITIAL EVACUATION Large Spill from a Tank (or from Many Containers, Drums, etc.) First Isolate in All Directions, feet	Then Evacuate in a Downwind Direction Width, miles	Then Evacuate in a Downwind Direction Length, miles
Oleum (1831)	280	580	1.5	2.2
Perchloromethylmercaptan (1670)	220	450	1.1	1.6
Phosgene (1076)	600	1250	3.3	5.2
Phosphorus trichloride (1809)	110	220	0.5	0.8
Pyrosulfuric acid (1831)	280	580	1.5	2.2
Sulfur dioxide (1079)	100	220	0.5	0.8
Sulfuric acid, fuming (1831)	280	580	1.5	2.2
Sulfuric anhydride (1829)				
Sulfur trioxide (1829)				
Titanium tetrachloride (1838)	30	60	0.2	0.2
Trimethylamine anhydrous (1083)	90	170	0.4	0.8

67

financially solvent to help prevent their deficiencies from falling back to the audited company. The audits must identify all deficiencies and nonconformities in addition to recommending corrective action procedures.

Many pollution-control wastes and emissions (e.g., those from incineration) may be regulated under the Resource Conservation and Recovery Act (RCRA). This act specifies that the generator is responsible for hazardous wastes from cradle to grave. This means that even though a company may pay someone to dispose of its waste, the company still remains responsible for it. Liability would only end if the waste material were used as a feed chemical to produce another chemical product or if the waste were chemically converted to a nonhazardous form.

Audits should evaluate

- compliance and adequacy of corporate policies and procedures
- internal management controls of all environmental contractors
- public and regulatory perceptions of all organizations involved
- equipment design and maintenance and operation procedures
- exposure risks from all emissions and disposed-of wastes
- future viability of environmental contractors from a financial perspective

These requirements imply that the composition of audit teams is critical to the successful completion of the assignment. The team must include persons knowledgeable with the regulations; at senior staff level; familiar with operating limitations of the equipment and the processes; capable of good communications; independent to function freely; and knowledgeable in legal contracts.

Files must be checked to assure that proper procedures are being followed and that records are kept as required for things such as

- manifests
- training
- contracts
- system performance and reliability
- third party contracts (i.e., subcontracts)
- laboratory analyses and procedures
- ultimate disposal

This includes the requirement that the audit team check the files of the regulatory/permitting agencies for their records on

- routine and special instructions
- reports
- construction and operating permits
- all permit applications

- citizen complaints
- enforcement actions
- reports
- risk assessment data
- public hearing records
- other information such as monitoring data

1.12 CHAPTER PROBLEMS

1.12.1 PARTICULATE MATTER

(Probability graph paper should be used for this problem.) A certain particulate emission is known to contain normally distributed spherical particulate matter with a $\bar{d}_M$ of 5.6 μm and σ_g of 2.3. Assume a control device could be built to remove 99% of the >3 μm particles, 97.5% of the 2–3 μm particles, 92% of the 1–2 μm particles, 85% of the 0.5–1 μm particles, and 15% of the <0.5 μm particles. Density of the particles is 1.5 g/cm^3.

a. Determine the new $\bar{d}_M$ and σ_g.
b. Estimate the original $\bar{d}_N$ and draw the log probability curve for distribution by number.
c. Approximate the control device's overall efficiency by mass.

1.12.2 FLUE GAS

A flue gas from a coal-fired boiler is being tested. This gas at 200°C contains 8% water, 12% CO_2, 5% O_2, and the balance nitrogen. Determine:

a. Average molecular weight of wet gas and of dry gas
b. Acoustic velocity (assume ideal gas)
c. Ratio of actual density to density at SC
d. Ratio of actual viscosity to viscosity at SC at 1 atmospheric pressure (assume air)

1.12.3 STACKS

The flue gas from problem #2 enters a 50-m-high rough concrete chimney at 200°C. During the fall (average temp. = 20°C), the cold face of the stack is 40°C when the hot face is 200°C, and the stack overall heat transfer coefficient based on inside area is 2,200 cal/(min m^2°C). An exit velocity of 20 m/sec is desired from the 3 meter ID stack exit. Barometric pressure is 29.92″ Hg.

Find:

a. Flue gas exit temperature
b. Gas flow rate in Nm^3/hr
c. Stack draft in cm of H_2O
d. Stack pressure loss in cm of H_2O

1.12.4 DISPERSION

Flue gases are released at a rate of 2.15 × 10^6 Nm^3/hr from a 7.2-m ID stack 65-m high at a temperature of 150°C. Assume it is a spring day with ambient temperatures of 25°C and with a 10-m wind speed of 2.5 m/sec NW. Concentration of SO_2 in the flue gases is 2,800 ppm.

Calculate:

a. Plume rise
b. Location of maximum ground-level concentration
c. Maximum ground-level SO_2 concentration in ppm
d. Compare this concentration to the allowable 24-hr max

REFERENCES

1 Rall, D. P. 1978. "Report of the Committee on Health and Environment Effects of Increased Coal Utilization," *Federal Register*, 43(10):2230–2240.

2 Rotty, R. M. and G. Marland. 1980. "Constraints on Carbon Dioxide Production from Fossil Fuel Use," US DOE Institute for Energy Analysis, Oak Ridge Associated Universities Research Memorandum ORAU/IEA-80-9(M).

3 Pettyjohn, E. S. and E. B. Christianson. 1948. "Effects of Particle Shape on Free Settling Rates of Isometric Particles," *Chem. Eng. Prog.*, 44(2):157–172.

4 Heiss, J. and J. Coull. 1952. In *Chem. Eng. Prog.*, 48:133.

5 Fuchs, N. A. 1964. *The Mechanics of Aerosols*. Elmsford, NY: Pergamon Press, Inc., p. 41.

6 Orr, C. 1966. *Particle Technology*. New York: MacMillan Publishing Co., Inc.

7 Hesketh, H. E. 1986. *Fine Particles in Gaseous Media*. Lewis Publishers, Inc.

8 Hougen, O. A., K. M. Watson, and R. A. Ragatz. 1965. *Chemical Process Principles, Part II—Thermodynamics*. New York: John Wiley & Sons, Inc.

9 Smith, J. M. and H. C. Van Ness. 1959. *Introduction of Chemical Engineering Thermodynamics, 2nd ed.* New York: McGraw-Hill Book Company.

10 Glasstone, S. 1946. *Textbook of Physical Chemistry*. New York: Van Nostrand Reinhold Company.

11 1988. "Ambient Monitoring—Noncriteria Pollutants and Air Toxics," *The Entropy Quarterly*, VIII(3).

12 Schroeder, W. H. et al. 1987. "Toxic Trace Elements Associated with Airborne Particulate Matter: A Review," *JAPCA*, 37(11):1267–1279.

13 1987. "Evaluation of Control Techniques for Hazardous Air Pollutants," EPA-650/6-011, A&B.

14 1976. "National Air Quality and Emissions Trends Report, 1975," Preliminary data, EPA-450/1-76-002.

15 1988. "National Air Quality and Emissions Trends Report, 1986," US EPA, EPA-450/4-88-001.

16 1984. "Compilation of Air Pollutant Emission Factors," As Appended, AP-42, US EPA.

17 1986. *Environmental Defense Fund Newsletter*, XVII(4):4.

18 Michaels, R. A. 1988. "Health Risk Assessment: WTE vs. Peanut Butter," *Solid Waste and Power*, II(5):22–27.

19 1980. "Air Quality Criteria for Particulate Matter and SO_2, External Draft #2," US EPA, Vol. II, Chap. 5.

20 Godish, T. 1989. Indoor Air Quality book w/Lewis Publishing.

21 Brown, G. A., J. J. Cramer, and D. Samela. 1988. "The Impact of the Proposed Clean Air Act Amendments," *CEP*, 84(12):41–49.

22 Pasquill, F. 1961. "The Estimation of Dispersion of Windborne Material," *Meteorol.*, 90(1063):63.

23 Touma, J. S. 1977. "Dependence of the Wind Profile Power Law on Stability for Various Locations," *J. Air Poll. Control Assoc.*, 27(9):863–866.

24 Gifford, F. A. 1961. "Uses of Routine Meteorological Observations for Estimations of Atmospheric Dispersion," *Nuclear Safety*, 2(4):47–51.

25 Sutton, O. G. 1932. "A Theory of Eddy Diffusion in the Atmosphere," *Proc. Roy. Soc.*, 135:145–165.

26 Ranchoux, R. J. P. 1976. "Determination of Maximum Ground Level Concentration," *J. Air Poll. Control Assoc.*, 26(11):1088, 1089.

27 1984. "Receptor Model Technical Series, Vol. V," EPA-450/4-84-020.

CHAPTER 2

Basic Energy Sources, Conservation Processes, and Emission Products

2.1 ENERGY AND CONSERVATION

CONCERNS about waste disposal have risen in recent years to the tops of the agendas of scientists everywhere, eclipsing the old environmental issues of energy use and conservation. Available waste-disposal space is rapidly being consumed, and it is becoming clear that every industrial project must now pay serious attention to questions of disposal techniques and procedures. Minimization of waste generation, recovery of energy from waste, optimizing and/or altering processes, changing the types of wastes produced—all of these are important both in protecting the environment and in operating a facility to provide an acceptable economic return.

The sun is the primary source of energy for the earth, and anything humans do to affect atmospheric retention of sunlight can have very adverse results. In addition to expending dwindling fossil fuel reserves, combustion processes can add pollutants, CO_2, water vapor, and heat to the atmosphere, and can also produce a residue which must be disposed of in concentrated form. In the past, these environmental costs were tolerated in the interest of producing useful energy. However, it is becoming increasingly clear that the presence of these emissions in the atmosphere can result indirectly in a greenhouse effect, with a subsequent heating-up (when sunlight is trapped more than normal) or cooling-down (when sunlight is released more than normal) of the Earth. These emissions also exacerbate the problem of acid rain, as was discussed in Chapter 1. The waste residues disposed of may produce leachate containing acids, heavy metals, or other substances which must be controlled. The energy produced is needed, yet the disadvantages associated with combustion processes must be minimized—and in a way which allows costs to be optimized. It is to this end that this book is directed.

Annual energy consumption in the U.S. in 1990 is estimated to be about 3.3×10^{19} calories (1.31×10^{17} Btu) and, as noted in Chapter 1, emissions due to combustion of fossil fuels are expected to increase for about forty-five years after that. Transportation, electric utilities, and industry, in order of energy expenditure, account for about 80% of the total energy consumption. About half of this energy is dissipated to the environment and is not utilized in any way. Thermal cycle limitations and mechanical losses result in rejection of about 75% of the transportation energy input and about 65% of the utility energy input. Even when this energy to be used is available as unit energy (work), mechanical losses account for the second highest energy rejection mechanism and amount to about 25% of the industrial energy waste due to motion such as friction, wear, and seal leakage.

The Electric Power Research Institute (EPRI) is a research and development arm of the electric power industry. EPRI recently correlated electricity generation output in the U.S. to real Gross National Product (GNP). This includes generation from electric utilities, industry, and other non-utility sources. Electric output in 10^9 KWh is essentially equal to 2.28 times GNP minus 850. In the U.S., energy consumption is about:

residential/commercial	34%
transportation	26%
industry	32%
others	8%

Energy in the U.S. is supplied by oil (48%), gas (25%), coal (19%), nuclear (3%) and others (4%), e.g., hydro, geothermal, solar, and waste. Waste-to-energy should increase in the future. EPRI [1] estimated the 1980 U.S. industrial energy consumption to be as shown in Table 2.1. Note that most of the electricity is produced by fossil fuels also.

Table 1.12 shows that the greatest amount of air pollution emissions come from transportation (about 50%), stationary fuel combustion which is primarily the electric utilities (about 24%) and industry (about 14%). This includes both controlled and uncontrolled emissions. This chapter discusses, as related to air pollution control, the resources primarily used as sources of energy and the related combustion processes which are most responsible for the largest energy consumption and air pollution emissions. Disposal of solid wastes and other important processes are included.

TABLE 2.1. Annual U.S. Industrial Energy
Consumption in 10^{15} Btu for 1980 [1].

	Electricity	Natural Gas	Oil & LPG	Coal & Coke	Total
Primary metals	1.70	0.95	1.85	0.45	3.30
Chemicals	1.40	1.50	0.30	0.25	3.45
Fabricated metals & machines	0.60	0.30	0.10	0.05	1.05
Paper	0.55	0.40	0.35	0.25	1.55
Food	0.45	0.40	0.20	0.10	1.25
Petroleum	0.30	0.95	0.05	0.05	1.35
Stone, clay, glass	0.30	0.35	0.25	0.30	1.20
Transportation equipment	0.30	0.15	0.05	0.55	1.05

2.2 COMBUSTION

Since early school days we have known that combustion requires a fuel–oxygen mixture at or above kindling temperature. However, there are additional constraints for *good* combustion in our energy/pollution-conscious environment. The three T's of good combustion are time, temperature, and turbulence. Combustion efficiency requires proper application of all three factors. However, it is approximately a function of: the cube of system turbulence; the square of the temperature; and directly related to time. It is the combination of all three that establishes the efficiency for a given fuel. One of these can change if compensated for by the other two so that combustion efficiency could remain the same for various combinations of conditions.

Pyrolysis (combustion under conditions of substoichiometric oxygen) and other forms of incomplete combustion do take place and serve desirable functions such as producing a more usable fuel. For this discussion, only complete combustion will be considered and thus proper design and operation of combustion facilities will produce the products CO_2 and water.

Combustion is a chemical reaction and all reactions are reversible to some extent; no reaction ever goes to absolute completion. A stoichiometric combustion equation has just the amounts of reactants required by the balanced chemical equation for theoretical complete combustion. A reaction can be made more complete by increasing the concentration of the reactants, so in combustion operations it is common to use more than the theoretical stoichiometric amount of oxygen. Combustion of gas, oil, and coal fuels having a generalized formula C_mH_n have the stoichiometric combustion equation with oxygen represented as

$$C_mH_n + \left(m + \frac{n}{4}\right)O_2 \rightarrow mCO_2 + (n/2)H_2O \qquad (2.1)$$

In actual combustion, many different steps may occur before reaching the final state of Equation (2.1). If moisture is present it is heated and converted to water vapor. Liquid fuels are atomized to provide large surface area for heating and evaporation. When solid fuels enter the combustion system, they are heated to drive off moisture and volatile matter. The volatiles ignite and burn with a hot luminous flame. Then the fixed carbon residue burns without a visible flame. All of this can occur in an oxidizing or reducing environment. The final state must be an oxidizing environment to achieve the results in Equation (2.1).

Liquid and solid particles attain surface temperatures of about 2000°F before significant pyrolysis or ignition occurs. Ignition most commonly occurs on the surface of solid particles and precedes rapid devolatilization. Combustion then takes place with the gaseous volatiles. The most common sequence of combustion is: heating and minor devolatilization; ignition; major devolatilization; and burning of the carbon residue. Solid fuels can be classified according to mode of ignition as surface ignition only, gas-phase ignition only, and hybrid fuels. Ignitability of solid fuels is determined by surface properties of both the original and the partially charred (carbonized) particles, not by the volatile content.

2.2.1 FLUE GAS COMPOSITION

An example can be given for the combustion of hydrocarbons such as the fossil fuels (gas, oil, and coal). Assume a properly designed and operated combustion system. The theoretical stoichiometric complete combustion equation of the simplest hydrocarbon (natural gas, methane) with air is

$$CH_4 + 2O_2 + 7.52N_2 \rightarrow CO_2 + 2H_2O + 7.52N_2 \qquad (2.2)$$

Air is considered to be 79% N_2 and 21% O_2. Mole fraction equals volume fraction, so the moles of N_2 equals (2 moles O_2) (79 moles N_2/21 moles O_2) or 7.52 moles. The presence of nitrogen and oxygen at elevated temperatures results in the formation of nitric oxide (NO) which converts to nitrogen dioxide (NO_2) in the atmosphere. The amount of NO is very small compared to the SO_2, H_2O, and N_2. It will be discussed with the specific fuels. The composition of the flue gas produced by Equation (2.2), if all components are gases, can be determined using the gas laws in Section 1.3.1.

$$\left[\frac{1 \text{ mole } CO_2}{(1 + 2 + 7.52) \text{ mole total}}\right](100) = \left(\frac{1}{10.52}\right)(100)$$

$$= 9.5 \text{ mole or volume } \% \ CO_2$$

$$\left(\frac{2}{10.52}\right)(100) = 19.0\% \ H_2O$$

$$\left(\frac{7.52}{10.52}\right)(100) = 71.5\% \ N_2$$

This is called a wet gas analysis. The molecular weight of this flue gas is then 27.6 g/mole (see Section 1.3.2.2) and the density at SC is

$$\left(\frac{27.6 \text{ g}}{22.4 \text{ l}}\right)\left(\frac{l}{1,000 \text{ cm}^3}\right)\left(\frac{273\,°C}{273 + 20\,°C}\right) = 1.15 \times 10^{-3} \text{ g/cm}^3$$

Continuing the methane combustion example, 5% excess air would be 105% theoretical air and the resultant balanced chemical equation would show (1.05)(2) moles of oxygen and (1.05)(7.52) moles of nitrogen

$$CH_4 + 2.1O_2 + 7.9N_2 \rightarrow CO_2 + 2H_2O + 0.1O_2 + 7.9N_2 \qquad (2.3)$$

The complete combustion example shows, as before, no CH_4 or CO in the products but the excess O_2 now is present. Theoretical composition of the combustion gas is 9.09% CO_2, 18.18% H_2O, 0.91% O_2, and 71.82% N_2. Dry gas composition is

$$\left(\frac{9.09}{100 - 18.18}\right)(100) = \left(\frac{9.09}{81.82}\right)(100) = 11.11\% \ CO_2$$

$$\left(\frac{0.91}{81.82}\right)(100) = 1.11\% \ O_2$$

$$\left(\frac{71.82}{81.82}\right)(100) = 87.78\% \ N_2$$

This type of analysis is obtained directly by measuring the actual flue gas composition with an Orsat gas analyzer.

This example can be expanded for other values of theoretical air. Figure 2.1 is produced in this manner. Note that at less than 100% air, any or all of CH_4, CO, and C must be present. Particulate matter as well as gases can be present; however, this is not significant for CH_4 combustion as only near and greater than 100% theoretical air is being considered. Nitrogen is the largest component of typical flue gases and is the major factor responsible for the large size of pollution control equipment. It also accounts for most of the gas pumping costs and for the greatest energy (heat) loss. Use of oxygen instead of air for combustion would reduce these wastes but the cost and energy of producing oxygen must be balanced against the savings and the losses in turbulence and dilution.

Another relationship shown by Figure 2.1 is the dilution effect. More fuel is actually converted to CO_2 and water as increased excess air is used but, because of dilution, the percentage of these products decreases.

2.2.2 EXCESS AIR

Excess air (EA) is mentioned several times in the discussion related to

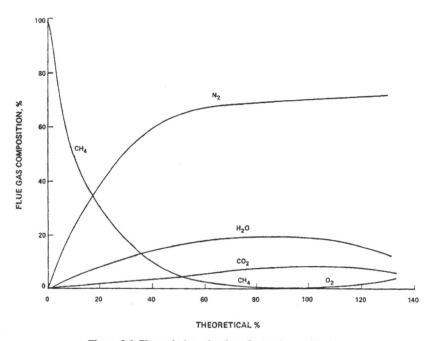

Figure 2.1 Theoretical combustion of natural gas with air.

combustion. The percent excess air can be determined by any of the following three equivalent relationships.

$$\% \ EA \ = \ \frac{Total \ Air \ - \ Theoretical \ Air}{Theoretical \ Air} \ (100)$$

$$= \ \frac{Excess \ Air}{Theoretical \ Air} \ (100)$$

$$= \ \frac{Excess \ Air}{Total \ Air \ - \ Excess \ Air} \ (100)$$

Theoretical air is that amount required to stoichiometrically convert all combustible species (mainly C, H, and S) to complete normal products of combustion (i.e., CO_2, H_2O, and SO_2). These relationships are stated as mole ratios of air which equal volume ratios. Although moles of air are specified in each term, the moles of oxygen can be used instead if applied consistently. Thus, it is proper to call this relationship either "excess air" or "excess oxygen." However, excess oxygen is also the percentage of unburned oxygen in the flue gases and can lead to confusion of terms if care is not exercised. This is demonstrated in the oil combustion Example Problem 2.2.

Various federal EPA and state regulations give procedures for calculating excess air based on dry gas (Orsat type) analyses. Two of these are

$$\% \ EA \ = \ \frac{\left(O_2 \ - \ \dfrac{CO}{2}\right) 100}{0.266 N_2 \ - \ \left(O_2 \ - \ \dfrac{CO}{2}\right)}$$

or as a crude approximation

$$= \ \frac{95 \ O_2}{21 \ - \ O_2}$$

where O_2, N_2, and CO represent dry gas values of oxygen, nitrogen, and carbon monoxide in percent. In most cases with good combustion, the amount of CO is negligible.

The relative magnitudes of oxygen and carbon dioxide as obtained in a dry gas—Orsat type—test can be validated and percent excess air can be estimated for various types of fuel using available data such as shown in the Figure 2.2 nomograph [2]. This quickly demonstrates whether O_2 and CO_2 pairs are acceptable, and if fuel type and one Orsat value is known, the other can be read from the curve.

To use the nomograph for validation purposes, draw a connecting line between the O_2 and CO_2 percentage values as measured with the Orsat and note where the line intersects with the fuel type line. If the point of intersection is within the brackets associated with the stated fuel type, then the Orsat values are acceptable based upon U.S. EPA criteria.

However, there are several circumstances that can cause the point of intersection for a set of Orsat values to be outside of the desired fuel type brackets even though the Orsat data are valid and reproducible:

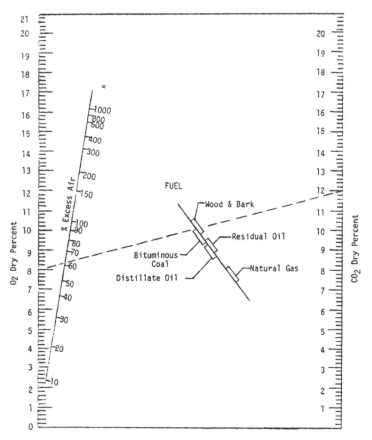

Figure 2.2 Orsat validation nomegraph (courtesy Walter Smith [2], President, Entropy Environmentalists, Inc.).

(1) The actual fuel is not representative of the normal fuel type, causing the intersection point to be outside of the brackets.
(2) The actual fuel is not the stated fuel or is a mixture of two or more fuels.
(3) If the intersection point on the fuel type line is above the bracket de-limiter (e.g., towards the wood fuel bracket), then the CO_2 value may be high because
   a. CO_2 is being manufactured in the process (e.g., from $CaCO_3$ in the fuel).
   b. Other acid gases are present in percentage quantities (HCl for example) which would be included by the Orsat analyzer in the CO_2 percent value.
(4) If the intersection point is low (i.e., in the example toward natural gas), CO_2 could be absorbed in a scrubber.

A value for waste fuel for incineration does not appear in Figure 2.2, but the data seem to indicate that this falls in the Natural Gas region.

In the example shown, 8% O_2, 12% CO_2, and 62% excess air are good values for a bark burner. A fuel mixture of natural gas and bark would lower the curve for good Orsat readings.

2.2.3 AIR–FUEL RATIO

Air–fuel ratio (AF) is the *mass* of air per unit *mass* of fuel. This can be obtained directly from the equation. For example, using the theoretical combustion of CH_4 with stoichiometric air, Equation (2.2), the air–fuel ratio is

$$\text{AF} = \frac{(2 \text{ mole } O_2)\left(\dfrac{100 \text{ moles air}}{21 \text{ moles } O_2}\right)\left(\dfrac{29 \text{ g air}}{\text{g mole}}\right)}{(1 \text{ mole } CH_4)\left(\dfrac{16 \text{ g}}{\text{g mole}}\right)} = 17.3$$

As the fuel becomes heavier, i.e., the ratio of moles of hydrogen per mole of carbon decreases, the air–fuel ratio decreases. For example, at stoichiometric conditions, gasoline ($\sim C_8H_{18}$) has an AF of 15 and for pure carbon it is 11.5.

2.2.4 FLUE GAS QUANTITY

Volume of flue gas can be related to the fuel consumption rate. If the combustion equation is known, this can be used. For example, using Equation (2.3) with 5% excess air, the volume of wet flue gas in actual cubic meters at 150°C per mole of CH_4 is

$$\left(\frac{1 + 2 + 0.1 + 7.9 \text{ moles total gas}}{\text{mole } CH_4}\right)\left(\frac{22.4 \text{ l}}{\text{g mole}}\right)$$

$$\times \left(\frac{273 + 150 \text{ K}}{273 \text{ K}}\right)\left(\frac{m^3}{1,000 \text{ l}}\right) = 0.382 \text{ Am}^3/\text{mole } CH_4$$

This may also be reported as the volume of dry gases at standard conditions by simply substracting the moles of water vapor in the product gases and proceeding from there. More likely, only the flue gas analysis is known. Using the analysis, the volume of dry flue gas at SC per mole CH_4 is

$$\left(\frac{100 - 18.18 \text{ moles dry gas}}{9.09 \text{ moles } CH_4}\right)\left(\frac{22.4 \text{ l}}{\text{g mole}}\right)\left(\frac{293 \text{ K}}{273 \text{ K}}\right)$$

$$\left(\frac{m^3}{1,000 \text{ l}}\right) = 0.216 \text{ Nm}^3/\text{mole } CH_4$$

2.2.5 ENERGY LOSSES

Maximum thermal cycle efficiencies in the mid-70 percentile are the limit for the best practicable systems. Combined efficiency for combustion-electrical generation increased from about 22% in 1948 to a peak of only 34% in 1965. The 34% peak is based on essentially complete combustion. Obviously, incomplete combustion products constitute a potentially significant loss of the fuel's available heat input. In practice, using adequate air and proper application of the 3 "T's" makes it possible to consistently attain nearly complete combustion.

One of the largest sources of energy loss in the combustion operation is the stack heat losses. This depends not only on the exhaust gas temperature but on the amount of excess air used. Specific heat, C_p, of the flue gas depends on composition of the flue gas and the temperature. For a gas such as produced by Equation (2.3) the specific heat is about 0.26 cal/g K over typical flue gas temperature ranges. The heat losses calculated for this gas are plotted as Figure 2.3 based on inlet reactant's temperature of 20°C. An energy loss of 5% is normal for flue gases.

Derating of a generating facility occurs if power to move air and gases is obtained from a self-contained loop. In these cases, if pollution control

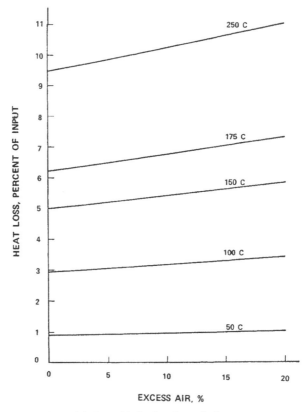

Figure 2.3 Typical boiler heat losses in flue gases.

devices are included, power losses can range up to 5% of the system-generated output. For example, for every 100 MW of generated output up to 5 MW could be lost as power for emission control. At an overall system efficiency of 34% input power would then be equivalent to about 300 MW of which up to 15 MW could be required to power the emission control systems.

Combustion of waste fuels usually results in up to a 10% energy loss. This accounts for the approximately 1% loss as non-combusted feed and 5% flue gas loss with the balance due to other factors such as control devices, possible inadequate insulation, and cyclic operation (this does not account for start-up/cool-down losses). Burning of wet waste fuels can significantly change the flue gas percentage composition and specific heat as large quantities of water can be added. The C_p of water vapor is 0.5 cal/

(g °C) or Btu/(lb °F) resulting in a higher C_p average for the flue gas, e.g., up to about 0.325.

2.3 COAL COMBUSTION

Coal is the most readily available energy resource in the U.S. at this time. In 1988, U.S. utilities burned 750 million tons of coal at an average cost of $32/ton. The major problems with coal as a fuel are (1) it is not easy to handle and use in smaller installations, and (2) it results in more *uncontrolled* air pollution emissions when burned. Even the predicted rate of consumption is low considering the potential availabilities of the various sources of energy. Technologies to produce clean gas and/or liquid fuels from coal may spur increased use of this energy source, which is predicted to be sufficient for the next 600 years.

2.3.1 COAL

Coal analyses are reported on several bases as shown in Table 2.2. This enables one to compare various types of coal on a common basis and also to determine handling weights on a heat value basis. The same sample of bituminous coal can also be reported on a more complete chemical analysis basis called an ultimate analysis. The "as received" analysis given in Table 2.2 is the "proximate" analysis. The ultimate analysis for this sample is 79.90% carbon, 4.85% hydrogen, 0.69% sulfur, 1.30% nitrogen, 6.50% ash and 6.76% oxygen by difference. Coal has a true specific gravity of 1.40–1.70 for anthracite and 1.25–1.45 for bituminous. The bulk density of loose-packed coal is about 0.64 g/cm^3.

Although we speak of anthracite and bituminous as hard and soft coal, respectively, there are a number of groups in each class as shown in Table 2.3. (The abbreviation MAF means moisture, ash-free.) The complete

TABLE 2.2. Coal Analyses for One Bituminous Coal Sample.

| Basis | Weight Percent | | | | Heat Value | |
	Moisture	Volatile Matter	Fixed Carbon	Ash	Cal/g	(Btu/lb)
As received	3.2	21.0	69.3	6.5	7,330	(13,200)
Moisture free (MF)		21.7	71.6	6.7	7,580	(13,640)
Moisture-ash free (MAF)		23.3	76.7		8,120	(14,620)

TABLE 2.3. Approximate Classification of Coal.

Class	Group	Weight % on MAF Basis		Calorific Value, cal/g, Ash-Free Basis	Agglomerating Character[a]
		Fixed Carbon	Volatile Matter		
Anthracite	Meta anthracite	≥98	≥2		
	Anthracite	92–98	2–8		non
	Semianthracite	86–92	8–14		
Bituminous	Low volatile	78–86	14–22		agglomerating
	Medium volatile	69–78	22–31		agglomerating
	High volatile A	<69	>31	≥7,780	agglomerating
	High volatile B			7,220–7,780	agglomerating
	High volatile C			6,390–7,220	agglomerating
Subbituminous	High volatile A			5,830–6,390	non
	High volatile B			5,280–5,830	
	High volatile C			4,610–5,280	
Lignite	High volatile A			3,500–4,610	
	High volatile B			<3,500	

[a]Non means nonagglomerating.

85

classification takes into account the proximate analysis, heat content (calorific value), and agglomerating character of the ash. Volatile matter is produced in the coal when it is heated. After heating, the fixed carbon and ash remain. Fixed carbon can be burned off leaving the ash.

The noncombustible mineral matter in coal produces the bottom ash and some of the fly ash particulate matter. Both inherent and extraneous mineral matter exist in coal. The inherent mineral matter is intimately mixed with coal and usually makes up less than 3% of the coal. It originated as part of the growing plant life which produced the coal and is uniformly distributed. Extraneous mineral matter is inorganic material typical of the strata surrounding the coal seam.

During the combustion, the coal is subject to either or both oxidizing and reducing conditions because of the presence of oxygen, carbon, and carbon monoxide in the bed. The mineral matter when heated reacts with the chemicals present, then sinters (becomes hard) or melts, forming the ash. Properties of the final ash vary depending on furnace conditions and the original composition of the mineral matter. Some typical ash properties are given in Table 2.4. (MF means moisture free.) Fly ash is mostly comprised of ash particles that escape collision until they are cool and no longer sticky. Wall slag and bottom ash are slags formed by molten ash.

The sulfur contained in coal can exist in three forms—pyrites, organics, or sulfates. Pyrites, or iron sulfides, usually contain over half the total sulfur. Organic sulfur consists of complex organic molecules. Sulfates represent a small portion of the total sulfur and are found in the fresh coal.

TABLE 2.4. Typical Ash Properties.

Property	High Volatile Bituminous	Subbituminous	Lignite
Ash weight (MF), %	17	6.5	12.8
Ash fusibility temperature, °C			
Reducing[a]	1,093	1,088	1,080
Oxidizing[a]	1,260	1,199	1,132
Softening temperature, °C			
Reducing[a]	1,238	1,193	1,166
Oxidizing[a]	1,332	1,216	1,199
Hemispherical temperature, °C			
Reducing[a]	1,249	1,232	1,177
Oxidizing[a]	1,343	1,227	1,210
Fluid temperature, °C			
Reducing[a]	1,271	1,254	1,227
Oxidizing[a]	1,432	1,260	1,254

[a]Initial deformation temperature for condition noted, i.e., reducing and oxidizing.

Conventional coal-washing processes can remove half, or more, of the pyrites and much of the mineral matter. Organic sulfur is not separated from the coal by this process. Typical washing processes may reject about 15–20% of the coal during the cleaning, and this cleaning process may effectively reduce the total sulfur content by about half. Rejection of over 30% of the coal is usually uneconomical [3].

Coal contains nearly every chemical element known. Reported analysis typically notes 10 major and minor elements plus trace elements. The trace elements usually occur in coal at about the same concentration as found in average rocks found in the earth's crust. However, boron, cadmium, and selenium occur in greater concentrations in coal. A representative coal from the U.S. Interior Province, Eastern Region (Illinois, Indiana, and western Kentucky) might contain elements as shown in Table 2.5.

2.3.2 COAL BURNERS

Coal is basically burned as fuel piled in a bed, in suspension, or in a fluidized state. In the U.S., it has been more economical to use beds for systems up to 100,000-lb/hr steam capacity. Suspension firing is used for larger systems. Coal-bed firing usually consists of one of three types of stokers. Stokers use a coarse coal and differ in the flow direction of fuel and air as shown in Figure 2.4. They are:

underfeed: air from below; fuel enters 90° from direction of bed and ash movement; uses ¼- to 2-in size coal and 20–50% excess air

crossfeed: air from below; fuel and ash move in continuous direction; can burn a variety of coal including low ash-fusion temperature coal; uses 15–50% excess air

overfeed: air from below; fuel from above; uses 30–60% excess air

Table 2.6 lists coal-burner combinations for both coal-bed and suspension-firing furnaces.

Suspension firing uses pulverized or crushed coal. Burner and furnace arrangements for this are shown in Figure 2.5. Pulverized coal with more than about 15% volatile matter is explosive and ignites easily. The primary air serves to pneumatically convey the pulverized coal to the burner in a fluidized state at about 15 m/sec in Figure 2.5(a–d). This air may equal 10–20% of the total combustion air. Secondary air enters near the fuel inlets. Total air normally amounts to 15–20% excess if the unit is completely water-cooled and up to 40% excess if partially water-cooled.

Coal crushed to 4-mesh size (4.7 mm) is used in the cyclone burner. The fine particles burn in suspension and the coarse particles are thrown by centrifugal force to the wall where they stick to the molten slag and burn. Excess air is normally 10–15%.

TABLE 2.5. Composition of Some Mid-Central U.S. Coal, Moisture-Free Basis [4].

10 Major and Minor Elements	Typical Weight % Range
Aluminum	0.60–2.77
Calcium	0.14–1.68
Chlorine	0.02–0.54
Iron	0.68–2.89
Magnesium	0.01–0.17
Potassium	0.04–0.24
Silicon	0.58–2.88
Sodium	0.009–0.145
Sulfur	0.50–6.00
Titanium	0.02–0.08

23 Trace Elements	Typical ppm by Weight
Antimony	0.4–5.2
Arsenic	4–93
Beryllium	0.5–3.2
Boron	12–170
Bromine	9–33
Cadmium	<0.1–20
Chromium	4–26
Cobalt	4–34
Copper	5–30
Fluorine	51–135
Gallium	2.1–4.9
Germanium	<1–22
Lead	6–218
Manganese	6–181
Mercury	0.04–0.60
Molybdenum	<1–29
Nickel	8–32
Phosphorus	<10–320
Selenium	0.4–3.3
Tin	<1–30
Vanadium	16–62
Zinc	10–3173
Zirconium	12–133

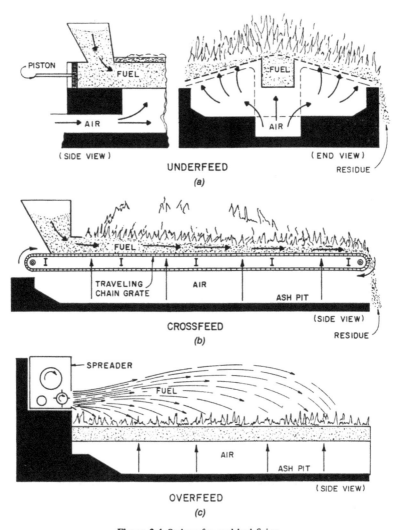

PISTON

FUEL

AIR

(SIDE VIEW)

FUEL

AIR

(END VIEW)

RESIDUE

UNDERFEED

(a)

FUEL

TRAVELING
CHAIN GRATE

AIR

ASH PIT

(SIDE VIEW)

RESIDUE

CROSSFEED

(b)

SPREADER

FUEL

AIR

ASH PIT

(SIDE VIEW)

OVERFEED

(c)

Figure 2.4 Stokers for coal-bed firing.

TABLE 2.6. Coal-Bed and Suspension Firing Furnace Combinations.

| | Coal-Bed Stokers | | | Suspension Boilers | |
Coal	Underfeed	Crossfeed	Overfeed	Pulverized Coal	Cyclone
Anthracite		X		X	
Bituminous					
17–25% volatile	X		X	X	X
25–35% volatile				X	X
coking	X	X	X	X	X
weak coking		X	X	X	X
Lignite		X	X	X	X

Fluidized-bed combustion is being developed. This process provides an intimately mixed fuel–air combustion zone at a low temperature (from 800–1,000°C) as shown in Figure 2.6. This produces less nitrogen oxides and reduces volatilization of sodium and potassium from the ash. Heat transfer rates to tubes immersed within the bed as shown are high. Blown-over carbon and other bed material is high. Figure 2.6 shows how a cy-

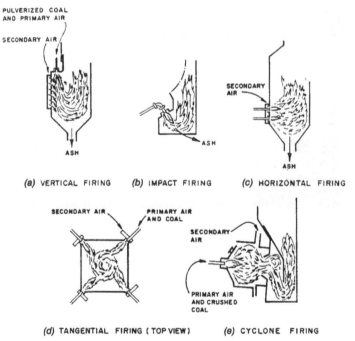

(a) VERTICAL FIRING   (b) IMPACT FIRING   (c) HORIZONTAL FIRING

(d) TANGENTIAL FIRING (TOP VIEW)   (e) CYCLONE FIRING

Figure 2.5 Suspension firing of pulverized and crushed coal.

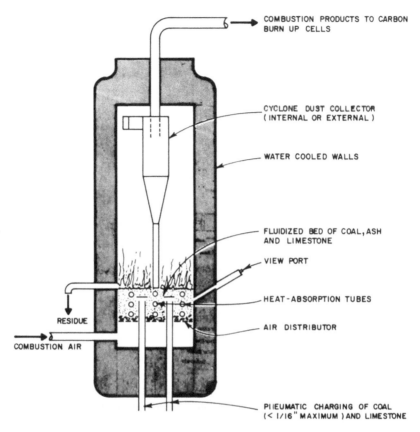

Figure 2.6 Fluidized bed combustor.

clone is required to return entrained material to the combustion bed. Carbon burn-up cells are needed after the fluid-bed combustion to achieve good fuel utilization.

High SO_2 removal can be obtained in a fluidized bed as shown by Henschel [5] in Figure 2.7 for coal of unspecified sulfur content using 420-to 500-μm limestone particles added to the bed. High SO_2 control requires high limestone stoichiometry resulting in much unused limestone in the bed discharge material. These experimental data were obtained using a 1.2-m fluidized bed depth, 1.8-m/sec gas velocity, 20% excess air, and 815°C temperature.

2.3.3 COMBUSTION PRODUCTS

The air pollution emissions from the combustion of coal include particulates, sulfur oxides, carbon monoxide, hydrocarbons, nitrogen oxides,

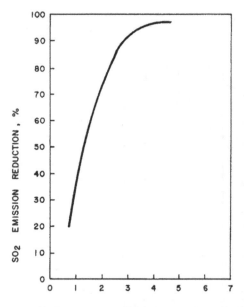

LIMESTONE TO SULFUR STOICHIOMETERY, MOLES Ca/S

Figure 2.7 Fluidized-bed SO_2 removal by limestone addition [5].

and aldehydes. Amounts of each are highly dependent on the type of coal, combustion process, and operation of the process. The particulates consist mostly of ash particles. They also include unburned coal blown over during the turbulent combustion. Typical pulverized coal-fired boiler flue gas contains 2–10 ppm CO, 3.9–5.1% O_2, and 14.3–15.3% CO_2 for new units.

Essentially all (95%) of the sulfur in the fuel is converted to SO_2 and normally leaves with the flue gases except for combustors such as fluidized beds which contain alkali material in the bed. Presence of rare metals such as platinum and vanadium in the ash can catalyze the conversion of SO_2 to SO_3. In the presence of excess air, up to 5% of the SO_2 can be converted to SO_3 [6]. Water vapor and SO_3 form acid mist. The acid dew point of flue gas containing as little as 15 ppm SO_3 could be expected to be as high as 125°C.

Nitric oxide is produced at the high temperatures of combustion. Increased nitrogen from either the air or the fuel increases the amount formed. Large coal-fired boilers emit more NO_x than any other combustion source. The combustion paradox is that increased air forms increased nitric oxide but decreased CO and hydrocarbons. The reverse is true for decreased air.

A compilation of air pollutant emission factors is given in AP-42 [7]. This listing assumes the boiler or furnace is at steady-state conditions, is properly designed and operated, and maintenance is good. Some of these values are given in Table 2.7. Inadequate air or inadequate time-temperature-turbulence results in increased carbon particulate matter, CO, and hydrocarbons, HC. Excess air can change the amounts of pollutants released. For example, overfire air in pulverized boilers can be optimized for best operation. NO_x emissions in the range of 189–301 ng NO_2/J (5.8–9.2 g NO_2/kg coal with 7,330 cal/g) are reported for typical operation [8]. Heat loss also increased 0.25% for every 10% decrease in theoretical air due to carbon deposits. Excessive air can result in increased particulates due to fly ash carryover. NO_x emissions are reduced by lowering excess air, reducing load, and increasing the separation between overfire air and burner nozzle tilt angles.

Note from Table 2.7 the relative advantages and disadvantages of various boiler types. For example, the cyclone-crushed coal units produce less fly ash, but, because they operate at higher temperatures, produce more NO_x. Exceptions from the NSPS (Table 1.11) are being considered for these units, in areas where oxidant concentrations are low, to permit them to burn gob pile (refuse) coal that cannot be burned in other types of boilers.

Coal combustion emissions also include trace quantities of toxic and hazardous air pollutants. This includes materials such as arsenic, beryllium, lead, and mercury plus radioactive material. Chlorides present in the coal are also emitted in the exhausted flue gases. All of these emissions are significant either as a potential health hazard or as a potential problem in the construction material for the air pollution control devices.

U.S. coal-fired power plants generate over 80 million tons of waste per year. Flue gas desulfurization accounts for 5 million tons per year, of which about 1% is utilized for gypsum and wallboard. Bottom ash amounts to over 20 million tons per year, and fly ash to over 50 million tons per year. Ash utilization breakdown by group is as follows.

	Utilization As, %	
	Bottom Ash	Fly Ash
Aggregate	1.3	0.7
Blasting grit	24.5	–
Cement addition	9.7	18.6
Concrete admixture	2.5	35.6
Concrete block	0.9	3.6
Dam building	–	1.8

TABLE 2.7. Typical Uncontrolled Emissions from Bituminous Coal Combustion [7].

Furnace			Uncontrolled Emissions, g/kg Coal					
Boiler Category	Heat Input, 10⁶ kcal/hr	Type	Particulate Matter[a]	Sulfur Oxides[b]	Carbon Monoxide	Hydro-carbons[c]	Nitrogen Oxides[d]	Aldehydes
Utility or large industrial	>25	Pulverized	8A	19S	0.5	0.15	9	0.0025
		Cyclone	1A	19S	0.5	0.15	27.5	0.0025
Commercial or industrial	2.5–25	Underfeed & crossfeed	2.5A	19S	1	0.5	7.5	0.0025
		Overfeed	6.5A	19S	1	0.5	7.5	0.0025
Small commercial or domestic	<2.5	Overfeed	1A	19S	5	5	3.0	0.0025
		Hand fired	20	19S	45	45	1.5	0.0025

[a]A = Weight percent ash in coal.
[b]S = Weight percent sulfur in coal; pollutant weight is reported as SO_2.
[c]Pollutant weight is reported as CH_4.
[d]Pollutant weight is reported as NO_2.

	Utilization As, %	
	Bottom Ash	Fly Ash
Fill	13.8	4.5
Grouting	–	6.2
Hazard waste stabilization	–	0.7
Road cinders	11.4	0.5
Road construction	11.8	2.6
Roofing particles	13.6	–
Miscellaneous	10.5	25.2
	100.0	100.0

As of 1985, 17% of the fly ash and 33% of the bottom ash were utilized. Most of those materials are disposed of in ponds with about 30% going to landfills. The Edison Electric Institute's data for 1984 can be used to show the waste disposition breakdown as follows.

	Disposition, %				
Waste Product	Sold	Paid to Remove	Land-filled	Ponded	Other
Fly Ash	9.4	12.8	35.6	32.9	9.3
Bottom Ash	4.5	28.2	34.7	23.7	8.9
FGD Sludge	0.7	23.3	34.2	41.8	0
Weighted Av.	5.7	19.5	35.1	33.3	6.5

Example Problem 2.1

Estimate the combustion gas composition and quantity and concentrations of uncontrolled emissions anticipated from new 150-MW pulverized coal and cyclone-type boilers and compare these values with the NSPS in Table 1.11. Assume the fuel is bituminous coal with the analysis given in Table 2.2 and Section 2.3.1.

Solution

Table 2.6 shows that both units can burn this coal, however, some assumptions must be made: assume both boilers operate at 15% excess air as discussed in Section 2.3.3, that it requires about 2.5×10^8 kcal/hr (10^9 Btu/hr) heat input per 100 MW, the stack temperature is 170°C and that the unit is operating at full load. Complete combustion is assumed.

Basis 100 g coal:

$$\frac{79.9 \text{ g C}}{12 \text{ g/atom}} = 6.66 \text{ atoms C}$$

$$\frac{4.85 \text{ g H}}{2 \text{ g/mole}} = 2.43 \text{ moles H}_2$$

$$\frac{0.69 \text{ g S}}{32 \text{ g/atom}} = 0.02 \text{ atoms S}$$

$$\frac{1.30 \text{ g N}}{28 \text{ g/mole}} = 0.05 \text{ moles N}_2$$

$$\frac{6.76 \text{ g O}}{32 \text{ g/mole}} = 0.21 \text{ moles O}_2$$

Stoichiometric combustion of these with air requires

$$6.66 + \tfrac{1}{2}(2.43) + 0.02 - 0.21 = 7.69 \text{ moles O}_2 \text{ or } (7.69)\,\frac{100}{21}$$

$$= 36.60 \text{ moles air}$$

At 15% excess air this equals 42.08 moles air input.
Combustion gas composition is then about

CO_2 = 6.66 moles	or 15.29 vol %
H_2O = 2.43 moles	or 5.58 vol %
SO_2 = 0.02 moles	or 0.05 vol %
N_2 = 0.05 + (79/100)(42.08)	
= 33.29 moles	or 76.44 vol %

$$O_2{}^* = \left(\frac{21}{100}\right)(42.08) + 0.21 - (7.69 + 0.21)$$

$$= \underline{1.15 \text{ moles}} \qquad \text{or } \underline{2.64 \text{ vol\%}}$$

Total = 43.55 moles per 100 g coal $\qquad$ 100 vol%

(This gives SO_2 concentration in flue gas to 4 significant places as 0.0459% or 459 ppm.)

*O_2 = balance of in-out.

Quantity of flue gases at full load are

$$\left(\frac{43.55 \text{ moles}}{100 \text{ g}}\right)\left(\frac{\text{g}}{7.33 \text{ kcal}}\right)\left(\frac{2.5 \times 10^8 \text{ kcal/hr}}{100 \text{ MW}}\right)(150 \text{ MW})\left(\frac{22.4 \text{ l}}{\text{mole}}\right)$$

$$\times \left(\frac{273 + 170}{273}\right)\left(\frac{\text{m}^3}{1,000 \text{ l}}\right) = 809,850 \text{ Am}^3/\text{hr}$$

Reference literature [8] suggests that the flue gas quantity equals about 340,000 Nm^3/hr per 100 MW *at 27% excess air* using 7.0 kcal/g coal. Under the conditions of this problem this would equal

$$\left(\frac{340,000 \text{ m}^3/\text{hr}}{100 \text{ MW}}\right)(150 \text{ MW})\left(\frac{273 + 170}{293}\right) = 771,100 \text{ Am}^3/\text{hr}$$

Uncontrolled emissions using data from Tables 2.2 and 2.7 are estimated and percent control needed to meet the 1979 NSPS as summarized in Table 2.8. To convert emissions in g/kg as fired coal to $g/10^6$ cal for this coal, use the multiplying factors

$$\left(\frac{\text{g coal}}{7,330 \text{ cal}}\right)\left(\frac{\text{kg}}{1,000 \text{ g}}\right)(10^6)$$

which equals 0.1364.

Note that a direct calculation of SO_2 assuming 95% combustion results in the same values as those predicted from Table 2.6 for this coal. This is

$$\left(\frac{0.69 \text{ g S}}{100 \text{ g coal}}\right)\left(\frac{64 \text{ g SO}_2}{32 \text{ g S}}\right)(0.95)\left(\frac{\text{g coal}}{7,330 \text{ cal}}\right)(10^6) = 1.79 \text{ g SO}_2/10^6 \text{ cal}$$

It is obvious from the preceding example that coal combustion emissions must be reduced significantly to achieve the NSPS. This can be done. However, even these standards may not be strict enough to prevent significant deterioration because of the increased use of coal anticipated. The Department of Energy predicted [9] U.S. emissions of total suspended particulates (TSP), SO_2, and NO_x as shown by the averages in Table 2.9. TSP are smaller particulates that do not settle from the air rapidly and are essentially those released from high-efficiency control devices. Total amount of TSP increases after 1985, and SO_2 and NO_x increase continuously. These are national averages so local areas downwind from generating facilities may experience significant increases in local ambient concentrations depending on the control technology used and the regulations.

TABLE 2.8. Example Problem 2.1. Estimated Emissions and Control Required Summary.

Boiler	Pollutant	Estimated Emissions in		NSPS Emission Limit $g/10^6$ cal	% Control Needed to Meet NSPS
		g/kg as Fired Coal	$g/10^6$ cal		
Pulverized Coal	Particulates	$(8)(6.5)$ = 52.0	7.09	0.054	99.24
	SO_2	$(19)(0.69)$ = 13.1	1.79	1.08	70 (min)
	NO_x as NO_2	9.0	1.23	0.9	0
	CO	0.5	0.07		
	HC as CH_4	0.15	0.02		
	Aldehydes	0.0025	0.0003		
Cyclone	Particulates	$(1)(6.5)$ = 6.5	0.89	0.054	93.93
	SO_x	13.1	1.79	1.08	70 (min)
	NO_x	27.5	3.75	1.26	66.4
	CO	0.5	0.07		
	HC as CH_4	0.15	0.02		
	Aldehydes	0.0025	0.0003		

TABLE 2.9. Predicted U.S. Emissions
Due to Increased Use of Coal Combustion [9].

Pollutant	Emissions During Year, in 10^9 kg		
	1975	1985	2000
TSP	14.1	8.6	12.0
SO_2	24.0	26.1	28.6
NO_x	15.8	19.9	25.9

2.4 OIL COMBUSTION

2.4.1 OIL

Oil is our most versatile energy source. It is easy to transport, burns relatively clean, is safe to use and has a high energy-to-volume ratio. In addition, it is a valuable lubricant and chemical feedstock. Production of domestic oil in the U.S. has not increased with the demand, and use of imported oil continues to rise. The American Petroleum Institute estimates show that as of 1989, the U.S. imported 8.0×10^6 barrels of crude oil per day compared to the U.S. production of 7.6×10^6 barrels per day. The Department of Energy estimates that, as of the end of 1988, proven crude reserves in the U.S. totaled 49 billion barrels. U.S. Geological Survey believes that there may be an additional 33–70 billion barrels of undiscovered recoverable petroleum reserves. It is also possible to produce synthetic oil from coal and this may ultimately be a significant source of oil.

Natural fuel oils are obtained from crude oil. Distillation towers are used to separate crude oil mixtures into components of various boiling ranges. As such, the distillate oil which comes from the upper sections of the tower has a lower boiling point. These compounds have shorter molecular chain lengths and more hydrogen. The residual oils are from the bottom sections of the still and are the longer chain, heavier compounds. Table 2.10 shows seven variations of oil. Diesel oil and #2 fuel (heating) oil are essentially the same thing. Heating value for this oil is 10.80 kcal/g or 19,440 Btu/lb (141,100 Btu/gal). Number 6 oil, also called Bunker C oil, is included, but it is not usually available. It is very viscous and usually must be heated in order to be pumped and atomized. Kerosene has an average molecular formula of $C_{12}H_{26}$.

2.4.2 OIL BURNERS

The immediate future use of oil combustion by utilities is of less sig-

TABLE 2.10. Typical Data for U.S. Fuel Oil at Standard Conditions.

Type	Grade	Specific Gravity	Heating Value (kcal/g)	Viscosity Range [g/(cm sec)]	Approx. Average Wt % of				
					Sulfur	Hydrogen	Carbon	Nitrogen	Ash
Distillate oil	Kerosene	0.81	11.39		0.04	15	85	<0.1	
Distillate oil	Diesel	0.84	11.19		0.2	13	87	<0.1	0.01
Distillate oil	#1	0.84	10.90		0.2	13	87	<0.1	0.01
Distillate oil	#2*	0.87	10.80	0.02–0.03	0.5	13	87	<0.1	0.01
Blend	#4	0.90	10.74	0.05–0.24	2	11.5	87	0.3	0.05
Residual oil	#5	0.92	10.71	0.29–1.41	2	11	87	0.3	0.05
Residual oil	#6	0.96	10.51	1.90–19.0	3	11	86	0.3	0.15

*In English units, #2 fuel oil has 141,000 Btu/gal or 19,440 Btu/lb.

nificance because of the federal mandate to convert to coal combustion. In 1988 only 5.5% of the utility power was produced by oil combustion. However, future production of synthetic oil from coal and/or refuse could reverse this.

Oil burners must atomize the oil to produce drops of about 50-μm diameter and mix these drops with combustion air. Burning time is proportional to the square of droplet diameter, so small droplet size is critical. Oil is sprayed into boilers through nozzles or from rotary spray cups. Air or hot steam can be used to assist in atomizing the oil. Steam also heats up the heavier grades of oil to about 100°C to reduce the viscosity. Low-pressure air systems use as much as 70% of the combustion air (about 7.5 m³/l of oil) for atomization at up to 25-cm Hg pressure. High-pressure air-assisted nozzles require 150- to 875-cm Hg air pressure and use up to 1.5 m³ air/l oil. Mechanical atomizers require oil at 375- to 1,500-cm Hg pressure depending on the viscosity. Oil burners can operate at 5–10% excess air.

2.4.3 COMBUSTION PRODUCTS

Properly designed and operated oil-fired boilers produce less uncontrolled air pollution emissions than coal-fired boilers but more than gas-fired boilers. Horizontal firing and tangential firing are the two basic arrangements used in oil combustion systems. The oil is directed toward the center of the furnace in a horizontally fired system resulting in a hotter flame and the formation of about twice the NO_x as is formed by tangentially fired oil furnaces. Large boilers burn hotter and produce more NO_x than small ones. About 0.65% of the sulfur dioxide converts to sulfur trioxide in an oil-fired boiler. Table 2.11 lists typical uncontrolled emissions for horizontally fired oil furnaces.

Example Problem 2.2

A heavy fuel oil with a specific gravity of 0.96 contains 86.5% by weight carbon, 12.38% hydrogen and 0.36% sulfur. The dry flue gas Orsat analysis shows 12.13% CO_2, 4.44% O_2 and 83.43% N_2. Determine the percentage excess air, concentration of SO_2 in the flue gases, and the wet gas analysis. Consider the air to be dry.

TABLE 2.11. Typical Uncontrolled Emissions from Oil Combustion [7].

				Uncontrolled Emissions, g/l of Fuel				
Boiler	Fuel Oil	Particulate Matter	Sulfur-Oxides[a]	Carbon Monoxide	Hydro-carbons[b]	Nitrogen Oxides[c]	Aldehydes
Utility or large industrial	Residual	1	19.2S	0.4	0.25	12.6	0.12
Commercial or industrial	Residual	2.75	19.2S	0.5	0.35	9.6	0.12
	Distillate	1.8	17.2S	0.5	0.35	9.6	0.25
Small commercial or domestic	Distillate	1.2	17.2S	0.6	0.35	1.5	0.25

[a]S = weight percent sulfur in oil; pollutant weight is reported as SO_2.
[b]Pollutant weight is reported as CH_4.
[c]Pollutant weight is reported as NO_2; use one-half this value for tangentially fired boilers.

Solution

Theoretical air needed for stoichiometric combustion

$$C + O_2 \rightarrow CO_2$$

$$H_2 + 1/2O_2 \rightarrow H_2O$$

$$S + O_2 \rightarrow SO_2$$

Basis: 100 moles dry flue gas

$$\% \ EA = \frac{excess \ oxygen}{theoretical \ oxygen}(100)$$

Amount of water vapor and SO_2 is found from a carbon balance and the combustion equations knowing that the 100 g of fuel contains

$$\frac{86.5}{12} = 7.21 \text{ atoms C}$$

$$\frac{12.38}{2} = 6.19 \text{ moles } H_2$$

$$\frac{0.36}{32} = 0.0113 \text{ atoms S}$$

$$(12.13 \text{ moles } CO_2)\left(\frac{1 \text{ atom C}}{1 \text{ mole } CO_2}\right)\left(\frac{6.19 \text{ moles } H_2}{7.21 \text{ atoms C}}\right)$$

$$= 10.41 \text{ moles } H_2 \text{ or } 10.41 \text{ moles } H_2O$$

$$(12.13 \text{ moles } CO_2)\left(\frac{1 \text{ atom C}}{1 \text{ mole } CO_2}\right)\left(\frac{0.0113 \text{ atoms S}}{7.21 \text{ atoms C}}\right)$$

$$\times \left(\frac{1 \text{ mole } SO_2}{\text{atom S}}\right) = 0.0190 \text{ moles } SO_2$$

$$\% \ EA = \frac{4.44}{12.13 + 10.41/2 + 0.0190}(100) = 25.6$$

$$\qquad\qquad \uparrow \qquad\quad \uparrow \qquad\quad \uparrow$$
$$\qquad\quad \text{for} \quad\ \text{for} \quad\ \text{for}$$
$$\qquad\quad CO_2 \quad\ H_2O \quad\ SO_2$$

(Note that this example shows 25.6% excess air, or excess oxygen, but the actual *percentage* of excess oxygen in the flue gases is 4.44% as given, which is a variation of roughly 5:1.)

Concentration of SO_2 was obtained using a carbon balance. On a dry gas basis this is 0.0190% or 190 ppm SO_2.

$$\left(\frac{12.13}{100 + 10.41}\right)(100) = \left(\frac{12.13}{110.41}\right)(100) = 10.99\% \ CO_2$$

$$\left(\frac{4.44}{110.41}\right)(100) = 4.02\% \ O_2$$

$$\left(\frac{83.43}{110.41}\right)(100) = 75.56\% \ N_2$$

$$\left(\frac{10.41}{110.41}\right)(100) = \underline{9.43\% \ H_2O}$$

$$\text{Total} = 100.00\%$$

$$SO_2 \text{ conc. is then} \left(\frac{190}{110.41}\right)(100) = 172 \text{ ppm}$$

2.5 GAS COMBUSTION

2.5.1 GASES

Gases are used to produce about 25% of the energy in the U.S. This ratio should continue to hold well into the 1990s. However, the distribution will vary with time as less gas is used by utilities and more by residential, commercial, and industrial consumers. Gases are the cleanest-burning fuels and are, therefore, in great demand for areas with high pollution levels. It is also the choice for incinerator secondary chamber auxiliary fuel.

Natural gas is the most available fuel. It can come directly from wells or it can be obtained at the refinery. In most cases it consists of almost pure methane (CH_4) with slight amounts of ethane (C_2H_4), nitrogen, helium, and carbon dioxide. It can contain trace amounts of substances such as hydrogen sulfide and mercaptans. The average gross heating value varies from 8,900–9,800 kcal/Nm^3. The density is about 7.5 × 10^{-4} g/cm^3 which gives a specific gravity of about 0.57 when referred to air.

Liquified petroleum gas (LPG) is a gasoline-refining byproduct and consists mainly of butane, propane, and trace amounts of propylene and butylene. Although it is distributed as a liquid under pressure, it reverts to a gaseous state when throttled and is burned as a gaseous fuel. LPG is commonly known as a bottled gas. Grade A is mostly butane and has a heating value of 6,480 kcal/l. Grade F is mostly propane with a heating value of 6,030 kcal/l (90,500 Btu/gal). Grades B–E are mixtures.

Table 2.12 lists properties of some gaseous fuels. Artificial gas is also known as manufactured or city gas and is a mixture of water gas, coal gas, and others. The heating value comes from the presence of the CH_4, CO, and H_2. Carbureted (water) gas and producer and coke-oven gases are made from coal or coke. The explosive limits in Table 2.12 are given as volume of fuel in air (oxygen plus nitrogen).

2.5.2 GAS BURNERS

The three major types of gas burners are the raw gas burner, the premix gas burner, and the forced-draft gas burner. In all cases, the gases (both the combustion gases and the air) are forced through nozzles at an exit velocity close to the flame ignition velocity so that the flame can be sustained. As a result, gas burners are noisy. The gas and air mixture leaving the burner nozzle or the gas mixed with air near the nozzle must remain in a combustion zone near the burner tip for 0.25–3 sec, depending on whether the temperature is "high" or "low" to achieve complete combustion. Gas burners typically operate at about 3–15% excess air.

The raw gas burner is similar to an old-style gas stove where gas exits through a number of holes and mixes with atmospheric air. A pilot light ignites the flame. Some raw gas burners use a venturi aspirator to induce some premix air (up to 40–60% stoichiometric air) into the pipe before the gas leaves through the gas orifice (or spuds or ring, if used). This air–gas mixture in the pipe must be kept above the lower explosive limits noted in Table 2.12.

A premix burner requires a metering system with a blower to proportion the air–fuel mixture. The mixture passes through a pipe to a gas orifice at a velocity high enough to prevent flashback. No further gas–air mixing is required. Gas mixing with the air in the atmospheric premix chamber is passed through a reducing valve to obtain a pressure of about 5- to 25-cm H_2O pressure. Both the raw gas burners and the atmospheric premix burners must operate with a draft break to prevent the stack draft from extinguishing the flame by pulling it out.

Forced-draft systems are usually the best burner systems and normally consist of a wind box, gas nozzle, and air nozzle. Air at about 14- to 24-cm

TABLE 2.12. Composition and Properties of Some Gaseous Fuels.

	Gas							
	Natural (Mid-U.S.)	Natural (PA)	LPG Grade A[b]	LPG Grade F[b]	Artificial	Producer	Coke Oven	Carbureted Blue
Typical composition:								
CH_4	97	67.6			20	0.5	25	10
C_2H_6		31.3			1		1	2
CO_2	0.8				3	6	2	5
CO					20	27	8	30
N_2	2.2	1.1			12	55	11	15
H_2					40	10	48	30
Sp gr at SC[a]	0.57	0.71	2.0	1.6	0.50	0.85	0.40	0.52
Heating values, kcal/Nm³	9,350	11,000	30,600	23,700	4,700	1,100	5,100	2,800
Stoichiometric combustion:								
Max. flame temp., °C	2,000	1,980	2,120	2,120	2,010	1,700	1,980	1,980
Max. CO_2 produced, %	11.7	12.3	12	11.6	14	18	11	20.5
Ignition velocity, cm/sec	30	30	25	45	60	20	60	60
Explosive limits gas in air, %:								
Lower	4.8	4.8	1.9	2.2	5.6	18.6	6.0	6.4
Upper	13.5	13.5	8.5	9.5	34.0	73.7	32.0	37.7

[a]Referred to air = 1.0 at SC.
[b]Although sold under pressure as a liquid, values are given for gaseous state at SC.

Hg pressure is brought into a mixing chamber tangentially or cocurrently with the gas. The mixed fuel is ignited as it leaves the surface of the holding block, which is normally made of a refractory.

Most current commercial and residential new gas furnaces operate at an overall efficiency of about 70%. Older units are 50–60% efficient. New systems, such as the condensing heat pipe, will be 90–95% efficient. These systems capture even the latent heat in water, but the water is acidic and the furnace must be constructed of corrosion-resistant parts. The pulse combustion furnace could have even higher efficiencies but may also be noisy.

2.5.3 COMBUSTION PRODUCTS

Gaseous fuels are normally the cleanest fuels available. The absence of sulfur results in very little SO_2 emissions and the ability to be combusted at low excess air results in less NO_x formation and maximized efficiency. However, presence of impurities in the gas and operation at deficiencies of air can result in very undesirable emissions. Note that for conventional premixed burner combustion of pure natural gas (i.e., with no nitrogen), the production of NO_x decreases with increased excess air. Personal data show dry NO_x concentration from burners up to 1.5×10^6 Btu/hr to be about:

Excess Air, %	NO_x, ppm @ 3% O_2
12	90
20	60
40	15
60	5
80	2
100	<1

Typical emissions resulting from gas combustion are given in Table 2.13.

2.6 TRANSPORTATION

Most of our transportation energy is derived from the combustion of a liquid petroleum fuel in an internal combustion engine. Therefore, the fuel, raw material, and emission products are similar to oil and gas combustion operations. Gasoline and diesel fuel, the two major transportation

TABLE 2.13. Typical Uncontrolled Emissions from Gas Combustion [7].

Boiler	Fuel	Uncontrolled Emissions[a]				
		Particulate Matter	Sulfur Oxides[b]	Carbon Monoxide	Hydro-carbons[c]	Nitrogen Oxides[d]
Utility or large industrial	Natural gas	80–240	20.9	272	16	11,200[e]
Commercial or industrial	Natural gas	80–240	20.9	272	48	1,920–3,680
	Butane (LPG)	0.22	0.01	0.19	0.036	1.45
	Propane (LPG)	0.20	0.01	0.18	0.036	1.35
Small commercial or domestic	Natural gas	80–240	20.9	320	128	1.280–1,920[f]
	Butane (LPG)	0.23	0.01	0.24	0.096	1.0–1.5[f]
	Propane (LPG)	0.22	0.01	0.23	0.084	0.8–1.3[f]

[a]Reported as g/10³ m³ natural gas or g/l LPG.
[b]Multiply SO₂ value listed times g sulfur/100 Nm³ natural gas or times g sulfur/100 m³ LPG vapor; pollutant weight reported as SO₂.
[c]Pollutant weight reported as CH₄.
[d]Pollutant weight reported as NO₂.
[e]Multiply NO₂ value by 0.151 exp (−0.0189 L) where L is percent load on boiler; use 4,800 g/10³ m³ for tangentially fired units.
[f]Lower value is for domestic and higher is for commercial heating systems.

fuels, are obtained from the refining of crude oil. Composition of these specific fuels depends on the crude oil from which it is obtained, the refining process used, and the amount of blending of various stocks.

It is unfortunate that the most energy-efficient means of transportation, the bicycle, is so often viewed as a nuisance by many drivers of internal combustion engine devices. For comparison, the bicycle, which ranks first in efficiency among traveling animals and machines, consumes only about 0.15 calories of energy per gram of body weight per kilometer traveled. Automobiles, jet transports and even walking all consume about five times that amount of energy. (Worldwide production of bicycles amounts to 40 million per year.)

In 1988, total U.S. oil consumption is estimated at 14.5 million barrels per day with about 44% being imported [10]. A little over 10% is used by electric utilities with the remainder going for transportation, heating, and chemical industrial use. About 50% of the crude oil is used for motor fuels, distributed as shown in Table 2.14. Note that about 73% of U.S. motor fuel is used on the highways and about 72.5% of all motor fuel used is in the form of gasoline. LPG, mainly butane, has very limited use as a motor fuel.

2.6.1 MOTOR FUELS

Crude oil is fractionated in a refinery to produce a number of petroleum

TABLE 2.14. Approximate U.S. Motor Fuel Usage Distribution.

		Percent of Fuel Used as				
Use	Vehicle Type	Gasoline	Diesel	Jet Fuel	LPG	Total
On Highway	Automobiles	51				51
	Trucks	14.9	6.1			21
	Buses & motorcycles	0.4	0.6			1
	Total	66.3	6.7			73
Off Highway	Aviation	1		12		13
	Industry & construction	1	3			4
	Agriculture	1.5	0.8		0.7	3
	Rail		3			3
	Marine	0.7	1.3			2
	Lawn & garden	2				2
	Total	6.2	8.1	12	0.7	27
Total		72.5	14.8	12	0.7	100

products. The amount of each product varies with the crude and the fractionating process. For example, a crude may be processed to obtain 56%
gasoline, 35% light oils (diesel oil, kerosene, heating oils, and jet fuels),
and 7% heavy oils (lubricating and asphalt oils) by continuous fraction in
a column such as shown in Figure 2.8. Adjustment of physical removal
location, temperatures, and rates can modify the type and amounts of each
cut.

Straight-run or natural gasoline separated directly from the crude oil
equals about 15%, which is not adequate to meet the requirements for gasoline. Most gasoline is obtained by cracking the gas-oil fraction of the
crude oil or polymerization of C_3–C_4 hydrocarbon gases. There is some
synthetic gasoline from coal, waste and other sources, but it does not represent a significant amount.

Cracking consists of thermally and catalytically breaking down the
longer hydrocarbon chains to produce shorter molecules of about C_6 to C_{11}
length. A furnace and catalytic cracker are shown in Figure 2.9. It is necessary to add hydrogen (hydrogenate) to obtain saturated molecules. Improved octane ratings can be achieved by adding lead and other additives,
but this is being prohibited so more aromatic and/or olefin unsaturated
hydrocarbons are being used to obtain the needed octane ratings. Gasoline
can literally contain hundreds of different hydrocarbon species. However,
an equivalent formula to represent the composition of average gasoline is
about $C_8H_{17.5}$. Sulfur compounds can be removed from petroleum products by "sweetening" processes. Also, the sulfur compounds are more

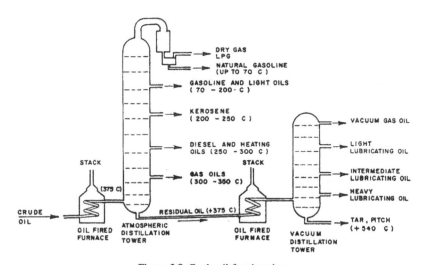

Figure 2.8 Crude oil fractionation.

Figure 2.9 Oil heating furnace and catalytic cracker at the Standard Oil Refinery, El Segundo, California.

likely to stay with the heavier oils. The average U.S. gasoline sulfur content is about 0.032%.

Diesel fuel can be obtained directly from fractionating towers. It is usually withdrawn from the atmospheric tower at about 270°C. Diesel fuel, being heavier hydrocarbons, would contain more sulfur than gasoline. Diesel fuel is essentially the same as Number 2 heating oil in Table 2.10.

2.6.2 MOTOR FUEL COMBUSTION

Most motor fuels are combusted by mixing the finely atomized-vaporized liquid fuel with stoichiometric amounts of air in an internal combustion chamber. This mixture is pressurized, then ignited by either a sparking device or by the cylinder-compression heat. This combustion procedure is used for about 86% of the motor fuel, which includes essentially all the gasoline fuel and the diesel fuel except for rail and some marine use. The rest of the fuel is combusted externally and the gases are then used to turn power turbines, turn generating turbines (i.e., diesel-electric), or to produce steam energy.

Stoichiometric combustion of gasoline and diesel fuels requires an air-fuel ratio of about 15 g air/g fuel. The calculation for average gasoline with air, assuming complete combustion, would be

$$C_8H_{17.5} + 12.375O_2 + 46.55N_2 \rightarrow 8CO_2 + 8.75H_2O + 46.55N_2$$

The air–fuel ratio (F) is then

$$AF = \frac{\text{mass air}}{\text{mass fuel}}$$

$$= \frac{(12.375 \text{ moles } O_2)\left(\dfrac{100 \text{ moles air}}{21 \text{ moles } O_2}\right)(28.9 \text{ g/mole air})}{113.5 \text{ g/mole fuel}} = 15.0$$

Combustion in the internal combustion engine (ICE) is intermittent and exists during only a very small fraction of the total cycle time. Spark ignition (SI) engines are one type of ICE and may use carburetors to pneumatically atomize-vaporize the gasoline fuel. Compression ignition (CI) engines are also ICE and use fuel injection to mechanically atomize the diesel fuel as it is injected directly into the combustion chambers. The diesel engines are more efficient, but, even so, maximum thermal efficiency is only about 40% and overall efficiency of the vehicle energy delivered is about 10%. In comparison, stationary boilers can operate at over 70% thermal efficiencies and 35% overall efficiency for electrical power generation. While both systems are far from ideal, the internal combustion engines could be improved significantly. In addition to improving the engine efficiency, fuel savings are obtained by reducing the mass and speed of the vehicle.

2.6.3 COMBUSTION PRODUCTS

As in the combustion of other fossil fuels, complete combustion is desired at stoichiometric amounts of air. Incomplete combustion results in the increased production of hydrocarbons and in the release of unburned or partially-burned hydrocarbon fuel. The unburned fuel emitted is not usually the same chemically as the original fuel as it undergoes thermal and pressure polymerization reactions. Increasing the amount of air helps reduce the amount of incomplete combustion, but it creates more nitric oxides. This is shown graphically in Figure 2.10. These curves indicate that at high air–fuel ratios there is dilution of the gases and a cooling of the combustion chamber. This cooling finally decreases NO_x concentration as shown.

Engine operation and modifications plus control systems are used on automobiles to meet certain emission levels. Emission standards for light duty vehicles as of 1989 are given in Table 2.15. The grams per mile and actual emission concentrations vary with engine size (volumetric output of gases). In a legal, but not a technical sense emissions in grams per mile can roughly be considered equivalent to:

Pollutant	g/mi	By Volume
HC	0.41	220 ppm
CO	3.4	1.2%
NO_x	1.0	225 ppm

PM-10 particulate emissions are also combustion products, but there are no emission standards for light duty vehicles. A study using receptor modelling and ambient data in a Colorado particulate non-attainment city re-

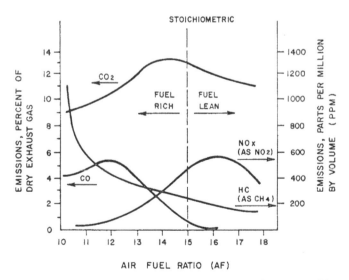

Figure 2.10 Typical effect of air–fuel ratio on automotive exhaust composition.

TABLE 2.15. **U.S. Automobile Emission Standards.**

	Grams per Mile for Model Year			
Pollutant	1977–1979	1980	1981	1983
HC	1.5	0.41	0.41	0.41
CO	15.0	7.0	3.4	3.4
NO$_x$	2.0	2.0	1.0	1.0
Particulates	—	—	—	0.60[a]

[a]For diesel engines only.

ported an average PM-10 filter mass of about 190 $\mu g/m^3$ [11]. The break-down of source contributions varied seasonally as

	Winter	Spring
Vehicle—tailpipe	3.9%	2.3%
Vehicle—fugitive	25.0%	72.8%
Fireplaces	4.5%	3.2%
Woodstoves	66.6%	21.7%

2.7 WASTE COMBUSTION

Copious quantities of waste materials are generated daily. Most of these materials have high heat values and can be readily combusted. Burning provides a mechanism to recover energy (if heat recovery is included in the process) as well as a means to reduce disposal requirements. Typically, a 10 to 1 volume reduction is achieved by incinerating solid wastes. Note that although incineration serves as a waste disposal technique, there are wastes and environmental emissions related to the process. These will be discussed in detail in the following sections.

The terms "combustion" and "incineration" are both used in this discussion and shall be considered synonymous. Combustion is a process of oxidation accompanied by the evolution of light and heat and incineration is the burning of a material to ashes.

Most wastes are solids with varying amounts of moisture, but some are liquids. There are many different types of wastes, different ways to classify these materials, and different techniques to incinerate them. A number of these are presented in this and the following sections. The Bureau of the Census, in their 1985 statistical abstract, report that in 1980 the U.S. solid waste production average in pounds per person per day is

Residential	3.74
Commercial	4.60
Total	8.34

This mainly represents combustible types of material.

The National Solid Waste Management Association (NSWMA) reports that residential and commercial solid waste generation rates differ for various geographic locations. For example, in pounds per person per day their estimate is

Los Angeles, CA	6.4
Philadelphia, PA	5.8
Chicago, IL	5.0
New York, NY	4.0
Tokyo, Japan	3.0
Paris, France	2.4
Toronto, Canada	2.4
Hamburg, W. Germany	1.9
Rome, Italy	1.5

Using average of NSWMA values, the total U.S. production of residential and commercial refuse in 1986 was about 160 million tons per year. About 10.7% of that is recycled and reused. By the year 2000, it is expected that the waste generation rate will increase to 200 million tons per year.

The U.S. EPA notes that disposal costs in 1988 averaged about $12/ton, but in the year 2000 is is expected to be from $100 to $150/ton. The NSWMA and U.S. EPA breakdown of residential refuse is shown in Table 2.16. This unrecycled waste is burned to form energy in many countries. Waste-to-energy (WTE) use in the U.S. currently accounts for about 5% of the waste. The rest is landfilled. WTE use in W. Germany is currently 30%, and in Japan it is 23%.

The Conservation Foundation, in a much more inclusive listing [12], estimates "all" U.S. waste production in 1987 in pounds per person per day and is on a very different basis:

Water pollutants—nonpoint source	63.0
Non-coal mining wastes	35.6
Agricultural wastes	11.0

TABLE 2.16. Estimated Breakdown of Residential and Commercial Refuse.

Component	Mass % before Resource Recovery in 1986	% Recycled	Mass % after Recycling in Year 1986	2000
Paper & paperboard	41.0	9.2	35.6	39.1
Glass	8.2	0.7	8.4	7.1
Metals	8.7	0.6	8.9	8.5
Plastics	6.5	0.1	7.3	9.2
Rubber & leather	2.5	0.1	2.8	2.3
Textiles	1.8	—	2.0	2.0
Wood	3.7	—	4.1	3.6
Food waste	7.9	—	8.9	7.3
Yard waste	17.9	—	20.1	19.0
Other	1.8	—	1.8	1.9

Hazardous wastes	6.5
Animal wastes	6.4
Silvicultural wastes	3.8
Municipal solid wastes	3.6
Air pollutants	3.4
Underground coal mining wastes	2.7
Industrial solid wastes	2.5
Air pollution control sludges	2.3
Brine wastes	2.3
Demolition wastes	1.9
Dredge wastes	1.5
Water pollution control sludges	0.8
Incinerator residues	0.6
Water pollutants—point sources	0.4
Radioactive wastes	0.3
Waste oils	0.2

All values are estimated dry weight except the hazardous wastes. Note that all of these are not combustible. These values are not totaled as the Conservation Foundation acknowledges that some may be double-counted.

Historically, wastes have been sent to landfills, but extreme problems have resulted from this. Groundwater contamination has resulted and toxic chemicals are released into the air. Chemicals volatilize from typical landfill situations at various rates. For example [13], time required for 75% volatilization is:

Dichloroethylene	4.2 months
Chloroform	11.0 months
Dichloroethane	11.7 months
Trichloroethane	2.0 years
Ethyl acetate	2.6 years
Benzene	41.0 years
Phenol	333.0 years

In addition to these problems, the National Solid Waste Management Association estimates that as of 1988, seven states had landfill capacities of less than five years. These included Ohio, Kentucky, West Virginia, Virginia, Pennsylvania, Connecticut, and Massachusetts. There were fifteen states that had an estimated five- to ten-year landfill capacity as of 1988 including Montana, Colorado, Oklahoma, Missouri, Minnesota, Illinois, Indiana, Michigan, Alabama, Florida, Delaware, New York, Vermont, New Hampshire, and Maine. By necessity, other disposal techniques, such as incineration, must be utilized to a greater extent.

2.7.1 WASTE AS FUEL

Municipal solid waste (residential and commercial refuse) is the major source of waste fuel. Industrial waste and bioinfectious waste (hospital infectious waste) are other good and consistent sources. Sewage sludge is also incinerated, but it can be used in land application for soil benefication, depending on heavy metal content. Hazardous wastes are good candidates for incineration, first of all in that they can be destroyed, and second of all, depending on concentration and heat value, in that they may provide enough energy for combustion with little if any supplemental fuel. Ash and residues from burned hazardous waste are also hazardous wastes. Therefore, current recommendations advise not combining hazardous waste with other combustible materials to prevent the ash and residue of the combined burning from being classified as hazardous. Hazardous wastes in very low concentration are encountered at some Superfund sites. Site cleanup can be achieved by passing the contaminated soil through an incinerator. This waste provides no heat value and requires large quantities of fuel, but often it is the most economical and practical cleanup technology.

Industrial wastes may be similar to municipal solid wastes or they may be completely different, e.g., liquid or radioactive. The same can be true for biomedical wastes. These materials require special handling. Industrial wastes could include organic liquids which can be fired as an oil.

Wastes do not need to be incinerated to recover useful energy from them. Solid wastes can be treated by various procedures to produce liquid and gaseous fuels. Pyrolysis, which is the heating of the wastes in a deficiency of oxygen, produces these types of fuels. Pure oxygen is used so that the gaseous emissions are nearly pure CO. Pressure is applied in some systems to compact the charge and aid in the release of hydrocarbon liquid fuels. The liquid can be processed to gasoline or used directly as a fuel after separating the water.

Solid waste in sanitary landfills can be used as a fuel source. The packed earth enclosure permits anaerobic decay to convert waste to a gaseous fuel in the oxygen-deficient cells. Each kilogram of typical refuse can produce about 0.00393 m^3 of gas with a heating value of about 4,450 kcal/m^3. This is equivalent to about 1.5 m^3 of gas production/m^3 of waste or 3,330–3,900 cal/g solid waste. Anaerobic decomposition products of waste are methane, carbon dioxide, water, organic acids, nitrogen, ammonia, and sulfides of iron, manganese, and hydrogen. Quantities vary with time after the solid waste deposit is made. This is shown in Table 2.17 for the major constituents of the gas. Perforated pipe headers can collect these gases and convey them to a processing plant where the gas is enriched, before use as a fuel. This may not be necessary as the gas can often be flared directly at the end of a stand pipe inserted into the landfill. A simple pipe flare is shown in Figure 2.11 where a pipe has been driven into a landfill cell. The combustible waste gases are burning as an invisible flame near the tip of the pipe. Location of this very hot methane flame can be obtained using the *long*-stem temperature sensor. It is much more desirable to collect and use these gases as noted.

TABLE 2.17. Typical Composition and Quantity of Landfill Gas Emissions [14].

Time Since Refuse Covered (days)	Average % by Volume			Cumulative Volume of Gas Produced, m^3/m^3 Waste
	N_2	CO_2	CH_4	
90	5.2	88	5	0.07
180	3.8	76	21	0.07
360	0.4	65	29	0.20
540	1.1	52	40	1.5
720	0.4	53	47	1.5
900	0.2	52	48	1.5
1,080	1.3	46	51	1.5
1,260	0.9	50	47	1.5
1,440	0.4	51	48	1.5

Figure 2.11 Landfill flare for combustion of anaerobic decomposition gases.

2.7.1.1 Waste Classification

There are 7 waste classifications commonly accepted by regulators and generators. They are:

Type 0—*Trash*

> 8500 Btu/lb. A mixture of highly combustible waste, such as paper, cardboard, wood, and floor sweepings from commercial and industrial activities. Contains up to 10% by weight of petrochemical waste, 10% moisture and 5% non-combustible solids.

Type 1—*Rubbish*

> 6500 Btu/lb. A mixture of combustible waste, such as paper, cardboard, wood foliage, and floor sweepings from domestic, commercial, and industrial activities. Contains up to 20% weight of restaurant waste, but little or no petrochemical wastes. Moisture content is up to 25% with 10% non-combustible solids.

Type 2—*Refuse*

> 4300 Btu/lb. An evenly distributed mixture of rubbish and garbage as usually found in as received municipal waste. Contains up to 50% moisture and 7% non-combustible solids.

Type 3—*Garbage*

> 2500 Btu/lb. Consists of animal and vegetable wastes from res-

taurants, cafeterias, hotels, hospitals, markets, and like installations. Contains up to 70% moisture and 5% non-combustible solids.

Type 4—*Human and Animal Remains*
1000 Btu/lb. Consists of carcasses, organs, and solid organic wastes from hospitals, laboratories, abattoirs, animal pounds, and similar sources. Contains up to 85% moisture and 5% non-combustible solids.

Type 5—*By-Product Waste*
Gaseous, liquid, or semi-liquid material such as tar, paints, solvents, sludges, fumes, etc. from industrial operations. Btu values must be determined from the individual materials to be destroyed.

Type 6—*Solid By-Product Waste*
Material such as rubber, plastic, wood waste, etc. from industrial operations. Btu values must be determined from the individual materials to be destroyed.

Typical properties of waste materials are given in Table 2.18. Actual values should be determined by laboratory analyses or test measurements. Table 2.19 lists proximate analyses of common wastes.

2.7.1.2 Municipal Solid Waste (MSW) and RDF

MSW is typical residential and commercial discarded materials. It contains discarded chemicals and other hazardous materials, but as of 1989, it is not classified as a hazardous waste. Normal materials such as discarded pool chlorine (calcium hypochlorite oxidizer), spent brake fluid (ignitable), old batteries (heavy metals such as cadmium, magnesium zinc, and lead), and thermometers (mercury) are just a few examples of toxic substances in typical MSW.

MSW is normally Type 2 waste with a specific gravity of about 0.38 and a heat content of 4300 Btu/lb. In the unprocessed form it contains about 64% combustible material, 26% uncombined water, and 10% non-combustible ash.

Mass-burn and modular facilities incinerate municipal waste almost as it is delivered to them, recovering and recycling only small amounts of aluminum, glass, and paper prior to burning. MSW can also be processed to make refuse-derived fuel (RDF). A minimum of processing by shredding and cleaning produces fluff RDF. Heating values of this material range from 5000 to 5500 Btu/lb and nearly all the MSW is recovered. Fluff RDF

TABLE 2.18. Properties of Waste Materials.

Waste	Heating Value, BTU/lb as Fired	Wt., lbs per cu ft (loose)	Content by Weight, %	
			Ash	Moisture
Type 0 waste	8,500	10	5	10
Type 1 waste	6,500	10	10	25
Type 2 waste	4,300	24	10	50
Type 3 waste	2,500	35	5	70
Type 4 waste	1,000	55	5	85
Newspaper	7,975	7	1.5	6
Brown paper	7,250	7	1.5	6
Magazines	5,250	35	22.5	5
Corrugated paper	7,040	7	5.0	5
Plastic coated paper	7,340	7	2.6	5
Coated milk cartons	11,330	5	1.0	3.5
Citrus rinds	1,700	40	.75	75
Shoe leather	7,240	20	21.0	7.5
Butyl sole composition	10,900	25	30.0	1.0
Polyethylene	20,000	40–60	0	0
Polyurethane (foamed)	13,000	2	0	0
Latex	10,000	15	0	0
Rubber waste	9,000–13,000	62–125	20–30	0
Carbon	14,093		0	0
Wax paraffin	18,621		0	0
Tar or asphalt	17,000	60	1	0
1/3 tar-2/3 paper	11,000	10–20	2	1
Wood sawdust	7,800–9,600	10–12	3	10
Wood bark	8,000–9,500	12–20	3	10
Corn cobs	8,000	10–15	3	5
Rags (linen or cotton)	7,200	10–15	2	5
Animal fats	17,000	50–60	0	0
Cotton seed hulls	8,600	25–30	2	10
Coffee grounds	10,000	25–30	2	20
Linoleum scrap	11,000	70–100	20–30	1

lasts about one week before significant breakdown (decomposition) occurs.

Additional processing of the raw MSW produces a medium-grade RDF with heating values of 5500 to 6000 Btu/lb. This requires a primary trommel, shredder, and shredder screener operations. Only about 80% of the raw MSW is recovered as processed fuel.

High-grade RDF with heating values of 6000 Btu/lb and higher can be made by additional processing. However, to reach 6000 Btu/lb only about 50% of the raw MSW is recovered as fuel.

TABLE 2.19. Proximate Analyses of Waste Materials.

Components	Moisture	Volatile Matter	Fixed Carbon	Non-Combustibles
Paper products				
Paper, mixed	10.24	75.94	8.44	5.38
Newsprint	5.97	81.12	11.48	1.43
Brown paper	5.83	83.92	9.24	1.01
Magazine	4.11	66.39	7.03	22.47
Corrugated boxes	5.20	77.47	12.27	5.06
Food waste				
Vegetable waste	78.29	17.10	3.55	1.06
Citrus rinds	78.70	16.55	4.01	0.74
Meat scraps	38.74	56.34	1.81	3.11
Vegetation				
Green logs	50.00	42.25	7.25	0.50
Furniture wood	6.00	80.92	11.74	1.34
Evergreen shrubs	69.00	25.18	5.01	0.81
Flowering plants	53.94	35.64	8.08	2.34
Lawn grass	75.24	18.64	4.50	1.62
Wood and bark	20.00	67.89	11.31	0.80
Domestic waste				
Tires	1.02	64.92	27.51	6.55
Leather	10.00	68.46	12.49	9.10
Leather shoe	7.46	57.12	14.26	21.16
Shoe heel and sole	1.15	67.03	2.08	29.74
Rubber	1.20	83.98	4.94	9.88
Mixed plastics	2.00	—	—	10.00
Polyethylene	0.20	98.54	0.07	1.19
Polystyrene	0.20	98.67	0.68	0.45
Polyurethane	0.20	87.12	8.30	4.38
Polyvinyl chloride	0.20	86.89	10.85	2.06
Linoleum	2.10	64.50	6.60	26.80
Municipal wastes, 3000 to 6000 Btu/lb	35–15	37–65	0.6–15.0	27.0–15.0

Densified RDF can be made if desired by pelletizing the RDF. This material has a longer product storage life and can be shipped more easily. It can also be burned in conventional fossil fuel combustion systems.

2.7.1.3 Bioinfectious Waste (Red Bag)

Hospital waste is typically a high-quality waste with heating values of 8500 to 9400 Btu/lb, with the higher value being more typical because of the substantial plastic content. Composition is about 85% combustibles, 10% moisture, and 5% ash. The combustibles are 60% paper/cardboard, 35% plastic, and 5% food and other. This waste is referred to as "red bag" waste and often contains infectious or biologically contaminated material which could be a health hazard. It contains paper, plastics, dressings, bandages, masks, gowns, needles, glassware, and all types of hospital material except body parts and fluids which are pathological wastes.

Waste generation rates in a hospital are 13 lbs/day per occupied bed and for a rest home it is 3. A laboratory would add 0.5 lb/day per patient and the cafeteria adds 2 lb/day per meal served. PVC plastic has a heat value of 9750 Btu/lb and stoichiometric air required to combust it is 8 lb air/lb waste. Respective values for other plastics normally found in hospital wastes are:

Polyurethane	11,200	9
Polystyrene	16,420	13
Polyethylene	19,690	16

2.7.2 SOLID WASTE COMBUSTION

Solid wastes can be combusted in conventional-type boilers or, as is more common, they can be incinerated. Gaseous and liquid waste fuels can also be combusted, the same as traditional gaseous and liquid fuels. This section is only a brief, generalized discussion relative to conventional boiler combustion of solid wastes. The reader is referred to Section 2.8 for a more comprehensive presentation.

Solid wastes are shredded and processed to produce enhanced-RDF fuel, as was noted. They can be burned alone or mixed with coal or oil, for example, to recover energy from the waste. This also serves to reduce undesirable emissions of sulfur oxides by reducing fossil fuel needs. Small, fluff waste can be pneumatically conveyed to boilers like pulverized coal, while larger, shredded waste can be handled by stoker-type boilers. Combustion reduces the weight of the refuse by 80–90% and the volume by 95% and generates a sterile residue that can have beneficient soil value. The char residue may have a density of about 2.2 metric ton/m³.

Fossil fuel is usually required at the startup of combustion operations. Large "cold" facilities must be brought to operating temperature slowly and may require a two-day startup period. A "hot" facility can be brought to operating temperature in one to three hours. The flue gas temperatures reach as high as 1,250°C (2,280°F) for waste-heat boiler systems.

Supplemental fuel is usually needed when metal slags are being produced from wastes, but this depends on the heat value of the waste. In such facilities, fossil fuel is used both as a means to keep slag taps clean and as pilot-burner fuel. It takes about 2.5 hours to reduce a typical engine block to slag in these systems. Air–fuel ratios of about 5.5 are required for some wastes, but combustion of material such as meat fat requires ratios as high as 12.

2.7.3 COMBUSTION PRODUCTS

The flue gases produced by direct flame combustion of solid waste are usually about 1,000–1,100°C (1830–2010°F) depending on the waste composition and water content and type of combustion. Water is often added during the processing stage for safety, health, and conveying purposes and drying of the waste could be required. A typical flue gas composition, on a dry basis, is given in Table 2.20.

If combustion air is inadequate, more CO will be produced and, more adversely, a reducing flame would result in the formation of phosgene from the chlorides in the plastic materials. The acids and moisture in the flue gases can be corrosive to the boiler if temperature drops.

Particulate concentrations range from about 0.1 to 5 g/Nm³ at 12% CO_2. The particulates are usually bimodal in size distribution with the burned ash being the smaller material. The larger material is entrained unburned

TABLE 2.20. Typical Flue Gas Composition from Direct Combustion of Solid Waste (Dry Basis).

CO_2	8%
CO	<0.1% (i.e., <1,000 ppm)
O_2	12.5%
Organic acids	150 ppm
NO_x	100 ppm
SO_2	100 ppm
HCl	150 ppm
HF	1 ppm
N_2	Balance

waste. Fly ash is the smaller at 30%, is less than 1 μm in size, and has a standard geometric deviation of about 2.3.

Pyrolysis systems generate smaller quantities of emissions, but these contain particulate (mainly oil mist) and gaseous pollutants. Typical gaseous pollution concentrations from pyrolysis of wastes contain about 70 ppm NO_x, 115 ppm SO_2, and 270 ppm CO.

Example Problem 2.3

Consider a municipal refuse with the ultimate analysis of: 28.16% moisture, 25.62% carbon, 21.21% oxygen, 3.45% hydrogen, 0.10% sulfur, 0.64% nitrogen, and 20.82% ash. The heating value is 2.35 kcal/g and the density is 0.38 g/cm^3. The flue gas analysis on a wet basis shows 7.2% CO_2, 10.9% O_2, 300 ppm CO, and 100 ppm NO_x. Determine (1) the percent excess air, (2) the Nm^3 of air that must be supplied by a blower for a 10-m^3/hr incinerator, and (3) the concentration of SO_2 in the flue gases.

Solution

(1) On the *basis of 100 g refuse*, the distribution by moles from the ultimate analysis is

$$\frac{28.16}{18} = 1.56 \text{ g moles } H_2O$$

$$\frac{25.62}{12} = 2.14 \text{ g atoms C}$$

$$\frac{21.21}{32} = 0.66 \text{ g moles } O_2$$

$$\frac{3.45}{2} = 1.73 \text{ g moles } H_2$$

$$\frac{0.10}{32} = 3.13 \times 10^{-3} \text{ g moles S}$$

$$\frac{0.64}{28} = 0.02 \text{ g moles } N_2$$

Based on stoichiometric combustion equations for C, H_2, and S, and accounting for the available O_2, the O_2 required from the air for complete combustion is

$$2.14 + 1.73/2 + 0.003 - 0.66 = 2.35 \text{ g moles}$$

The moles of excess O_2 from the flue gas analysis and a carbon balance equals

$$\left(\frac{10.9 \text{ moles } O_2}{7.2 \text{ moles } CO_2}\right)\left(\frac{1 \text{ mole } CO_2}{1 \text{ atom C}}\right)(2.14 \text{ atoms C}) = 3.24 \text{ moles } O_2$$

[Note, the first term in the previous line should be

$$\frac{(10.9 - 0.03 \text{ for CO}) \text{ moles } O_2}{7.2 + 0.03 \text{ moles } CO_2 + CO}$$

but accuracy of data and these values do not justify this.]
Therefore, the air supplied is $2.35 + 3.24 = 5.59$ moles O_2

$$\text{Percent excess air} = \left(\frac{3.24}{2.35 + 0.66}\right)(100) = 108\%$$

(2) The moles of air supplied

$$= (5.59 \text{ moles } O_2)\left(\frac{100 \text{ moles air}}{21 \text{ moles } O_2}\right)$$

$$= 26.6 \text{ g moles air/100 g refuse}$$

and the volume of air supplied by the blower is

$$\left(\frac{26.6 \text{ moles}}{100 \text{ g}}\right)\left(\frac{22{,}400 \text{ cm}^3}{\text{g mole}}\right)\left(\frac{293}{273}\right)\left(\frac{0.38 \text{ g}}{\text{cm}^3}\right)\left(\frac{10 \text{ m}^3}{\text{hr}}\right)$$

$$= 2.43 \times 10^4 \text{ Nm}^3 \text{ air/hr}$$

(3) The moles of N_2 in the flue gas comes from the air and the fuel.

N_2 combined as NO_x =

$$\left(\frac{0.01 \text{ mole } NO_x}{7.2 \text{ moles } CO_2}\right)\left(\frac{1 \text{ mole } CO_2}{\text{atom C}}\right)\left(\frac{2.14 \text{ atoms C}}{100 \text{ g refuse}}\right)$$

$$= \frac{0.003 \text{ moles } NO_x}{100 \text{ g refuse}}$$

N_2 in flue gas =

$$(5.59)\left(\frac{79 \text{ moles } N_2}{21 \text{ moles } O_2}\right) + 0.02 - 0.003 \text{ NO}_x = 21.05 \frac{\text{moles } N_2}{100 \text{ g refuse}}$$

Therefore total moles flue gas/100 g refuse

$$= (CO_2 + CO) + O_2 + H_2O + SO_2 + N_2 + NO_x$$

$$= (2.14) + 3.24 + (1.73 + 1.56) + 0.003 + 21.05 + 0.003$$

$$= 29.73$$

The SO_2 concentration on a wet gas basis is

$$\left(\frac{3.13 \times 10^{-3}}{29.73}\right)(10^6) = 105 \text{ ppm}$$

and on a dry basis is

$$\left[\frac{3.13 \times 10^{-3}}{29.73 - (1.73 + 1.56)}\right](10^6) = 118 \text{ ppm}$$

Note

In checking the final flue gas concentration, this gives

$$\left(\frac{2.14 \text{ moles } CO_2 + CO}{29.73 \text{ moles Total}}\right)\left(\frac{7.2 \text{ moles } CO_2}{7.2 + 0.03 \text{ moles } CO_2 + CO}\right)$$

$$\times (100) = 7.2\% \text{ } CO_2$$

$$\left(\frac{3.24}{29.73}\right)(100) = 10.9\% \text{ } O_2$$

which are the reported flue gas analysis values. Normally, there would be air leaks, fuel composition variations, operating variations, analysis errors, etc., which would result in differences.

Values of CO are often specified in regard to some base such as 12% CO_2 (dry basis) or 50% excess air (see Section 2.12). Correcting the given value of 300-ppm CO could be done simply by

$$\text{Concentration of CO corrected to 12\% CO}_2 = (300)\left(\frac{12}{8.1}\right)$$

$$= 444 \text{ ppm CO}$$

where

$$\text{dry \% CO}_2 = \left[\frac{2.14}{29.73 - (1.73 + 1.56)}\right]\left(\frac{7.2}{7.2 + 0.03}\right)(100) = 8.1\%$$

Concentration of CO corrected to 50% excess air

$$= (300)\left(\frac{208}{150}\right) = 416 \text{ ppm CO}$$

2.8 INCINERATION

Incineration is becoming an increasingly important tool for disposing of hazardous wastes, for converting waste to energy, and for solving a whole host of other industrial/environmental problems. For instance, incineration can also be used for the purpose of odor control—disposing of low-concentration odorous materials that have essentially no heat value as fuel. The incinerator feed could also be a low-concentration toxic waste such as PCB in soil with a highly negative heat value (i.e., much energy is required to heat up the feed charged to the incinerator for the purpose of "cleaning up" the soil).

Incinerators, as well as other combustion processes which promote oxidation reactions, must be designed to provide the following: adequate time for complete oxidation to occur; a sufficiently high temperature to enable oxidation within the given time; and enough turbulence so that all material to be oxidized is in contact with the oxygen at the time and temperature provided. These are, as was noted, the basic "3 T's" of combustion processes—time, temperature, and turbulence.

Classical incineration can be accomplished by direct-flame, thermal, or catalytic processes. Flame incineration is like other combustion operations. Temperatures from about 1600°F (870°C) up to the actual flame temperatures are required for about 0.3 sec or more. Direct incineration is most common for the disposal of solid or liquid wastes.

Solid combustible wastes are always charged to the primary combustion chamber to assure that the required DRE (destruction-removal efficiency) occurs. Liquid wastes can be successfully charged to either the primary or secondary, depending upon the system. In either case, they behave like liquid fuels. For organic liquids, the burning time is proportional to the 'square of the droplet diameter; and for aqueous liquids, droplet evaporation is proportional to the square of the droplet diameter. In both cases, good atomization is essential—especially when the liquids are charged to the secondary. In order to properly atomize liquids (and remembering these can be liquid wastes) the liquids must be filtered to remove solids. This is required to prevent the atomizers from plugging and failing.

Catalytic incineration requires the presence of a catalyst, usually a noble metal, to speed up the oxidation reaction. This can permit oxidation at temperatures as low as 570–930°F (300–500°C), depending on the type of fuel. Losses of the catalyst must be minimized because most of the noble metal catalysts are both hazardous and expensive. Catalytic incineration is often used in odor control processes. Catalysts are easily blinded or poisoned by contaminants in the fuel or air and usually are economically attractive only when these impurities are controlled.

2.8.1 INCINERATION TERMINOLOGY

Incineration is becoming such an important field that readers should be familiar with the most commonly used terms. Many of these terms are also common general combustion expressions.

Burn Rate—Total quantity of combined carbon and hydrogen that is converted to CO_2 and H_2O vapor. Usually expressed in pounds per hour.

Charge Rate—Quantity of waste material loaded into an incinerator, but not necessarily burned. Usually expressed in lb/hr.

Stuff and Burn—Where charging rate of burnable material is greater than the actual burn rate.

Controlled Air—Controlling air flow to attain desired rate of combustion. Refers to a system with two or more combustion chambers.

Primary Combustion Chamber—The combustion zone where solid fuels are fed to so they can heat, devolatize, and ignite. Often operated in a reducing atmosphere.

Secondary Combustion Chamber—The final combustion zone of a two-stage incinerator, designed to completely combust gases and all volatile material. Operated at excess air. May be called an afterburner.

Stoichiometric—Theoretical air required for complete combustion to CO_2, H_2O vapor, and other oxides.

Pyrolysis—Chemical destruction of organic materials in the presence of heat and the absence of oxygen.

Starved Air—Controlled burning at less than stoichiometric air requirements. Usually in the primary chamber.

Excess Air—Controlled burning at greater than stoichiometric air requirements.

Exothermic—Chemical reactions that liberate heat to surroundings; e.g.,

$$C_{(s)} + O_{2(g)} \rightarrow CO_{2(g)}$$
$$(\Delta H = -94,052 \text{ cal/g mole or } -169,294 \text{ Btu/lb mole})$$

$$C_{(s)} + \tfrac{1}{2}O_{2(g)} \rightarrow CO_{(g)}$$
$$(\Delta H = -26,416 \text{ cal/g mole or } -47,549 \text{ Btu/lb mole})$$

$$H_{2(g)} + \tfrac{1}{2}O_{2(g)} \rightarrow H_2O^{(g)}$$
$$(\Delta H = -57,798 \text{ cal/g mole or } -104,036 \text{ Btu/lb mole})$$

where ΔH is change in enthalpy at 298 K.

Endothermic—Chemical reactions that absorb heat from surroundings; e.g.,

$$C_{(s)} + H_2O \text{ (Steam)} \rightarrow CO_{(g)} + H_{2(g)}$$
$$(\Delta H = +31,382 \text{ cal/g mole or } +56,488 \text{ Btu/lb mole})$$

Heat Release—Total energy released from combustion; i.e., heating value (Btu/lb) times burn rate (lb/hr).

Burn Out—Amount of combustible material in fuel converted to CO_2 and H_2O.

Grate Heat Release Rate—A measure of incinerator capacity in Btu/ (hr ft² plan area).

Volumetric Heat Release Rate—A measure of incinerator capacity in Btu/ (hr ft³ volume).

Heating Value—Net energy available from chemical combustion. Expressed in Btu/lb.

Proximate Analysis—Determination of volatile matter, fixed carbon, moisture, and ash (non-combustible matter) in any given waste material.

Volatile Matter—That portion of waste material which can be liberated with the application of heat only. In two-stage controlled-air incineration, the volatile matter is released in the primary and burned in the secondary combustion chamber.

Fixed Carbon—The non-volatile portion of waste which must be burned at higher temperature.

Moisture—Both formed and contained which must be evaporated from the waste material by the heat released from the waste material or the supplemental fuel.

Pathological Waste-Material—From the study and treatment of diseases.

Pathogenic Waste-Material—Capable of causing disease.

Particulate Emissions—Fine solid and/or liquid matter suspended in combustion gases carried to the atmosphere. May be expressed in GR/DSCF corrected to 7% O_2 or other appropriate base.

Afterburner Gas Retention Time—Amount of time volatile matter is exposed to turbulent mixing, elevated temperature, and excess air for final combustion.

Boiler—Heat transfer portion of system where water is turned to steam.

Bottom Ash—Residue from the furnace.

Grate—Section of furnace used to support the waste material.

Tipping Fee—Cost charged to dispose of waste at incinerator or other disposal area.

Tipping Floor—Area where solid waste is delivered at incinerator site.

Trommel—Perforated, rotating horizontal cylinder used to separate solid waste and to remove cans from bottom ash.

Waterwall Facility—Combustor with vertical water tube walls to remove heat when waste is burned.

2.8.2 TYPES OF INCINERATORS

Direct-flame thermal and catalytic incinerators have been introduced with the note that direct flame is the most useful type for waste destruction. Direct-flame incineration includes open (trench) burning, single-chamber, multiple chamber, and controlled-air incinerators. Some areas have had regulations banning the use of open-burning and single-chamber incinerators since as early as 1957 [15].

The common terminology used to describe incinerators that burn waste is: waste-to-energy (WTE) facilities, controlled-air incinerators, rotary kilns, and fluidized beds. Fluidized beds are replacing multiple-hearth incinerators for sludge burning. Some of these will be discussed in the following subsections. In addition, trench or curtain-of-air destructors may be found as portable field units to clean up trees and brush from cleared areas and dispose of rubber and other refuse. They consist of a pit and an air blower with plenum chamber nozzles to direct air into the pit combustion zone. The cost, as well as the efficiency, of these units is low.

Domestic incinerators are sold for in-house use and may contain an auxiliary burner. Residential dwellings may use a flue feeding system. Pathological incinerators are designed to dispose of animal remains and other high-moisture organic material in a biologically clean manner. They are usually gas-fired batch-fed units with the waste placed on a hearth in the combustion chamber. Afterburners should be used to ensure complete combustion.

Transportable and mobile incinerators are used mainly for cleanup of hazardous materials at contaminated sites (see Reference [16] for details on these).

2.8.2.1 Waste-to-Energy (WTE) Facilities

These are municipal WTE units to recover energy from the MSW (municipal solid waste). As of late 1988, there were 111 facilities existing and 91 planned in the U.S. [17]. Breakdown by type is given in Table 2.21. WTE facilities disposal capacity is expected to increase from 60,000 tons per day (TPD) in 1988 to 250,000 TPD by the year 2000. Most of the systems burn the MSW as received (i.e., mass burn). All the modular facilities are under 500 tons/day. Newer systems are larger waterwall units with some larger than 1,000 tons/day. An inclined grate, waterwall mass burn system schematic is shown in Figure 2.12. RDF (refuse derived fuel) incineration systems would be similar except that fuel processing equipment would be used before combusting. Modular units are more of the controlled air type. These, as well as rotary kilns and fluidized beds are discussed in the next subsections.

Large MSW-fired boilers are able to obtain 70% thermal efficiencies. They operate at about 1600°F (870°C). Typical heat losses are: 14.2% to

TABLE 2.21. Waste-to-Energy Recovery Plants as of 1989 [16].

Facility Technology	Percent of	
	111 Existing	91 Planned
Mass burning—Water wall incineration	18.9	51.6
Modular	48.6	16.5
Rotary kiln	3.6	12.1
With sludge incineration	1.8	3.3
Refractory furnace	4.5	2.2
RDF burning—Shredded	11.7	9.9
Fluidized bed	1.8	3.3
Co-fired with coal	3.6	4.5
Other technologies	0.9	1.1

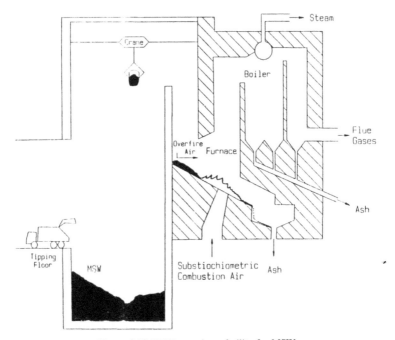

Figure 2.12 WTE mass burn facility for MSW.

flue gas heat loss; 4.6% to fuel moisture; 8.4% to water as vapor; 1.5% to incomplete burnout; and 1.8% to other losses. Burnout is a very important factor to control relative to thermal efficiency.

WTE incinerator ash has the potential to be a toxic waste. So far, the EPA exempts the ash when energy is recovered. Table 2.22 is a summary of some inorganic elements in MSW combined bottom plus fly ash as reported by the EPA. Oxygen, chlorine, and sulfate are excluded.

2.8.2.2 Controlled-Air Incinerators

A two-chamber unit is shown in Figure 2.13. This is used for all municipal, industrial, and bioinfectious wastes. The lower chamber is the primary and is usually operated at 60 to 80% stoichiometric air. Air enters under and/or around the grates. Some models use heat exchange surfaces to preheat the combustion air. Waste is charged onto the hearth either before firing or during the burn depending on the type of system. Excess air is added to the gases entering the secondary (the upper chamber). These gases may also be preheated using heat exchangers. Supplemental fuel may or may not be added to either chamber. The smoke from the smoldering primary fuel is completely combusted in the secondary.

2.8.2.3 Rotary Kiln Incinerator

A rotary kiln is a refractory lined horizontal cylinder fed with waste at the upper end and discharging ash residue at the lower end. This cylinder is sloped about 3° and rotates slowly at 1–5 rpm. Air and gas flow may be cocurrent with the waste flow or countercurrent. The feed is augered or ram fed. Length can vary from a few feet to several hundred feet. Figure 2.14 shows one type of system.

Rotary kilns are excess air incinerators with a hearth rather than a grate. Kilns typically operate at 1400–2300°F (760–1260°C) with from 50–250% excess air. Kiln afterburners operate at 2000–2500°F (1090–1370°C) at 120–200% excess air for up to 2 seconds. In the kiln, at 1–5 rpm, residence time may reach 1 1/2 hours. Heating value of fuel plus waste should average 8,000 Btu/lb (4,440 cal/g) or more. Average U.S. kiln heat release capacity is 61.4 × 10⁶ Btu/hr (15.5 × 10⁶ kcal/hr). Maximum feed rate should be such that it does not exceed 20% of kiln volume.

The advantages of a kiln are that it is a simple device, combustion and feed control are easy, feed can be large in size, and many different types

TABLE 2.22. **Significant Concentrations of Inorganic Elements in Typical MSW Combined Ash.**

Elements	Pounds/Ton of Ash		
Barium	0.16	to	5.40
Chromium	0.02	to	3.00
Lead	0.06	to	73.20
Mercury	below detection	to	0.04
Aluminum	10.00	to	120.00
Calcium	8.2	to	170.00
Copper	0.08	to	11.8
Iron	1.38	to	267.00
Magnesium	1.40	to	32.00
Manganese	0.03	to	6.26
Nickel	0.03	to	25.82
Phosphorus	0.58	to	10.00
Potassium	0.58	to	24.00
Silicon	2.76	to	392.14
Sodium	2.20	to	66.60
Titanium	2.00	to	56.00
Zinc	0.18	to	92.00

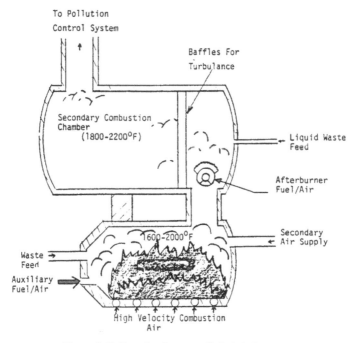

Figure 2.13 Two chamber controlled air incinerator.

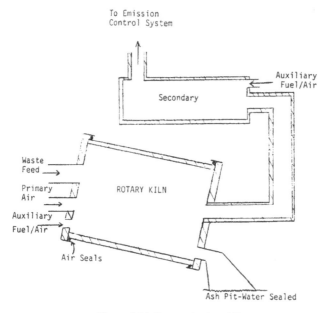

Figure 2.14 Cocurrent rotary kiln.

of material can be fed. Air inleak is a problem and seal leaks increase as temperature increases.

Rotary kiln incinerators can be designed on either a non-slagging or slagging basis. The non-slagging design is to avoid melting the inorganic materials in the kiln. Slagging kilns remove the ash as a liquid.

2.8.2.4 Fluidized Bed Incinerators

This is the newest technology, and it is extremely promising. Fluidized bed combustion (FBC) can be used for both incineration and for fossil fuel combustion as discussed in Section 2.3.2. This procedure uses a bed of sand and/or ash and limestone as a heat sink and the combustion medium. Combustion air rises through a bed support plate. The pressure drop across this plate is required to assure uniform air flow across the fluidized bed. The fuel/waste can be fed any or all of: below the bed, into the bed, or above the bed. Aqueous wastes are usually fed above the bed to provide time for the moisture to evaporate before the material reaches the bed.

Gas velocity through the bed must be adequate to produce fluidization of the bed. Beyond that, two types of operation can be achieved—bubbling mode or circulating mode. The bubbling mode is when gas velocities are rising at 2 to 10 ft per second (0.6–31 m/sec). Under these conditions, most of the bed material does not leave the reaction chamber. Figure 2.15 is a schematic of the reactor. Figure 2.5 is another variation. Another new design incorporates a concave downward grate to increase bed temperatures to about 2,000°F.

In the circulating mode, gas velocities are from 10 to 20 ft per second (3.1–6.1 m/sec), and most of the bed is in constant circulation out of the reactor to the hot cyclone and back to the reactor. The circulating fluidized bed combustor is known as a CBC, and a system is shown in Figure 2.16. Return from the hot cyclone can be mechanical or pneumatic. Bed material size and density actually determine at what gas velocity the different fluidization modes occur. Remember that stoker coal boilers operate near fluidization velocities and 76% of the ash is blown out with the flue gases. FBC ash removal can be accomplished from either the bed, the hot cyclone, or elsewhere downstream.

FBC systems have several advantages. Section 2.2 notes that combustion efficiency is most affected by turbulence. FBCs are extremely turbulent, and they also provide good holdup/contact time for the air and the fuel. Therefore, temperature can be reduced while still achieving excellent combustion efficiencies. Optimum FBC temperature for low NO_x production and high acid gas sorption is about 1550°F (840°C). Low NO_x is a benefit of FBC, and values of 50 ppm can be achieved.

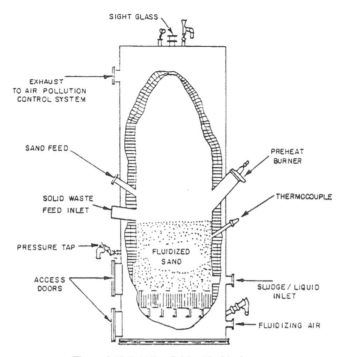

Figure 2.15 Bubbling fluidized bed incinerator.

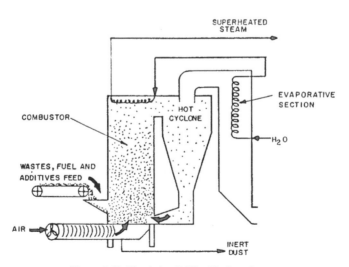

Figure 2.16 Circulating fluidized bed combustor.

137

It was noted that the bed could contain limestone ($CaCO_3$). This calcines to lime (CaO) with the release of CO_2. The lime combines with acid gases such as SO_2, HCl, and other fluorides and produces an alkali salt. Acid gas control as high as 90% can be achieved in a FBC at alkali to acid gas stoichiometrics of about 2.0.

Another advantage of FBCs is that they provide a high heat release because of the turbulence and good contact with heat exchange surfaces. Fluidized beds can also combust extremely low heating value wastes with no supplemental fuel required. Material with heat content as low as 3,000 Btu/lb (1,700 cal/g) can be self-combusted. Feed to FBCs must be 2 inches or smaller in size. FBCs are more difficult to control and, for example, are sensitive to feed variations. If the fuel feed is interrupted the flame could extinguish, as the amount of fuel in the bed at any time is very small. Feed cannot normally be interrupted and reinitiated without use of start-up fuel. Inorganics in the waste or otherwise slagging fuel can result in clinkers which cause fluidization to stop and shut down the system.

Sewage sludge incineration systems are converting to FBC systems [18]. Typical FBC sludge combustors are bubbling type, 8 to 22 feet in diameter by 35 feet high, used to process up to 10 tons wet sludge per hour. The bed is sand, but when auxiliary fuel is required, crushed coal is normally used so ash is also a major component. Coal is usually fed above the bed and the liquid sludge is fed into the bed.

2.8.3 DESIGN FACTORS

In addition to the specifics given for each incinerator, Table 2.23 summarizes design factors for several incinerators and for some conventional fossil fuel combustion systems.

2.8.4 COMBUSTION PRODUCTS

Sections 1.5 through 1.8 serve as indicators to forewarn us of potential problems that can be associated with incinerator emissions. In order to make a complete system balance, samples need to be taken and analyzed for all of waste feed, fuel, water, chemical additives, ash, and effluents. There exists the possibility that original chemicals can pass through the incinerator and that products of incomplete combustion (PICs) can be formed in addition to the traditional combustion emission products. The presence of halogens, especially the large quantities of chlorides, can result in chlorinated products being formed. Hydrogen chloride gas is one

TABLE 2.23. Typical Combustion System Design Parameters.

Furnace Type	Excess Air, %	Superficial Gas Flow Rate, ft/sec	Normal Temperature Range, °F	Retention Time	Grate Heat Release, Btu/(hr ft² Plan Area)	Volumetric Heat Release, Btu/(hr ft³ Volume)
Stoker	20–100	—	1800–2600	>2 sec (gas)	750,000	17,500
Pulverized coal (PC)	15–50	—	1900–2500	>2 sec (gas)	1,250,000	17,500
FBC, bubbling	20–25	2–10	1450–1700	>2 sec (gas)	725,000	22,500–35,000
FBC, circulating	15–20	10–20	1450–1700	>2 sec (gas)	1,500,000	20,000–30,000
Rotary kiln	50–250	10–15	1400–2900 (<2000 non-slagging) (>2000 slagging)	min-hr (solids)	—	15,000
Liquid injection (conventional)	20–60	12–15	1800–2400	>2 sec (gas)	—	25,000

product, but chlorinated organic compounds can be even worse. Dioxin or PCDD (polychlorinated dibenzo dioxins) and furans or PDCF (polychlorinated dibenzo furans), of which there are several hundred family variations, can be formed. Some of these materials are reported to be the most toxic chemicals known to humans.

Combustion of solid waste is reported [19] to produce small amounts of dioxin (<200 ng/dscm corrected to 7% O_2) and furan (<500 ng/dscm @ 7% O_2). Note that a nanogram (ng) is 10^{-9} grams. This report indicates that dioxin production is a result of some different operating conditions than is furan, but that both are formed in incinerators at temperatures $>1250°F$ ($680°C$) and that more is formed when excess oxygen is $<5\%$. Furans are formed more easily and are harder to destroy. Amount of chlorine in the feed seems to have little if any effect on their formulation although some chlorine must be present.

Uncontrolled particulate and gaseous emissions from FBC incineration of RDF solid waste are listed in Table 2.24. The presence of SO_2, HCl, and other substances can be estimated directly knowing the composition and quantity of sulfur compounds, plastics, and other substances in the refuse. The quantity of SO_2 from domestic wastes is usually about 0.25 kg/metric ton refuse and, from municipal waste, it is about 0.75 kg/metric ton [6]. Note that amounts of particulates and nitrogen oxides especially will vary with type of incinerator, operation, and refuse. Controlled emissions from a survey of 15 MSW combustion facilities are reported in Table 2.25 [20]. These systems all have heat recovery and control devices for particulates and/or gases. System sizes range from 50 to 1,000 ton MSW/day mass burn. Most systems are waterwall with reciprocating grates.

TABLE 2.24. Typical Uncontrolled Emission
Factors for RDF Waste Fuel in a FBC.

Emissions	Emission Rate, lb/ton Waste
Particulates	260
SO_2	8
NO_x	4.7
HCl	9.96
CO	3.26
Total HC	0.56
Pb	0.66
Hg	0.00168
HF	0.041

TABLE 2.25. Average Controlled
Emissions from MSW Facilities.

Dry Gas Analysis, %:	
O_2	11.3
CO_2	8.5
Moisture, %	14.3
Particulates, gr/dscf @ 7% O_2	0.015
Gases, ppm @ 7% O_2:	
CO	126
Total HC	2
SO_2	53
HCl	89
HF	1.2
NO_x (as NO_2)	212
Metals, ng/dscm @ 7% O_2:	
arsenic	0.25
beryllium	2
cadmium	5
chromium	337
lead	53
mercury	167
nickel	158
PCDD & PCDF, ng/dscm	136
TCDD & TCDF, ng/dscm	7
Aldehyde, mg/dscm	0.7

The following listing serves as an example to identify the fate of various constituents in waste feeds:

Starting Material	Product
moisture	water vapor
	water condensate
hydrogen	water vapor
nitrogen	NO gas
carbon/HC	CO_2 gas
	CO gas
	C particles
	PICs
sulfur	SO_2 gas
chlorine	HCl vapor
	chlorinated organics
other halogens	same as Cl
"ash"	"fly ash"
asbestos	asbestos
mercury	mercury vapor
lead	metal oxide vapors
antimony	metal oxide vapors
beryllium	BeO particles
arsenic	metal oxide particles
cadmium	metal oxide particles
chromium	metal oxide particles
nickel	metal oxide particles

Example Problem 2.4

Determine the amount and type of uncontrolled emissions and make a comparison for the incineration of an odorous gas stream from meat processing by direct incineration in an oil-fired industrial single-chamber incinerator and in an industrial boiler. The processing plant produces 60 metric tons of smoked meat in an 8-hr shift with the release of 9 kg of odorous organic matter in 360 Nm³/hr of exhaust gases. Assume complete combustion of the odorous organic matter. The incinerator and the boiler burn #2 fuel oil and the odorous gas stream is essentially air.

TABLE 2.26. Typical Uncontrolled Emissions from Refuse Incinerators [7].

	Uncontrolled Emissions (kg/metric ton)			
Incinerator	Particulates	Carbon Monoxide	Hydrocarbons	Nitrogen Oxides
Municipal				
multiple chamber	15	17.5	0.75	1.5
Industrial and commercial				
multiple chamber	3.5	5	1.5	1.5
single chamber	7.5	10	7.5	1.0
Trench				
wood	6.5	NA	NA	2.0
rubber	69	NA	NA	NA
municipal refuse	18.5	NA	NA	NA
Domestic single chamber				
without primary burner	17.5	150	50	0.5
with primary burner	3.5	~0	1	1.0
flue-fed	15	10	7.5	1.5
Pathological	4	~0	~0	1.5

Solution

If properly designed, emissions from either system should be the same. However, if an incinerator with no reheat is used, all emissions would result from disposing of the 9 kg of organic waste. In the boiler, all emissions would result from energy production and no additional emissions would be produced. It is required that the boiler air flow be at least 360 Nm^3/hr and maintenance checks be made to prevent deposits of organic material.

Emissions from incineration can be obtained from Table 2.11 for an industrial boiler using distillate oil and from Table 2.10.

Based on 100 g typical #2 grade oil

$$\frac{13}{1} = 13 \text{ atoms H}$$

$$\frac{87}{12} = 7.25 \text{ atoms C}$$

$$\frac{0.5}{32} = 0.016 \text{ atoms S}$$

Stoichiometric combustion requires

$$13/4 + 7.25 + 0.016 = 10.52 \text{ moles } O_2$$

And at 5% excess air, this requires

$$(1.05)(10.52)\left(\frac{100 \text{ moles air}}{21 \text{ moles } O_2}\right)\left(\frac{0.022400 \text{ m}^3}{\text{g mole}}\right)\left(\frac{293}{273}\right)$$

$$= 1.26 \text{ Nm}^3 \text{ air to burn } 100 \text{ g oil}$$

The amount of oil burned per 8-hr shift is

$$\left(\frac{360 \text{ Nm}^3}{\text{hr}}\right)(8 \text{ hr})\left(\frac{100 \text{ g}}{1.26 \text{ Nm}^3}\right)\left(\frac{\text{cm}^3}{0.87 \text{ g}}\right)\left(\frac{1}{1,000 \text{ cm}^3}\right) = 262 \text{ liters of oil}$$

Emissions from a horizontally fired oil boiler are then

Particulates	= (1.8)(262)	= 472 g
SO_2	= (17.2)(0.5)(262)	= 2,253
CO	= (0.5)(262)	= 131
HC	= (0.35)(262)	= 92
NO_x	= (9.6)(262)	= 2,515
Aldehydes	= (0.25)(262)	= 66
	Total	= 5,529 g

$$\text{or} \sim 5.5 \text{ kg}$$

Conclusion: from incineration only with no recovery of heat, 5.5 kg of pollutants are generated to dispose of 9 kg of organic waste at an energy penalty of 2.5 million kcal. It would be much more desirable to charge this waste gas stream to the industrial boiler if possible.

2.9 OTHER PROCESSES

Methods are being studied to convert existing fuels into more easily handled and combusted fuels which generate fewer pollutants. Some of these fuels include landfill matter to methane, MSW to RDF, agricultural crops to gasoline, and converting coal or coal-like fuel, at a conversion facility, to a liquid or gaseous fuel which is then sent to conventional combustion facilities. The liquid and gaseous synthetic fuels may or may not be similar to existing fuels and hence could require modifications of the combustion facilities. The synthetic fuel production processes in themselves are subject to air pollution emission and control requirements.

2.9.1 GASIFICATION

Low-, medium- and high-heat content gases are being produced from coal reserves in small-scale evaluation studies. Heat-content values are relative, but can be compared with the values in Table 2.12. Tests have been conducted on gases with heating values ranging from 3,700 to 11,200 kcal/Nm³ to compare synthetic gas to natural gas in the areas of flame stability, flame length, flame emissivity, furnace efficiency, and NO_x emissions. Modifications to the boiler/burner system are usually necessary when medium and low ($< 7,500$ kcal/Nm³) heating-value gases are used [21].

The coal-gasification processes can usually be varied to produce different gas mixtures for specific end uses. One process, at high pressures of about 70 atmospheres, produces synthetic natural gas. At intermediate pressures of about 30 atmospheres a turbine fuel is produced and at low pressures a synthesis gas is produced. Synthesis gas is a mixture of varying amounts of CO and H_2 (see Table 2.12).

2.9.2 LIQUEFACTION

There were at least 14 different coal-liquefaction processes being studied. Demonstration-size facilities are being planned, but it will be years before such processes will be commercially available.

2.10 HEATING VALUE

It is important to know the heating value of the fuel, fuel/waste combinations, and waste used as fuel in a combustion process. Two different heating values are listed in the literature—higher heating value (HHV) and lower heating value (LHV). The HHV is reported considering the moisture condensed while the LHV assumes the moisture remains in gaseous forms. These values are usually reported for pure, dry fuel material. Both the HHV and LHV as reported in thermodynamic data tables are for the enthalpy change between the fuel and the completely reacted products. This change is for fuel, air, and products all at the same temperature (usually 20°C). Tables of thermodynamic data can also be used to calculate reactant-product combustion heat releases (e.g., for carbon to CO_2 and hydrogen to water) accounting for enthalpy and bond-cleavage energy requirements.

Waste fuels may have HHV and LHV which are relatively close in numeric magnitude. For example, pure dry cellulose ($C_6H_{10}O_5$) has a HHV of

$$C_6H_{10}O_5 + 6O_2 \rightarrow 6CO_2(g) + 5H_2O(\ell) + 7,500 \text{ Btu/lb cellulose}$$

A mass balance shows that 1 lb cellulose produces 0.56 lb water. Using 1,000 Btu/lb as the latent heat of evaporization for water the LHV is 560 Btu less or

$$C_6H_{10}O_5 + 6O_2 \rightarrow 6CO_2(g) + 5H_2O(g) + 6,940 \text{ Btu/lb cellulose}$$

Many waste fuels are wet and contain inert material. In order to assign a heating value to this material, a correction must be made similar to the one used in converting an MAF coal heating value to an as-fired value. This resultant value is the net heating value (NHV). It is a type of HHV which assumes reactants and products at the same temperature, assumes the water as liquid, and accounts for noncombustible components. NHV can be calculated knowing pure component HHV and the waste composition. Heating values for coal, oil, gas, and wastes are given earlier in this chapter.

HHV of pure polyethylene (C_2H_4)$_n$ is 20,000 Btu/lb. A waste containing 80% polyethylene, 5% inerts and 15% water would have an NHV of (20,000)(0.8) or 16,000 Btu/lb. Note that in a combustion process, and in particular when wastes containing water and inerts are combusted, much of the heat is required to evaporate the free water and only the remainder of this energy would be available to raise the temperature of the inert material in the furnace. A bomb calorimeter sample is dried first. Therefore, the NHV of a material can be calculated from the calorimeter HHV by correcting this HHV to Btu per lb of original wet weight of fuel.

NHV for a sample can also be estimated using DuLong's formula and knowing the chemical composition. The formula was developed for coal but works well on organic-type fuels. This formula is

$$NHV = 14,544C + 62,208(H - O/8) + 4050S \qquad (2.4)$$

where

NHV = heat release in Btu/lb fuel as fired
C, H, O, S = ultimate analysis weight fraction of combustible carbon, hydrogen, oxgyen, and sulfur, respectively

Use of this is included in Example Problem 2.5. This formula must be corrected if other exothermic materials are present.

Example Problem 2.5

Find the NHV and estimate the chemical composition of a waste mixture consisting of 40% cellulosic material (containing 12% water and 3% inerts), 40% wood (containing 25% moisture and 5% inerts), and 20% polyethylene (containing 6% water and 10% inerts). Check NHV using DuLong's formula.

Data can be found to show the following properties:

Pure, Dry Chemical	Composition, Wt Fraction						HHV Btu/lb
	C	H	O	N	Ash	Total	
cellulose	0.444	0.062	0.494	—	—	1.000	7,500
wood	0.496	0.061	0.438	0.001	0.004	1.000	8,500
polyethylene	0.857	0.143	—	—	—	1.000	20,000

These values must be corrected to reflect the composition of the waste mixture. Doing this for each component gives

Mixture Components		Corrected HHV, Btu/lb
% Material	Corrected Mass Fraction	
40 Cellulose	$C = 0.444(0.85) = 0.377$	$(7,500)(0.85)$
	$H = 0.062(0.85) = 0.053$	$= 6,380$
	$O = 0.494(0.85) = 0.420$	
	Moisture $= 0.120$	
	Inerts $= 0.030$	
	Total $= 1.000$	
40 Wood	$C = 0.496(0.70) = 0.347$	$(8,500)(0.70)$
	$H = 0.061(0.70) = 0.043$	$= 5,950$
	$O = 0.438(0.70) = 0.306$	
	$N = 0.001(0.70) = 0.001$	
	Ash $= 0.004(0.70) = 0.003\ \}\,0.053$	
	Inerts $= 0.050$	
	Moisture $= 0.250$	
	Total $= 1.000$	

Mixture Components		Corrected HHV,
% Material	Corrected Mass Fraction	Btu/lb
20 Poly-ethylene	C = 0.857(0.84) = 0.720 H = 0.143(0.84) = 0.120 Moisture = 0.060 Inerts = 0.100	(20,000)(0.84) = 16,800
100	Total 1.000	

NHV of
the resulting
waste mixture $= (0.40)(6,380) + (0.40)(5,950) + (0.20)(16,800)$

$= 8,290$ Btu/lb

Composition of the waste mixture in mass fractions is then

$$
\begin{aligned}
C &= (.40)(.377) + (.40)(.347) + (.20)(.720) = 0.434 \\
H &= (.40)(.053) + (.40)(.043) + (.20)(.120) = 0.062 \\
O &= (.40)(.420) + (.40)(.306) \qquad\qquad = 0.290 \\
N &= (.40)(.001) \qquad\qquad\qquad\qquad\quad = \ - \\
\text{Ash + inerts} &= (.40)(.030) + (.40)(.053) + (.20)(.100) = 0.053 \\
\text{Moisture} &= (.40)(.120) + (.40)(.250) + (.20)(.060) = 0.160 \\
&\qquad\qquad\qquad\qquad\qquad\quad \text{Total} = 1.000
\end{aligned}
$$

Using DuLong's formula

$$NHV = 14,544(0.434) + 62,208(0.062 - 0.290/8)$$

$$= 7,914 \text{ Btu/lb}$$

This is within 4.5% of original estimated NHV.

2.11 COMBUSTION TEMPERATURE

It is important to be able to estimate combustion temperature, especially for incineration systems. These can be done using fuel chemistry, fuel and air rates, thermodynamic data and mass and heat balances. Certain simplifying procedures can be used. It is necessary to know composition of fuel/

waste feed, amount of air, degree of burnout, and system heat loss. NHV and excess air can then be estimated using procedures already discussed.

Simplifying assumptions that can be used are:

(1) Reference temperature is 68°F (20°C).

(2) Latent heat of vaporization of water is 1,000 Btu/lb.

(3) Specific heat of dry combustion gas from 68 to 2000°F in normal flue gas concentrations is 0.27 Btu/(lb °F).

(4) Specific heat of water vapor from 68 to 2000°F in normal flue gas concentrations is 0.52 Btu/(lb °F).

(5) All ash exits the system at combustion temperature with a specific heat of 0.4 Btu/(lb °F).

(6) Fuel burnout is assumed to be 99% so fuel heat loss is 1%.

(7) System heat loss is 7% of input heat.

The energy balance is: heat in equals heat out, at steady state conditions. Heat in equals combined fuel NHV. Set NHV equal to heat out as follows

$$NHV = Q_1 + Q_2 + Q_3 + Q_4 + Q_5 \qquad (2.5)$$

where for an incinerator per pound of fuel/waste mixture

Q_1 = heat loss = 0.07 NHV
Q_2 = latent heat of vaporization of water in feed plus formed water
    = 1,000 (lb water per lb fuel/waste)
Q_3 = sensible heat loss in ash and unburned fuel
    = $(0.4)(t - 68)(0.01 +$ lb ash per lb fuel/waste)
Q_4 = sensible heat of dry combustion gas
    = $(0.27)(t - 68)$ (lb dry gas per lb fuel/waste)
Q_5 = sensible heat of formed water plus water contained in feed
    = $(0.52)(t - 68)$ (lb water vapor per lb fuel/waste)
t = combustion temperature, °F
(Note: water mass of Q_2 and Q_5 are the same.)

Combustion temperature is greatly reduced by excess air, water in the fuel, the presence of inert material, and by incomplete burnout. A suggested empirical method of approximating combustion temperature is

$$t = \frac{4300(AHV) - 7.4 \times 10^6}{(1 + EA)^{0.9}(AHV)} + 10^{(1+EA)} \qquad (2.6)$$

where

t = combustion temperature, °F
AHV = available heating value, Btu/lb fuel
EA = excess air, fraction

The value of AHV accounts for heat loss due to water vapor in the flue gas and inerts/ash in the feed. It is estimated by

$$AHV = NHV - [1,000(w) + 800(I)]$$ (2.7)

where

w = lb flue gas water/lb fuel
I = mass fraction of ash plus inerts in fuel

This appears to be good for AHV from about 3,300 to 15,000 Btu/lb.

A different approach is suggested for site clean-up incineration where large quantities of soil with no heat value are incinerated with fuel required to provide all the heat. These systems frequently consist of a rotary kiln primary with an afterburner secondary. The following equations are developed specifically for a kiln system with 70°F inlet, 50% excess air, 1800°F, no heat value for waste feed, and using number 2 fuel oil. Under these conditions, there would be 50,800 Btu of heat available to heat soil per gal of oil.

Fuel oil required for the primary, FO_p, in gal/hr would be

$$FO_p = [(0.7)(W') + 15.14](T')$$ (2.8)

where

W' = % water in waste feed
T' = waste feed rate, ton/hr

The total acfm of flue gases from the primary at 1800°F, Q_p, is

$$Q_p = [(103.57)(W') + 2,251](T')$$ (2.9)

Similarly, for the secondary of this system operating at 20% excess air, inlet gases at 1800°F, outlet gases at 2200°F using number 2 fuel oil for heat, there would be 49,300 Btu heat/gal oil available to heat the gases. Secondary fuel oil requirements, FO_s, in gal/hr for this are

$$FO_s = [(0.485)(W') + 8.16](T')$$ (2.10)

Total gas flow from the secondary, Q_s, in acfm at 2200°F is

$$Q_s = [(189.9)(W') + 3,794](T') \qquad (2.11)$$

Equations (2.8 through 2.11) were developed by the author from empirical data. A 10% heat loss is assumed for each combustion chamber.

Example Problem 2.6

Use the data from Example Problem 2.3 and calculate the combustion temperature of the waste mixture. Check the calculated value using Equation (2.7).

Given is the NHV of 2.35 kcal/g or 4,230 Btu/lb using Appendix B conversion factors. Flue gas composition is:

7.2%	CO_2
10.9%	O_2
0.03%	CO
0.01%	SO_2
0.01%	NO_x
70.8%	N_2

$$\frac{1.73 + 1.56}{29.73} \, 100 = 11.1\% \; H_2O$$

There are (29.73 lb moles)/100 of wet flue gas produced per lb of waste.

$$Q_1 = (0.07)(4,230) = 296 \; \text{Btu/lb waste}$$

$$Q_2 = (1,000)\left(\frac{1.73 + 1.56 \; \text{moles} \; H_2O}{100}\right)(18) = 592 \; \text{Btu/lb waste}$$

$$Q_3 = (0.4)(t - 68)(0.01 + 0.2082) = 0.087(t - 68) \; \text{Btu/lb waste}$$

$$Q_4 = (0.27)(t - 68)(0.2973)[(0.072)(44) + (0.109)(32)$$

$$+ \; (0.708)(28)]$$

$$= 2.126(t - 68) \; \text{Btu/lb waste neglecting CO, } SO_2 \text{ and } NO_x$$

$$Q_5 = (0.52)(t - 68)(0.2973)(0.111)(18)$$

$$= 0.309(t - 68) \; \text{Btu/lb waste}$$

The heat balance by Equation (2.5) is then

$$4230 = 296 + 592 + (0.087 + 2.126 + 0.309)(t - 68)$$

so: $t = 1393°F$.

Note: Supplemental fuel must be used to incinerate this properly because of the high moisture content (28.16%) of the wet feed. Check combustion temperature using Equation (2.6):

$$w = \frac{1.73 + 1.56}{100} \quad (18)$$

$$= 0.592 \text{ lb flue gas water/lb waste fuel}$$

$$I = 0.2082 \text{ lb ash/lb waste fuel}$$

From Equation (2.7)

$$AHV = 4{,}230 - [1{,}000(0.592) + 800(0.2082)]$$

$$= 3{,}471 \text{ Btu/lb}$$

$$EA = 1.08$$

$$t = \frac{(4{,}300)(3{,}471) - 7.4 \times 10^6}{(1 + 1.08)^{0.9}(3{,}471)} + 10^{2.08} = 1242°F$$

2.12 CORRECTION FACTORS

Frequently, it is necessary to adjust to a standardized or common basis. Some of this can be done using procedures already presented in Chapters 1 and 2, yet many others are unique and confusing. This section attempts to summarize accepted techniques relative to combustion and gas measurements data.

Section 2.2.2 gives two expressions for calculating excess air based on Orsat-type dry gas values. The user should be cautioned that unless otherwise specified, all CO_2 values must represent combustion CO_2 and should not include CO_2 from calcining of limestone, e.g., as in a fluidized bed.

In addition, gas analyses made after a wet scrubber must be corrected to account for the CO_2 absorbed by the scrubber. The adjusted value of dry % CO_2 at the wet scrubber outlet, $(CO_2)_A$, is

$$(CO_2)_A = (CO_2)_I \frac{100 + (EA)_I}{100 + (EA)_0} \tag{2.12}$$

where

$(CO_2)_I$ = % dry gas CO_2 at scrubber inlet
$(EA)_I$ = % excess air at scrubber inlet
$(EA)_0$ = % excess air at scrubber outlet

Emission concentrations may be required to be corrected to an equivalent 12% CO_2 base. The corrected emission concentration $(C)_{12}$ would be

$$(C)_{12} = (C)_T \frac{12}{CO_2} \tag{2.13}$$

where

$(C)_T$ = emission concentration obtained from test data
CO_2 = % CO_2 from dry flue gas measurement

Units of $(C)_{12}$ are the same as those for $(C)_T$. In the case where waste is burned, it may be necessary in Equation (2.13) to use only the dry gas concentration of % CO_2 in flue gas from the combusted waste $(CO_2)_W$. This is found by

$$(CO_2)_W = \frac{(CO_2)_T - (CO_2)_F}{Q_T} \, 100 \tag{2.14}$$

where

$(CO_2)_T$ = volume of CO_2 emitted during the test at standard conditions (SC)
$(CO_2)_F$ = volume of CO_2 calculated to be emitted during test at SC from stoichiometric combustion of fuel
Q_T = total volume of flue gas in test period at dry SC

The most common emission correction is to a base of 7% O_2. Concentrations corrected to 7% O_2 $(C)_7$ are found by

$$(C)_7 = (C)_T \frac{14}{21 - O_2} \tag{2.15}$$

where

$(C)_T$ = emission concentration from test data
O_2 = dry gas % O_2 value

Units of $(C)_7$ would be the same as units of $(C)_T$.

There may be occasion to report CO concentration corrected to 50% excess air, $(CO)_{50}$. To do this, use

$$(CO)_{50} = (CO)_T \frac{100 + EA}{150} \qquad (2.16)$$

where

$(CO)_T$ = test value of CO
EA = % excess air

Units of $(CO)_{50}$ would be the same as $(CO)_T$, usually dry ppm.

Combustion efficiency, CE, can be measured using dry gas analyses if all the CO_2 and CO come from a combusted fuel. This is

$$CE = \frac{CO_2}{CO_2 + CO} 100 \qquad (2.17)$$

where CO_2 and CO are % by volume dry gas readings and CO and CO_2 are released in the combustion chamber only from the combustion of organic fuel (e.g., no limestone is present).

Both English and metric units are in common use, and both types will be seen. Two conversion relationships useful for particles in flue gases are

$$mg/Nm^3 = 4.366 \times 10^{-4} \text{ grains/SCF}$$

$$\mu g/dscm = \frac{g/sec}{Q} \frac{2.119 \times 10^{11}}{100 - H_2O}$$

where Q is total volumetric gas flow rate in acfm and H_2O is percent water vapor in the flue gas.

2.13 EMISSION MINIMIZATION FROM COMBUSTION PROCESSES

The primary requirement is that the combustion facility be properly designed and constructed for the particular fuel to be burned. Assuming this to be true, the next requisite is that it be properly operated. In addition to these factors, it is possible to reduce emissions by initiating precombustion operating practices, such as fuel preparation. The final step is postcombustion emission control as discussed in the following chapters of this book.

2.13.1 PRIMARY COMBUSTION CHAMBERS

Section 2.2 discusses basic combustion. The following discussion relates primarily to incineration systems. The purpose of the primary combustion chamber is to produce evaporization and volatilization of the feed material and to assure good burnout of the fuel charged. Mass and heat transfer occur during these physical changes. Even though the primary is substoichiometric in air, some chemical destruction can occur. However, most of this takes place in the secondary.

The most important primary chamber variable is the quantity of flue gas produced. This dictates sizing for both primary and secondary. Fuel composition as well as rate is critical in determining the amount of flue gas produced. For example, compare methane, CH_4, and paper, $C_6H_{10}O_5$, as fuels, each fuel fed at a rate of 1,000 lb/hr and each containing 20% ash and 10% water. Assuming complete combustion for each, the production of wet flue gas is 44,120 lb/hr for the CH_4 and 9,730 lb/hr for the $C_6H_{10}O_5$. The CH_4 gas rate is over 4.5 times that of the $C_6H_{10}O_5$ on a mass basis. The fuel heating values must also be adequate to maintain a temperature of greater than 1600°F (except for FBC systems).

Liquids must be atomized properly and fed at a rate so that the air–fuel velocity equals the flame velocity. Otherwise, the flame could extinguish and an explosion could occur. Gas fuel feed velocities are critical for the same reason. The combustor volume should be about 10% greater than the flame envelope at maximum rates. This is effected by fuel flow rate, pressure, viscosity, impurities content, and heating value.

Solid fuel volume and gas velocities in primary chambers should also be restricted. For example, in a rotary kiln, the solid volume should not exceed 15% of the kiln volume, and gas velocities should be from 7 to 25 ft/sec.

2.13.2 SECONDARY COMBUSTION CHAMBERS

The purpose of the secondary combustion chamber is to ensure complete combustion of the gases whether they come from the primary or from liquid fuels and/or gases injected into the secondary. The secondary should be operated at from 1800 to 2200°F (980–1200°C). The maximum gas temperature in the secondary is typically 100 to 300°F above the exit temperature. Excess air in the secondary helps assure both good mixing and good destruction. Kephart [22] recommends that excess air be kept at from 5 to 9%.

2.13.3 WASTE RECYCLING

In addition to improving fuel (e.g., producing RDF from MSW) and reducing wear and corrosion of furnace parts, stack emissions can be reduced by processing fuel to recover and recycle useful materials. This discussion goes beyond the cleaning of fuel, e.g., coal cleaning as previously discussed, and is addressed at municipal waste type fuels which are homogeneous mixtures of plastics, textiles, rubber, yard material, paper, metals, and other household discards. Acid gases are produced from refuse containing chlorine, sulfur, nitrogen, and fluorine. Halogenated polycyclic organics are also formed. Metals from batteries and other sources are either vaporized, oxidized, or remain with the bottom ash. The vaporized material can condense as particulates or on fly ash or can leave in vapor form with flue gases.

These materials can come from surprising sources. Some examples in solid waste combustion are:

> cadmium—26% from plastic and 60% from metals (mostly batteries)
> chromium—42% from rubber/leather and 43% from metals
> mercury—10% from plastics, 13% from paper and 60% from metals (mostly alkaline batteries)

In addition, tires are a source of many chemicals:

> zinc oxide—comprises 5% by wt of rubber (used as curing activator and pigment); also present in cardboard
> sulfur—represents 1.5 to 2% of tire (as vulcanizing compound)

antimony and arsenic—used for mold and algae retardants
cobalt and boron—adhesion-enhancement addition
other trace elements—e.g., borium, lead, copper, calcium, sodium, and potassium

Resource recovery can reduce many incineration system emissions before the pollution control system. Half of the municipal systems in Japan use plastic-separation procedures. Sweden limits mercury so mercury batteries are collected and recycled. Completely recovered metals are valuable. Currently, metal values are [23]: Cu $0.675/lb, Cd $1.20/lb, Cr $1.25/lb, Ni $3.45/lb and Zn $0.44/lb. Japan has also instituted large-scale recycling of batteries, thermometers, and cans. A material recovery system at the Gallatin, TN, waste-to-energy plant resulted in a reduction of certain emissions:

lead—52%
chromium—64%
cadmium—73%
complex HC—75%
CO—63%

The Gallatin system produced half the ash of a normal MSW incinerator, which reduced plant disposal costs.

2.13.4 MINIMIZING NO$_x$

It has already been noted that combustion NO$_x$ can be reduced by use of fuel with low available nitrogen, low O$_2$ concentrations, and low temperatures (e.g., staged combustion and flue gas recirculation). In an EPA/EPRI study [24], of the fourteen independent variables evaluated, temperature reduction was most significant. This is significant for waste incinerators which operate with high secondary temperatures of 1800–2200°F and possibly up to 2600°F.

2.13.5 OPERATION AND INSTRUMENTATION

Operation, operator training, and instrumentation may be mandated, and if so, the regulations must be complied with. Specific operating procedures in addition to those already noted should be incorporated for waste incineration systems to further reduce emissions. One of these is frequent cleaning of boiler tubes. The reason for this is that dioxins may be formed when fly ash is held on the cooler boiler tubes. It is believed that more dioxin is produced as residence time at 570°F (300°C) increases when exposed to incinerator flue gases [25]. Other operating recommendations are summarized:

- modulating burners and combustion air flow rates
- continuous temperature recorders for each chamber
- O$_2$ and CO monitor and recorder in flue gas stream
- combustion air flow recorder and totalizer
- auxiliary fuel flow recorder and totalizer
- waste feed rate recorder and chemical analysis

2.14 COMBUSTION FLASHBACK

For safety and to reduce release of unwanted chemicals to the atmosphere, incinerator burner flame flashback and blowoff is important. A flame will remain stationary in space when the rate of consumption of unburned combustible material mixture equals the rate at which combustible material is fed to the flame. When combustible material rate exceeds the feed rate, the flame is blown off. The phenomena of flashback can occur in an explosive gas–air mixture if the tube diameter is large enough to permit flame propagation. Flashback velocity, U_F, in ft/sec at ambient temperatures can be calculated by

$$U_F = 0.2015G_L D'$$
(2.18)

where G_L is critical boundary velocity gradient, second^{-1} and D' is diameter of hole or orifice, ft. Flashback velocities through large holes (0.5 to 5 inches) are much greater than the fundamental flame-burning velocities. It is flashback velocities plus a safety factor of four (i.e., $4 \times U_F$) that must be used to prevent flashback in an incinerator burner tube or orifice.

Values for G_L are given by Grumer et al. [26] and vary with the fuel gas–air composition. Typical maximum values of G_L at 20°C occur at slightly greater than stoichiometric air–fuel mixtures (with temperature extrapolations) and are indicated below

methane 425/sec @ 300°K or $(5.22T-1140)$
nat. gas 480/sec @ 300°K or $(5.22T-1075)$
Av. paraffins 500/sec @ 300°K or $(5.22T-1055)$
propane 600/sec @ 300°K or $(5.22T- 955)$
ethane 630/sec @ 300°K or $(5.22T- 925)$
hydrogen 10,000/sec @ 300°K

The temperature extrapolations assume linear escalation where T is °K.

2.15 UNCONFINED SOURCES

Emission sources can contribute to atmospheric pollutants by either emissions, by secondary reaction products, or by reemission of pollutants that settle from the original plume. In addition, mechanical operations and natural events can generate and release more pollutants. As an example, Section 2.6.3 listed specific pollutants released by automobiles. This did not include those resulting from mechanical use and natural events which include suspension of pollutants due to

- pavement wear and deposition
- vehicle wear and deposition
- settled dust
- litter
- carry out of dirt by vehicle
- carry out of dirt by erosion
- ice control compounds

As a result, an additional 3.4 g per vehicle mile occurs when driving over paved roads and over 50% of this material is <5 μm in size [27].

Other natural unconfined emissions release millions of tons per year of particulate and gaseous pollutants in the U.S. Considering a forest as fuel, forest fires are reported to release about 4% of their mass as particulates during forest fires with most of this mass in the 0.1 to 1 μm size range [28]. Although CO and hydrocarbons are also released, negligible amounts of NO and SO_2 are released.

Most of our pollutants result from combustion sources used to convert a fuel to energy. In addition to the direct emissions, these two examples show that secondary effects of reentrainment, wear, and erosion and natural effects should be considered and controlled with equal concern. These emissions can equal or even exceed in quantity the direct process emissions.

2.16 CHAPTER SUMMARY

It is continuously necessary to tie together and balance energy needs with the current availability of energy sources, technological conversion capabilities, and the control requirements needed to protect man and the environment. All of these factors change with time. It is predicted [29] that by the year 2010 coal will surpass petroleum as the world's most utilized fuel, and that by 2050 global coal production will be three times the current rate. If this is to occur in a way that is not environmentally catastrophic, great attention will have to be paid to the following: precombustion processing (e.g., coal cleaning); during-combustion design and operation; and postcombustion controls to minimize emissions and maximize energy recovery.

Current utilities can operate coal combustion systems to achieve particulate control of over 99%, SO_2 control in the high 90s, and NO_x control of over 50%. The environmental-control costs for utilities continue to increase as control efficiencies increase. Currently, about one-third of the total power plant cost is taken up by the total environmental controls [29].

These trends may not continue forever. Looking to the long-term future, it is predicted [30] that by the 22nd century most of the world's energy will be derived from waste incineration and from nuclear and direct solar power. But even if this is so, many of the environmental concerns will be the same, and there could also be many new emissions forms to consider. Whatever the energy source, techniques will have to be developed to minimize the adverse environmental impact in a way that is economically feasible. A few basic examples of techniques in current use have been discussed in this chapter, but most will be given detailed presentation in Chapters 5 and 6.

2.17 CHAPTER PROBLEMS

2.17.1 FLUE GASES

A natural gas, assumed to be pure methane, is burned to produce a flue gas which is analyzed as 1% CH_4, 8.96% CO_2, 1% O_2, 17.91% H_2O and the balance nitrogen. Flue gas temperature is 120°C.

a. What general statement can be made regarding this combustion process?
b. Determine the percent theoretical air used.
c. Write the balanced stoichiometric equation.
d. Write the actual equation (this may not be easy; think about it and go on).
e. Calculate the actual air–fuel ratio and compare with the stoichiometric AF.

2.17.2 COAL COMBUSTION

A new 40,000 lb/hr industrial crossfeed stoker-fired boiler with no fly ash reinjection is designed to burn coal at 18% excess air. The as-received coal analysis is 22.1% moisture, 33.4% volatile matter, 40.1% fixed carbon, and 4.4% ash. The ultimate analysis of the wet coal is 6.3% H, 54.3% C, 1.3% N., 32.4% O and 1.3% S. Heating value is 5,340 cal/g. This steam boiler is 75% efficient and at normal loads consumes 2.5 metric ton coal/hr.

a. Determine the coal analysis on a moisture-free basis.
b. Determine the coal analysis on a moisture- and ash-free basis.

c. Classify the coal using Table 2.3.

d. Estimate the flue gas composition assuming complete combustion.

e. Calculate the flue gas volumetric flow rate in Nm³/hr.

f. Determine the particulate and SO_2 emissions and determine the amount of control required to meet the NSPS.

2.17.3 OIL COMBUSTION

A household oil furnace burning number 2 distillate oil has an Orsat dry flue gas analysis of 14.42% CO_2, 1.48% O_2, and the balance nitrogen. It is determined that the wet gas actually contains 11.45% water. The maximum home heat losses are 1.5×10^8 cal/hr and the heating system is 80% efficient. Flue gas temperature is 190°C. Assume inlet combustion air is dry and oil analysis is as in Table 2.10.

a. Determine the actual wet flue gas analysis.

b. Find the excess oxygen percentage and the percent excess air.

c. If the maximum stack velocity is to be about 4.0 m/sec, determine the diameter of stack needed.

d. Estimate the SO_2 and particulate emission rates in g/hr under these conditions.

2.17.4 INCINERATION

A 300 metric ton/day municipal incinerator is being operated at 70% load. The dry flue gas analysis shows 9.2% CO_2, 375 ppm CO, 12% O_2, and the balance nitrogen. The flue gases contain 5.5% water. Assume 50% of the water is produced by combustion of hydrogen in the fuel and that the oxygen in the fuel is just adequate to combust the hydrogen in the fuel.

a. Estimate the amount of uncontrolled emissions in kg/hr.

b. If the emission limit is 500 ppm CO corrected to 12% CO_2, would these emissions be acceptable?

c. Give the CO concentration corrected to 50% excess air.

2.17.5 INCINERATION OF WET WASTE

A 70×10^6 Btu/hr heat input incinerator is used to destruct 3.5 ton/hr of wet sludge. The NHV of the sludge is 1500 Btu/lb. The incinerator is to operate at 1800°F. Natural gas is the supplemental fuel. The sludge contains 30% water, 40% carbon, 7% hydrogen, 4% chlorine, 6% oxygen, 3% nitrogen, and 10% inerts. Operation is at 25% excess air. Heat capacity, C_p, of the inerts is 0.5 Btu/(lb °F). Use thermodynamic data and

a. Estimate the quantity of natural gas required in ft³/min.
b. Calculate the composition of the flue gas.

2.17.6 INCINERATOR EMISSIONS

An incinerator burns 3 ton/hr of a waste containing 6% chlorine. Flue gas emissions are 85,000 acfm at 350°F and contain 9.5% CO_2, 40 ppm CO, 12% O_2 (dry basis), and 15% water. HCl control required is 99.9% and particulates are limited to 0.1 grains/dscf corrected to 7% oxygen.
a. Calculate emission rates of HCl and particulates in lb/hr.
b. Estimate combustion efficiency.

2.17.7 COMBUSTION TEMPERATURE

A waste contains 40% cellulosive material (80% cellulose, 15% water, and 5% inerts), 30% motor oil (19,300 Btu/lb), 10% water, and 20% ash. The flue gases analyze 10.5% CO_2, 50 ppm CO, and 11.5% O_2 on a dry basis.
a. Find NHV.
b. Find EA.
c. Estimate combustion temperature.

2.17.8 FBC MASS-HEAT BALANCE

An atmospheric FBC operating at 1600°F burns hazardous waste having a heating value of 4,500 Btu/lb and containing 20% water, 40% carbon, 5% hydrogen, 4% oxygen, 10% chlorine, and 21% inert material (C_p = 0.5). Natural gas can be used as a supplemental fuel. Limestone is charged with the feed at an alkali to chlorine stoichiometry of 2:1. Operation is in circulating mode. The flue gas analysis is 11% CO_2, 10.5% O_2, 40 ppm CO, and 10 ppm HCl on a dry basis. State assumptions and
a. Specify if supplemental fuel is required after the bed reaches operating temperature.
b. Find the combustion efficiency.
c. Calculate volumetric flow rate of flue gases at 400°F if waste charge rate is 5 ton/hr.
d. Determine how much HCl leaves the system in lb/hr.

Use 1,000 Btu/ft³ for natural gas and consider all CH_4

C_p mean from 68 to 1600°F in Btu/lb °F is

0.87 for CH_4
0.27 for CO_2, CO, air, and N_2
0.52 for H_2O
0.25 for O_2
0.20 for HCl
3.55 for H_2

REFERENCES

1 Whitaker, R. 1984. "Electricity: Lever on Industrial Productivity," *EPRI Journal*, 9(8):6–15.

2 Smith, W. S. 1985. "Orsat Excess Air Nomograph," *The Entropy Quarterly*, 6(3).

3 Simon, J. A. 1977. "1977 Coal, Illinois' Major Fuel Resource," Illinois Energy Resources Commission.

4 Ruch, R. R., H. J. Gluskoter, and N. F. Shimp. 1974. "Occurrence and Distribution of Potentially Volatile Trace Elements in Coal," *Illinois State Geological Survey*, No. 72.

5 Henschel, O. B. 1978. "Emissions from FBS Boilers," *Environ. Sci. and Tech.*, 12(5):534–538.

6 Danielson, J. A., ed. 1973. *Air Pollution Engineering Manual, 2nd ed.* U.S. Environmental Protection Agency.

7 1973–1978. "Compilation of Air Pollutant Emission Factors," U.S. EPA, AP-42, plus supplements 1–8.

8 Smith, W. S. and C. W. Gruber. 1966. "Atmospheric Emissions from Coal Combustion—An Inventory Guide," U.S. Department of Health, Education and Welfare, Publication #999-AP-24.

9 1977. "Annual Environmental Analysis Report," U.S. Department of Energy.

10 1990. Private communiqué, The American Petroleum Institute, Washington, DC, phone (202)682-8000, Statistics Dept.

11 Dresser, A. L. 1988. "A Dispersion Model Analysis of a Western Community's PM-10 Problem," *JAPCA*, 38(11):1419–1421.

12 1987. "State of Emission, a New Report," Conservation Foundation Letter, No. 5.

13 Scheible, M. et al. 1982. "An Assessment of the Volatile and Toxic Organic Emissions from Hazardous Water Disposal in California," State of California Air Resources Board.

14 Brunner, D. R. and D. J. Keller. 1972. "Sanitary Landfill Design and Operation," EPA Publication SW-65ts.

15 Cross, F. L. and H. E. Hesketh. 1985. *Controlled Air Incineration*. Lancaster, PA: Technomic Publishing Co., Inc.

16 Hesketh, H. E., F. L. Cross, and T. Tessitore. 1990. *Incineration for Site Clean-Up and Destruction of Hazardous Wastes*. Lancaster, PA: Technomic Publishing Co., Inc.

17 1988. "Waste-to-Energy at a Glance," *Waste Alternations*, 1(2):6–7.

18 Brinkman, W. K. and R. F. Forbers. 1988. "Sewage Sludge Incineration: An Overview of the Technology," *Solid Waste and Power*, 2(5):12–20.

19 Visalli, J. R. 1987. "A Comparison of Dioxin, Furan and Combustion Gas Data from Test Programs at Three MSW Incinerators," *JAPCA*, 37(12):1451–1463.

20 Hesketh, H. E. 1988. CTA Files.

21 Lachapelle, D. G., Project Officer. 1977. "Burner Design Criteria for NO_x Control from Low Btu Gas Combustion; Vol. I. Ambient Fuel Temperature," EPA-600/7-77-094a.

22 Kephart, W. L. 1987. "Incineration System Control with RCRA Constraints," *Proceedings of the National Symposium on Incineration of Industrial Wastes*, St. Louis, MO: Randolph, Breyer and Grove.

23 1987. *The Hazardous Waste Consultant*, 5(4):1–22.

24 Croom, J. M. et al. 1987. "NO_x Formation in a Cement Kiln: Regression Analysis," *Proceedings of the 1987 Symposium on Stationary Combustion Nitrogen Oxide Control*, Vol. 2, Section 49, New Orleans.

25 Clarke, M. 1987. "How Plant Operators Can Minimize Emissions," *Waste Age*, pp. 156–164.

26 Grummer, J., M. E. Harris, and V. R. Rowe. 1956. "Fundamental Flashback, Blowoff, and Yellow-Tip Limits of Fuel Gas–Air Mixtures," ROI 5225, U.S. Dept. of Interior.

27 1977. "Compilation of Air Pollutant Emission Factors," EPA AP-42, Draft Supplement Section 11.2.5.

28 Radke, L. F., J. L. Stith, D. A. Hegg and P. V. Hobbs. 1978. "Airborne Studies of Particulates and Gases from Forest Fires," *J. Air Poll. Control Assoc.*, 28(1):30.

29 Douglas, J. 1989. "Quickening the Pace in Clean Coal Technology," *EPRI Journal*, 14(1):4–15.

30 Hanlon, S. 1989. "The Future's so Bright . . . Engineering the 22nd Century," *AIChExtra*, p. 1.

CHAPTER 3

Particulate Control Mechanisms

3.1 INTRODUCTION

A number of basic relationships and definitions related to airborne particulate matter are given in Chapter 1 of this book. If you are not familiar with the following material you may wish to return to Chapter 1 and review the sections on: aerodynamic diameter, equivalent diameter, sedimentation or Stokes diameter, cut diameter, dynamic shape factor, and principles of relating size and size distribution.

Primary particles are those which are emitted directly into the atmosphere and have the normal distribution range of both large and small particles as discussed in Chapter 1. Secondary particles are formed in the atmosphere as the result of chemical reactions and are basically smaller in size. Promulgation of the PM-10 National Ambient Air Standard makes it necessary to give more consideration to secondary particles and to be able to predict the behavior of all types.

Secondary particles can form when acid gases react with other materials present in the atmosphere. Important atmospheric gases which react to form secondary particles include sulfur dioxide, nitrogen oxides, ammonia, hydrogen sulfide, organics, and acid vapors including nitric, sulfuric, hydrochloric, and hydrofluoric acids.

Formation of homogeneous secondary particles can be by either of two distinct mechanisms: gas–gas or gas–particle reactions. Both of these reactions can be highly complex and take place in a sequence of steps. In gas phase reactions, two or more gases interact to form another vapor product which nucleates as small particles about 0.001 μm in diameter. These particles can increase in size by diffusing to and/or condensing on another particle surface, or by coagulating into larger particles. Also,

165

either or both the reactant and product gases can adsorb onto particle surfaces. Normally, the particles grow to about 0.2 to 2 μm in size. Heterogeneous secondary particles result when molecules of supersaturated vapor condense at the surface of a foreign particle.

Secondary particles are important contributors to PM-10 ambient particulate loading. Once formed, they behave as other particles and their behavior can be predicted. The complexing factor is that they do not appear as particles at the time of the emission.

In addition to these types of particles, several other terms commonly used when discussing particles are:

>   Acicular—Needlelike or long slender crystal
>   Amorphous—No apparent crystal structure
>   Angular—With sharp edges and corners
>   Cenosphere—Hollow sphere
>   Chainlike—A number of particles joined in a line
>   Fibrous—A particle with length $\geq$ 20 times diameter
>   Floc—A fine aggregate of particles
>   Rosetta—Symmetrical growth pattern resembling a rose
>   Vitreous—Having the luster of broken glass

Minerals account for the largest amount of primary atmospheric particulate matter—about 65% by mass. Most of these originate as natural or fugitive emissions. Combustion products are next, amounting to about 25%. Other primary particulates include biological material (pollen, spores, cellulose, and plant tissue) and miscellaneous particles such as iron and other metals and rubber. Secondary particles are then added to these.

The most predictable behavior for particles in a dynamic state is when the particles are all the same size (monodisperse), spherical in shape, and in steady-state motion in a straight line. This assumes that the particles in question are in motion in a gas medium at the velocity of the gas and that they are not influenced by the presence of other particles and forces. This chapter will cover this situation first, then variations are presented to include other factors. After discussing the procedures for predicting particle behavior, methods are given for applying these principles to particle collection by use of impaction, interception, diffusion, and phoretic forces. Note that although these discussions are for solid-gas systems, the same procedures are applicable to other fluid (e.g., solids-liquid) systems. See Section 5.11 for the special case of sweeping impaction.

3.2. DIMENSIONLESS NUMBERS

Dimensional analysis is a useful flag for determining when an equation is set up incorrectly. That is, it does not necessarily show that the equation setup is correct, but it certainly shows when the setup is wrong. Dimensional analysis can also be used to relate analogous groups as ratios of various forces or effects in a given system. The resultant dimensionless ratios are useful for estimating when one or the other effect described by the ratio of terms becomes significant. From empirical evaluations, it is possible to establish cutoff values, based on these dimensionless numbers, that permit us with acceptable accuracy to estimate behavior when it is dependent on some given significant effect.

Some important dimensionless numbers useful for predicting particle behavior are given here. Note that these include numbers related to gas behavior because air pollution control is concerned with evaluating behavior of particles in gaseous media. Some dimensionless numbers such as dynamic shape factor and Mach number were given in Chapter 1.

3.2.1 FLOW REYNOLDS NUMBER (Re_f)

This dimensionless group is the ratio of inertial-to-viscous forces of a flowing fluid. Viscous forces dominate when Re_f is $<2,100$. The flow is turbulent above about 4,000. Flow Reynolds number, Equation (1.25), is rewritten

$$Re_f = \frac{Dv_g \varrho_g}{\mu_g} \qquad (3.1)$$

where

D = diameter (or equivalent) of containing device
v_g = velocity of gas or other fluid
ϱ_g = density of gas or other fluid
μ_g = viscosity of gas or other fluid

3.2.2 PARTICLE REYNOLDS NUMBER (Re_p)

Motion of a particle is characterized by the particle Reynolds number. This includes motion in an unconfined fluid medium (such as in the atmosphere) or in a system where the walls of the device have no influence on the particle. Particle Reynolds number is

$$Re_p = \frac{d(v_p - v_g)\varrho_g}{\mu_g}$$ (3.2)

where

d = diameter of particle
v_p = velocity of particle

Note that density and viscosity are fluid properties, not particle properties, even though the particle could be a liquid. Frequently particle Reynolds numbers of 0.1 and 1 characterize limits of effects which can be used to predict particle behavior. Values up to a maximum of about 400 will be used, but this is still greatly less than the normal magnitude of flow Reynolds numbers.

3.2.3 KNUDSEN NUMBER (Kn)

Knudsen number is the ratio of gas molecule mean free path distance to particle size and is

$$Kn = \frac{2\lambda_g}{d}$$ (3.3)

The value of gas mean free path, λ_g, can be obtained from Equation (1.19). For air at SC it is about 6.53×10^{-6} cm.

Extremely small particles view the fluid medium as being somewhat discontinuous. That is, when particles decrease below a certain size for a given gas molecule mean free path they begin to escape being contacted by the gas molecules. When this occurs, the particles behave differently. This occurs when Kn becomes >0.1.

Four size regimes are used, based on values of Kn, to classify particulates when the Mach number is much less than 1. In order from large particles to small particles, these size regimes as given in Table 3.1 are Continuum, Slip Flow, Transition, and Free Molecule Regimes. The approximate diameters shown are calculated for air at SC.

3.2.4 CUNNINGHAM CORRECTION FACTOR (C)

When Kn is greater than 0.1, behavior of particles can be predicted if a correction for slip is made using the Cunningham slip correction factor, C.

TABLE 3.1. Particle-Size Regimes Characterized
by Approximate Knudsen Number and Particle
Diameter for Air at SC.

Size Regimes	Approximate Value	
	Kn	d_p, μm
Continuum (Stokes)	<0.1	>1.3
Slip Flow (Cunningham)	≤0.3	≥0.4
Transition	10–0.3	0.01–0.4
Free Molecule	>10	<0.01

This equation includes thermal and momentum accommodation factors based on the Millikan oil-drop studies and is empirically adjusted to fit a wide range of Kn values. The form suggested by Davies [1] is

$$C = 1 + Kn[1.257 + 0.400 \exp - (1.10/Kn)] \qquad (3.4)$$

This equation will underestimate values of C if used under conditions of high temperature and pressure as discussed in Section 3.7.9 using the predictions of Rao [2].

A simplified equation is given by Calvert [3] for use in air at near normal pressures

$$C = 1 + \frac{(6.21 \times 10^{-4})T}{d_p} \qquad (3.5)$$

where T is in units of °K and d_p is particle diameter in μm. Usually the Cunningham corrections are neglected for particles larger than 1 μm in air at SC.

3.2.5 STOKES' NUMBER (St)—IMPACTION PARAMETER (K_I)

Stokes' number is the ratio of particle stopping distance to a characteristic distance the particle must travel to be captured. In this book, we will consider the impaction parameter to equal Stokes' number; however, there are several other commonly used definitions for K_I and one must check the specific definition used. In this work it is

$$St = K_I = \frac{2X_s}{D_c} \qquad (3.6)$$

where X_s is particle stopping distance and D_c is collector diameter. This impaction parameter is the ratio of drag-to-viscous forces.

3.3 LINEAR–STEADY-STATE MOTION

Particle dynamics are influenced by the particle size regime and the effects of the forces acting on the particle. The forces to be considered are external forces, medium resistance forces, and interaction forces. External forces commonly include gravitational, centrifugal, electrical, magnetic, and thermal. Sonic, nuclear, and others may also be present. The medium resistance forces are considered separately and are always present in air pollution control work.

A force balance can be made on a single spherical particle in motion in a continuous gaseous medium when $Ma \ll 1$. Using Stokes' expressions for form and friction drag, and recognizing that the total force, F, acting on the spherical particle is the sum of buoyant plus drag forces, F_D:

$$F = 1/6\pi d^3 \varrho_g g + F_D \tag{3.7}$$

where g is gravitational acceleration (980 cm/sec^2 at sea level).

3.3.1 UNDER INFLUENCE OF GRAVITY

3.3.1.1 Continuum Regime

Behavior of the larger particles that are characterized as shown in Table 3.1 was studied extensively by Stokes. This led to Stokes' expression for form and friction drag of

$$F_D = 3\pi\mu_g d(v_p - v_g) \tag{3.8}$$

Combining Equations (3.7) and (3.8) with the gravitational force, F_G

$$F_G = 1/6\pi d^3 \varrho_p g \tag{3.9}$$

completes a force balance on a freely falling particle.

The solution of this force balance equation gives terminal settling velocity, v_s (where v_s is terminal v_p),

$$(v_s - v_g) = \frac{d^2 g(\varrho_p - \varrho_g)}{18\mu_g} \tag{3.10}$$

Stokes' relationship is within 1% accuracy for $0.01 < Re_p < 0.1$ and within 10% accuracy for smaller and larger Re_p values to $10^{-5} < Re_p < 1.0$ assuming spherical particles in air at SC. These equations can be simplified by assuming gas density to be insignificant compared to particle density and by noting that in still air v_g equals zero.

Oseen [4] modified Stokes' drag relationship which improves the accuracy of determining v, for slightly larger particles. This modification is

$$v_s - v_g = \frac{d^2 g(\varrho_P - \varrho_g)}{18\mu_g(1 + 3/16 Re_p)} \tag{3.11}$$

and requires trial-and-error solution. Accuracy is within 1% for particles in air at SC with Re_p from 0.1 to about 0.5.

Terminal settling velocity of even larger particles in air at SC with Re_p of 3–400 can be estimated using Klyachko's [5] modification of Stokes' drag relationship. This gives

$$v_s - v_g = \sqrt{\frac{4d(\varrho_P - \varrho_g)g}{3 C_D \varrho_g}} \tag{3.12}$$

where C_D is the dimensionless drag coefficient which is sometimes called the particle friction factor, f. C_D is related to Re_p by Stokes for Re_p 0.01 to 0.1

$$C_D = \frac{24}{Re_p} \tag{3.13}$$

and by Klyachko

$$C_D = \frac{24}{Re_p} + \frac{4}{Re_p^{1/3}} \tag{3.14}$$

This means that Equation (3.12) is solved by trial and error using either Equation (3.13) or (3.14). Equation (3.14) is better for the "larger" values of Re_p up to 400. These expressions are used to obtain the plot of C_D vs Re_p given as Figure 3.1. A word of caution: Do not extrapolate this figure for larger Re_p.

3.3.1.2 Slip Flow-Transition Regimes

Table 3.1 notes that these particles are smaller than the Continuum Re-

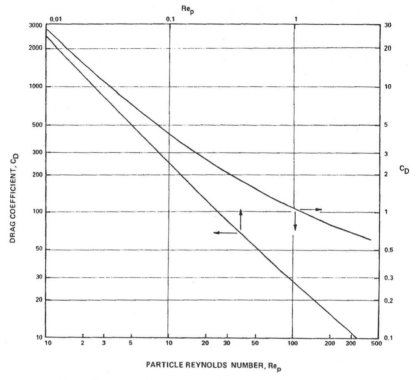

Figure 3.1 Drag coefficient as a function of particle Reynolds number.

gime particles. The table also calls the Slip Flow size particles Cunningham particles. These are smaller than the Stokes particles. Reasonably accurate estimates of particle behavior can be obtained by dividing Stokes' drag relationship by the Cunningham slip correction factor, C. When the resulting force balance is solved for terminal settling velocity the expression becomes

$$v_s - v_g = \frac{d^2 g (\varrho_p - \varrho_g) C}{18 \mu_g} \qquad (3.15)$$

This gives within 1% accuracy for particles in air at SC with Re_p of 10^{-7} to 10^{-6}. Accuracy within 10% is expected when Re_p values reach a low of 4×10^{-11} and a high of 10^{-5}. This 10% accuracy for the smaller particle groups corresponds to Kn values from about 0.3 to 13 which includes the Transition Regime. Predictions for behavior of particles smaller than this can be made but accuracy decreases.

3.3.1.3 Free Molecular Regime

Particles with $Kn > 10$ would have diameters smaller than 0.01 μm in air at SC. These size particles would tend to remain suspended in still air because of gas convective motion and the presence of other naturally occurring external forces, even though they may be extremely small.

3.3.1.4 Summary

Table 3.2 summarizes the various equations presented for estimating behavior of spherical particles in still air at SC under the influence of gravitational force. The particle diameters and Knudsen numbers would change from those noted as particle density varies from unity (1 g/cm^3). These can be calculated using the noted values of Re_p. Values of v_s for particles of $0.5 < Re_p < 3$ are obtained using either Equation (3.11) or (3.12). Accuracy will be less than indicated.

3.3.2 NONSPHERICAL PARTICLES

Most of the equations to predict particle motion presented in the previous section are based on Stokes' law or some variation of that relationship. As such, all these equations require use of the Stokes or sedimentation diameter, d_s. Sedimentation diameter is defined as the diameter of a sphere having the same terminal settling velocity and density as the particle under consideration. If the particle is nonspherical and d_s is not known, the equivalent diameter, d_e, should be calculated using the particle volume and Equation (1.1). A value of d_s can be obtained directly from d_e and the dynamic shape factor, χ, Equation (1.3). Otherwise, use d_e and divide the behavioral equations by χ for the equations containing d^2 in the numerator.

This method of accounting for nonspherical shape should be applied to all behavior-predicting equations whether the motion is linear and steady-state or otherwise.

Example Problem 3.1

Determine the terminal settling velocity and particle Reynolds number and estimate the accuracy of these values for a cylindrical particle with a 2:1 length-to-width axis, falling in still air at SC. The particle length is 0.5 μm and density is 1 g/cm^3.

Solution

Dynamic shape factor, χ, from Figure 1.1 is about 1.14.

TABLE 3.2. Summary of Applicable Terminal Settling Expressions for Unit Density Spherical Particles in Still Air at SC.

Applicable Range of Re_p	For Unit Density Particles			Equation		Notes
	d_p, μm	Kn	Regime	No.	Name	
3–400	130–1,500	$<10^{-3}$	Continuum	3.12	Klyachko-Stokes	Plus Equations (3.13) or (3.14)
0.1–0.5	38–70	3.4–1.9×10^{-3}	Continuum	3.11	Oseen-Stokes	Within 1% accuracy
0.01–0.1	16–38	8–3.4×10^{-3}	Continuum	3.10	Stokes	Within 1% accuracy
10^{-5}–0.01	1.6–16	0.08–0.008	Continuum	3.10	Stokes	Within 1–10% accuracy
10^{-6}–10^{-5}	0.8–1.6	0.16–0.08	Continuum	3.15	Cunningham-Stokes	Within 10% accuracy
10^{-7}–10^{-6}	0.36–0.8	0.36–0.16	Slip Flow	3.15	Cunningham-Stokes	Within 1% accuracy
4×10^{-11}–10^{-7}	0.01–0.36	13–0.36	Transition	3.15	Cunningham-Stokes	Within 10% accuracy
4×10^{-13}–4×10^{-11}	0.001–0.01	130–13	Free Molecular	3.15	Cunningham-Stokes	Accuracy decreased

Equivalent diameter using Equation (1.1) is then

$$d_s = \left(\frac{6V}{\pi}\right)^{1/3} = \left[\frac{(6)\pi(0.25)^2(0.5)}{\pi 4}\right]^{1/3} = 0.3606 \ \mu m$$

Sedimentation diameter from Equation (1.3a) is

$$d_s = \frac{ds}{\sqrt{\chi}} = \frac{0.3606}{\sqrt{1.14}} = 0.3377 \ \mu m$$

Terminal settling velocity in still air at SC is found by Equation (3.15) using $C = 1.55$ from Figure 3.14 or Equation (3.4).

$$v_s = \frac{(0.3377 \times 10^{-4} \ cm)^2 (980 \ cm/sec^2)(1 \ g/cm^3)(1.55)}{18(1.83 \times 10^{-4} \ g/cm \ sec)}$$

$$= 4.74 \times 10^{-4} \ cm/sec$$

Accuracy of this would be between 1 and 10%.

Particle Reynolds number from Equation (3.2) is

$$Re_p = \frac{(0.3047 \times 10^{-4} \ cm)(4.28 \times 10^{-4} \ cm/sec)(1.2 \times 10^{-3} \ g/cm^3)}{(1.83 \times 10^{-4} \ g/cm \ sec)}$$

$$= 9.5 \times 10^{-8}$$

3.3.3 AERODYNAMIC DIAMETER

Aerodynamic diameter, d_a, is defined in Section 1.2.1 as the diameter of a sphere of unit density that attains the same terminal (steady-state) settling velocity, v_s, as the particle under consideration. This concept is used in the sizing of particles by dynamic techniques that evaluate the in-motion character of the particle. For this reason, d_a relates to the particle mass and shape that govern particle motion. Aerodynamic diameter is normally used when noting collection efficiency of a device as function of particle size. Note that aerodynamic diameter equals true diameter only for spherical unit density particles.

Aerodynamic diameter is an extremely useful concept when dealing with particles in the Cunningham and Stokes size ranges. When Re_p is greater than about 10^{-11}, Equation (3.15) can be used in the form

$$v_s - v_g = \frac{d_a^2 g \varrho_o C_a}{18 \mu_g} \tag{3.16}$$

where

ϱ_o = unit density, 1 g/cm³
C_a = Cunningham correction applied to d_a

The relationship between aerodynamic and true or sedimentation diameter is obtained by setting Equation (3.15) equal to (3.16) and solving for

$$d_a = d \sqrt{\frac{C\varrho_p}{C_a}} \qquad (3.17)$$

This shows that $d_a = d$ when $\varrho_p = 1$ g/cm³, d_a is larger when $\varrho_p > 1.0$ and d_a is smaller when $\varrho_p < 1.0$.

Solution of Equation (3.17) is by trial and error. As an aid, Figure 3.2 can be used. The solid lines on this figure are Equation (3.17) and the

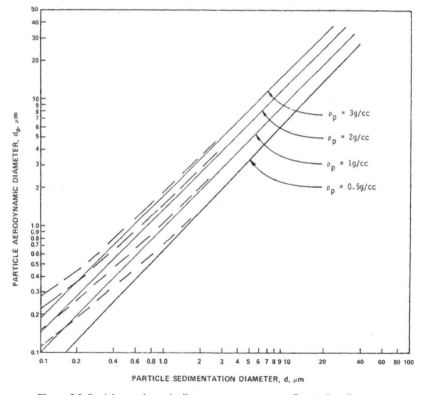

Figure 3.2 Particle aerodynamic diameter versus true or sedimentation diameter.

dashed lines are for $d_a = d\sqrt{C\varrho_p}$, which is obtained when one defines aerodynamic diameter so that it includes the square root of the Cunningham correction factor. As can be seen from Figure 3.2, this difference can be significant for the submicron particles.

3.3.4 PARTICLE DIFFUSION

Particles as well as gas molecules are in constant motion as a result of thermal energy. The average kinetic energy of particles and gas molecules are equal at the same temperature signifying that the particles, with their much heavier mass, are moving more slowly. Smaller particles move faster as shown by the root mean square thermal velocity, v_D

$$v_D = \sqrt{\frac{8D_{PM}}{\tau}}$$ (3.18)

Diffusivity of a single particle through a stationary medium, D_{PM}, in cm²/sec is given by the Stokes-Einstein equation corrected for slip

$$D_{PM} = \frac{CkT}{3\pi\mu_g d}$$ (3.19)

where

k = Boltzmann's constant, 1.38×10^{-16} g cm²/(sec² particle K)
T = absolute temperature, K

Particle diffusivity is plotted in Figure 3.3 for various size particles in air at SC. Relaxation time, τ, is a characteristic time defined as

$$\tau = \frac{d^2\varrho_p}{18\mu_g}$$ (3.20)

Diffusional displacement of small particles becomes much more significant than displacement as a result of settling. Although particle diffusion is random and in all directions, a net linear displacement occurs when $t \gg \tau$. The mean displacement due to diffusion, $\overline{\Delta X_d}$, is

$$\overline{\Delta X_d} = \sqrt{\frac{4D_{PM}t}{\pi}}$$ (3.21)

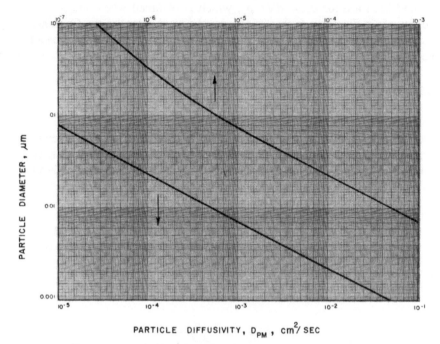

Figure 3.3 Particle Brownian diffusivity in air at standard conditions.

The thermal mobility, B, of a particle is the net steady-state velocity resulting because of diffusion and in units of sec/g is

$$B = \frac{C}{3\pi\mu_g d} \tag{3.22}$$

B is also known as mechanical or diffusional mobility and as Brownian diffusion.

3.4 CENTRIFUGAL–STEADY-STATE MOTION

Centrifugal force is a simple mechanism used to obtain particle-gas separation forces greater than those obtained by gravity alone. Centrifugal separation is used in cyclones and systems that cause spinning or swirling, but it is also the dominant mechanism for inertial impaction and interception. However, these occur in nonsteady-state interactions.

Terminal centrifugal velocity in the normal direction, v_n, is obtained for spherical particles in the Continuum Regime by

$$v_n = \frac{2d^2 \varrho_p v_t^2}{18 \mu_g D_l} \tag{3.23}$$

where

v_t = tangential velocity of gas
D_l = diameter of streamline

Particles in the Slip Flow Regime would use the same equation but with a Cunningham correction factor, C, in the numerator.

The magnitude of the centrifugal force related to the gravitational force is found by dividing terminal centrifugal velocity, v_n, by terminal settling velocity, v_s. This is called the separation factor, SF, and equals

$$SF = \frac{2v_t^2}{gD_l} \tag{3.24}$$

Tangential velocity, v_t, at the diameter of the cylcone, D_c, approximately equals inlet gas velocity, v_i, when no appreciable expansion occurs upon entering the cyclone. Tangential velocities at other diameters, D, can be estimated in the expression noted by Strauss [6]

$$v_t = v_i \left(\frac{D}{D_c}\right)^n \tag{3.25}$$

where n is a constant from 0.5 to 0.7.

Strauss also shows that the time, t, for a spherical particle to drift from an initial diameter, D_i, to a final diameter, D_f, is

$$t = \frac{9}{4} \left(\frac{\mu_g}{\varrho_p - \varrho_g}\right) \left(\frac{D_f}{v_{tf} d}\right)^2 \left[1 - \left(\frac{D_i}{D_f}\right)^4\right] \tag{3.26}$$

where v_{tf} is the tangential velocity at the final location.

3.5 NONSTEADY-STATE MOTION

3.5.1 RELAXATION TIME

In Sections 3.3 and 3.4 particles were assumed to be in steady-state motion. However, this is actually quasisteady-state as the particles may start out or end up at some different velocity. Over long periods of time the initial and final changes may be insignificant and neglected. The time required for a particle to change from an initial steady-state condition to a final steady-state condition or vice versa is a significant property characteristic of the particle. This characteristic property, called relaxation time, τ, is the time during which most of the motion change occurs when a force is applied or removed. A particle accelerated from rest reaches two-thirds the terminal velocity when time $t = \tau$. A decelerating particle reaches one-third the initial velocity at time $t = \tau$, after removal of the force of motion. Thus, only relatively small change in motion occurs when $t > \tau$.

Relaxation time of Continuum Regime size particles is defined by Equation (3.20). Smaller particles would require the use of the Cunningham factor which would be for spherical particles

$$\tau = \frac{d^2 \varrho_p C}{18 \mu_g} = \frac{m_p C}{3 \pi \mu_g d} \tag{3.27}$$

where m_p is particle mass.

3.5.2 ACCELERATING PARTICLES

A force balance on accelerating spherical particles from rest by the application of a single force of acceleration results in a particle velocity, v_p, for particles in the Continuum Regime equal to

$$v_p = \tau a[1 - \exp - (t/\tau)] \tag{3.28}$$

where a is acceleration in units of cm/sec^2. This is applicable when $v_p \gg v_g$. When the accelerating force is gravity, $a = g$. It may be convenient to note from a comparison of Equations (3.10) and (3.20) that $v_s = \tau g$.

Acceleration of smaller spherical particles under the same constraints gives

$$v_p = \tau Ca[1 - \exp - (t/\tau C)] \tag{3.29}$$

Acceleration of larger particles can be accounted for using Newton's equation for drag forces, F_D, in place of Stokes' law. Newton's equation is

$$F_D = \frac{C_D \varrho_g (v_p - v_g)^2 \pi d^2}{8} \tag{3.30}$$

When $v_p \gg v_g$, and when final Re_p is from 0.1 to 400 (for unit density particles in still air), the solution as presented by Fuchs [7] for spherical particles falling from rest is

$$\int_{Re_p=0}^{Re_{pf}} \frac{dRe_p}{C_1 - C_D Re_p^2} = \frac{3\mu_g t}{4d^2 \varrho_p} \tag{3.31}$$

where

Re_{pf} = the final particle Reynolds number
t = time from start of free-fall
$C_1 = 4g \varrho_p \varrho_g d^3 / 3\mu_g^2$

If the particle is in motion, use initial Re_p instead of $Re_p = 0$ and $t = t_f - t_o$. If acceleration is other than due to gravity, use the appropriate acceleration in place of g.

This expression can be solved graphically by obtaining an area under the curve $1/(C_1 - C_D Re_p^2)$ vs Re_p equal to the numeric value of the right side of Equation (3.31). Trial and error values of Re_p are used for this. The value of v_p is found from the final Re_p, remembering that Re_p equals $dv_p \varrho_g / \mu_g$.

3.5.3 DECELERATING PARTICLES

Particles moving in a gas stream at an initial velocity, v_i, decelerate because of aerodynamic drag forces when the force of motion is removed. For a spherical particle of the Continuum Regime size in still air, the particle velocity, v_p, at time, t, can be determined from a force balance to be

$$v_p = v_i \exp - (t/\tau) \tag{3.32}$$

For smaller particles this would be

$$v_p = v_i \exp - (t/\tau C) \tag{3.33}$$

Larger particles, with initial Re_p from 0.1 to 400, have a velocity at time, t, that can be found from the final Re_p using a procedure similar to that for accelerating particles. The expression to use is

$$\int_{Re_{pi}}^{Re_p=0} \frac{dRe_p}{C_D Re_p^2} = - \frac{3\mu_g t}{4d^2 \varrho_p} \tag{3.34}$$

where Re_{pi} is the initial Re_p and t is stopping time. If area under the curve $1/(C_D Re_p^2)$ vs Re_p is obtained, it is negative when the integration is from right to left (i.e., when Re_{pi} is larger than Re_{pf}).

3.5.4 DISTANCE TRAVELED

Distance during steady-state motion is relatively simple to obtain. Distance traveled during acceleration and especially during deceleration is important in control of particulate pollutants and must be estimated by use of the nonsteady-state velocity equations.

Distance traveled, x, by a spherical particle in still air starting at rest while being accelerated is

$$x = \tau(at - v_p) \qquad (3.35)$$

for particles in the Continuum Regime. When acceleration is due to gravitational forces, use g for the acceleration. Distance traveled by smaller particles would be

$$x = \tau C(at - v_p) \qquad (3.36)$$

The nonsteady-state distance traveled by larger particles during acceleration is found using

$$\int_{Re_p=0}^{Re_{pf}} \frac{Re_p \, dRe_p}{C_1 - C_D Re_p^2} = \frac{3\varrho_g x}{4d\varrho_p} \qquad (3.37)$$

Terms are defined with Equation (3.31) and the solution would be similar.

Distance traveled by decelerating particles in the Continuum Regime is

$$x = v_i \tau[1 - \exp - (t/\tau)] \qquad (3.38)$$

and for smaller particles is

$$x = v_i \tau C[1 - \exp - (t/\tau C)] \qquad (3.39)$$

where v_i is initial velocity. At large values of t, the exponential term becomes negligible in both cases and stopping distance, x_s, becomes

$$x_s = v_i \tau \quad \text{or} \quad = v_i \tau C \qquad (3.40a,b)$$

Distance traveled by large particles during deceleration is

$$\int_{Re_{pi}}^{Re_p=0} \frac{dRe_p}{C_D Re_p} = -\frac{3\,\varrho_g x_s}{4d^2 \varrho_p} \tag{3.41}$$

The solution of this would be similar to that for Equation (3.34).

3.6 FORCES AND THEIR EFFECTS

The previous three sections have dealt with linear steady-state motion, centrifugal steady-state motion, and nonsteady-state motion. Each of these sections are subdivided into several specific areas. All possible areas are not included in those sections and, in fact, some areas not included are very significant to air pollution control. Rather, it is felt that some of these subject areas would be more easily understood and utilized in this section under forces and their effects. The following material, some of which has been covered to some extent in the previous three sections and some of which has not been discussed, is presented in this section:

(1) Gravitational settling and diffusive deposition (may be linear steady-state or nonsteady-state motion)
(2) Impaction and interception (nonsteady-state centrifugal motion)
(3) Electrostatic and phoretic motion (may be either steady-state or non-steady-state, but usually is considered quasisteady-state motion)

3.6.1 GRAVITATIONAL SETTLING

A check of relaxation time for a specific size, shape, and density of particle will show whether steady-state, nonsteady-state, or a combination of the two must be used. This section considers time $\gg \tau$ (therefore quasi-steady-state motion), spherical particles, negligible particle slippage in the gas, and horizontal flow of the gas stream. Any variations would need to be appropriately integrated into the final expression.

Gas flow is usually laminar at $Re_f < 1,200$ in a system with smooth walls and at a location sufficiently removed from obstacles, bends, or openings. The fraction of particles of specific size that settle due to gravitational force in a laminar system can be considered as a fractional individual collection efficiency, ϵ_i. For flow between flat horizontal plates

$$\epsilon_i = \frac{v_s L}{\bar{u} h} \qquad (3.42)$$

where L is the length of plates and h is distance between plates. The average gas bulk velocity, $\bar{u}$, under these conditions would equal two-thirds the maximum velocity.

The critical plate length, L_{cr}, required to permit complete settling of a specific size particle would be

$$L_{cr} = \frac{h\bar{u}}{v_s} \qquad (3.43)$$

Settling efficiency of particles from a gas in laminar flow in a horizontal tube is given by Thomas [8] as

$$\epsilon_i = \frac{2}{\pi}\left(2 C_2 \sqrt{1 - C_2^{2/3}} + \sin^{-1} C_2^{1/3} - C_2^{1/3} \sqrt{1 - C_2^{2/3}}\right) \qquad (3.44)$$

where $C_2 = (3v_s L)/(4D\bar{u})$ and arc sine is from 0 to $\pi/2$ radians. The diameter of the tube is D and average gas bulk velocity, $\bar{u}$, in a tube equals one-half maximum velocity.

The critical length for all particles to settle in the tube is

$$L_{cr} = \frac{4D\bar{u}}{3v_s} \qquad (3.45)$$

3.6.2 DIFFUSIVE DEPOSITION

Diffusive deposition becomes more significant than gravitational settling as a mechanism for removing particles from a gas stream as particle size decreases. Diffusive deposition in a stationary medium dominates when $t \ll 4D_{PM}/v_s^2$. Particles deposited from a stationary gas medium on the outside of a long cylindrical tube have been predicted by Slinn's modification of Pitch's procedure [9] as a function of Fourier number, Fo. These values are given in Table 3.3. Multiplying the values shown times container volume times initial number concentration of particles equals total number of deposited particles per unit length of cylinder over the time con-

TABLE 3.3. Relative Number of Particles per Unit Length of External Cylinder Surface as a Function of Fourier Number, *Fo* [9].

Fo	Relative No./Length
0.001	0.072
0.005	0.165
0.01	0.235
0.02	0.338
0.05	0.552
0.1	0.806
0.2	1.186
0.4	1.762

sidered. Fourier number is defined for this purpose as

$$Fo = \frac{4D_{PM}t}{D^2} \tag{3.46}$$

where D is the diameter of the cylindrical tube.

Amount of diffusive deposition from flowing gas streams becomes less significant as the flowrate increases. Deposition from laminar gas flow systems is estimated as shown in Figure 3.4 by Davies [10] and Gormley [11]. Value of the functional constant, C_3, in Figure 3.4 for cylinders is

$$C_3 = \frac{D_{PM}L}{D\bar{u}} \tag{3.47}$$

and for flat surfaces is

$$C_3 = \frac{16D_{PM}L}{h^2\bar{u}} \tag{3.48}$$

where h is distance between flat surfaces. Values to use for $\bar{u}$ are given in Section 3.6.1.

3.6.3 INERTIAL FORCE

Particles in a gas stream are normally considered to be moving with the gas with little or no slippage. When the gas stream changes direction, the

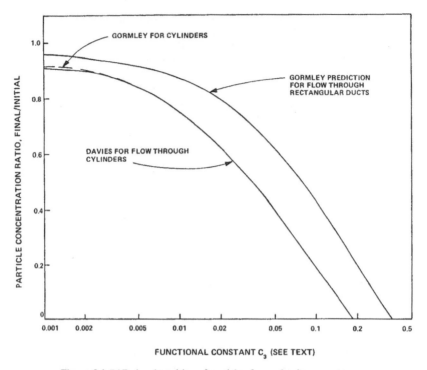

Figure 3.4 Diffusive deposition of particles from a laminar gas stream.

resulting centrifugal force is often too strong to permit the heavier particles to change direction significantly because of their greater inertia. A
stationary or slow-moving obstacle placed in a gas stream then becomes a
target upon which particles can be deposited by inertial impaction. Figure
3.5 depicts a stationary object in a moving gas stream. Gas flow streamlines above and below the stagnation streamlines move above and below
the target. The heavier particles with their greater mass and inertia tend to
deviate less than the gas streamlines as depicted. Dependent upon velocity
and location with respect to the target mass, particles may escape the target
as shown by the small particle, 1, and the particle located far from the stagnation streamline, 2. Some particles such as 3 will impact upon the target
and some particles such as 4 and 5 will just escape impaction upon the
target but will intercept the target and remain deposited. Do not assume
that the particle movement indicated in Figure 3.5 is the only possible
movement as settling and diffusion occur. Large particles can settle on the
target as a result of gravitational force and small particles can become attached to the target as a result of diffusion deposition as discussed in the

previous two sections. In addition, deposition can occur as a result of some of the other forces as discussed in the following two sections.

3.6.3.1 Inertial Impaction

Impaction that occurs as a result of the inertial force created when a particle moves around a target is a significant mechanism for separating particles from a gas stream. Impaction of particles on targets is related to the ratio of stopping distance, x_s, to collector dimension. As has been noted, this is called the Stokes' number, St, which is also defined in this book as the impaction parameter, K_I. Assuming quasisteady-state motion and spherical particles, this gives from Equations (3.40) and (3.20) for Continuum Regime size particles

$$K_I = St = \frac{d^2 \varrho_p (v_p - v_g)}{9 \mu_g D_c} \qquad (3.49)$$

and for Slip Flow Regime size particles

$$K_I = St = \frac{d^2 \varrho_p (v_p - v_g) C}{9 \mu_g D_c} \qquad (3.50)$$

where D_c is either collector diameter or a characteristic dimension of the collection device.

Particle removal from a gas stream becomes more efficient as the value of the impaction parameter increases. Collection by impaction occurs only when St is greater than the critical Stokes' number, St_{cr}. Experimentally

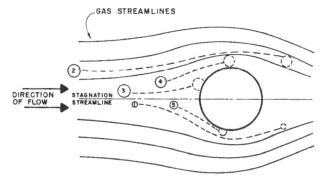

Figure 3.5 Possible particle movement near a target in a moving gas stream.

observed values of St_{cr} are

> 0.30 for a stream approaching a cylinder
> 0.20 for a stream approaching a circular disc
> 0.16 for a jet perpendicular to a plane
> 0.10 for a stream approaching a sphere

3.6.3.2 Impingement

Inertial impaction and inertial impingement mean essentially the same thing. The distinction used here is that impaction relates to the collection of particles on small targets. Relatively speaking, the gas stream carrying the particles is large compared to the target size. These targets could be droplets of atomized liquid or filter fibers and hairs. Impingement as used here relates to the collection of particles on large surfaces. This includes impingement of the particles into a liquid. In this case, the gas stream carrying the particles is small compared to the target size. Impaction is best discussed in relation to specific types of particle collection systems and is covered in Section 3.7.3.

Various impinger devices can be obtained commercially in which flat plane surfaces, large cylindrical surfaces, or large spherical surfaces are used as the targets for particle deposition. A Greensburg-Smith impinger with strike plate is shown in Figure 3.6. The nozzle width, W, would be used for the D_c in Equations (3.49) or (3.50) in determining Stokes' number. Standard strike plate spacing and exit jet velocity are given when flowrate is 1 cfm. Gas streamlines and the stagnation streamlines are depicted leaving the nozzle in Figure 3.6.

Impaction efficiencies for the various types of impingers have been determined experimentally. Ranz and Wong data [12] for the fractional impaction efficiency of specific size particles on flat surfaces can be represented by

$$\epsilon_i = 1.15St - 0.24 \tag{3.51}$$

The Ranz and Wong data for impaction on large spherical surfaces is

$$\epsilon_i = 0.203(5.04)^{St} \tag{3.52}$$

Lundgren data [13] for fractional impaction efficiency on large cylindrical surfaces is found by

$$\epsilon_i = 3.2 \times 10^{-4} \exp (10.5St) \tag{3.53}$$

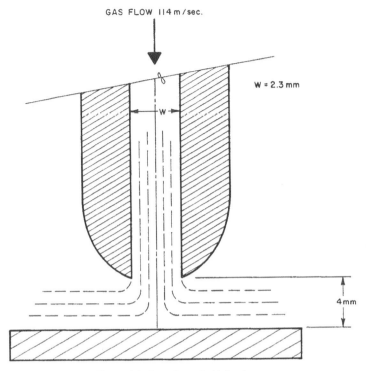

Figure 3.6 Greensburg-Smith impinger.

when $St \leq 0.7$ and by

$$\epsilon_i = 0.094(9.72)^{St} \tag{3.54}$$

when $0.7 < St < 1.04$.

3.6.4 ELECTROSTATIC FORCE

Particles in the atmosphere may have a positive, negative, or no charge because air is a poor conductor of electricity. Natural or uncharged particles will contain some small charge distribution as a result of contact with naturally occurring charged gaseous ions. The particle charge occurs when charged ions diffuse to and contact with the particle. This is called ion diffusion charging. Gaseous ions are created from corona discharges, radiation, and other energy releases. The charge on particles consists of specific whole numbers of electrons (elementary charges). Average charges are reported for various size particles by Whitby and Liu [14] as shown in Figure 3.7. This is achieved by some particles having zero charge, some having one charge, etc., as shown in Figure 3.8. Note the number of charges increases as particle diameter increases. Figure 3.8 also shows that for most particles the median number of natural charges is one electron.

Introduction of a high-energy electrical field in a gas can result in the formation of large concentrations of charged ions and, consequently, large numbers of charges on the particles present in this gas. Charged ions are formed in the electrically active region (the corona region) around a negative discharge electrode. Electrons are stripped from neutral gas molecules in the corona glow region resulting in an avalanche of electrons moving toward the collector electrode as shown in Figure 3.9. The positive ions return to the discharge electrode. Electrons moving toward the collector form negative ions from the gas molecules, G. The ions contact dust particles resulting in particles with large electron charges.

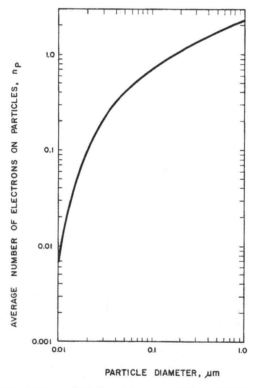

Figure 3.7 Average charge on spherical particles in equilibrium with natural atmosphere.

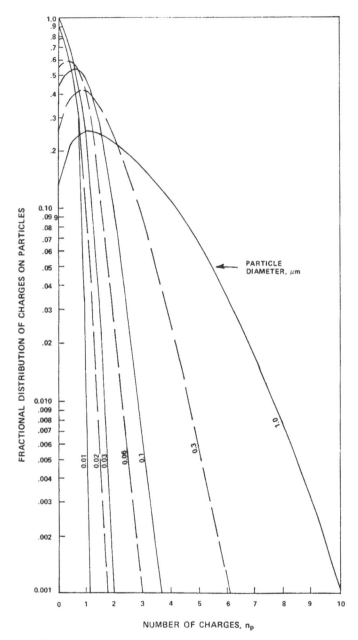

Figure 3.8 Fractional distribution of natural charges on particles.

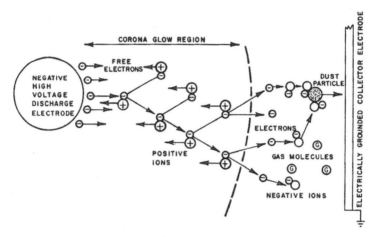

Figure 3.9 Particle charging process by corona discharge.

As the electrical voltage is increased on a given discharge electrode the amount of charge per particle increases toward a maximum. This maximum is limited by the electrical breakdown capacity of the gas. For dry air, this is about 10,000 volt/cm or 8 esu/cm² or 1.67×10^{10} electrons/cm² (one electron = 4.8×10^{-10} electrostatic units). Figure 3.10 shows the estimated maximum charge possible for particles in dry air. Normally, observed charges are about one-tenth the value noted in Figure 3.10.

Particle charging in an electrical field consists of field charging as shown in Figure 3.9 and diffusional charging. Diffusional charging occurs as a result of random motion of either or both the charged ions and the particles. As would be expected, diffusional charging is only significant for the small particles and dominates for the <0.2-μm particles. Field charging dominates for the >0.8-μm particles. Both methods contribute toward charging of the intermediate size particles from 0.2 to 0.8 μm in diameter, but a minimum particle charge in an electrical field occurs for particles about 0.3 μm.

A particle containing a charge in an electrical field will move at an electrical migration velocity, v_e, that can be shown by a force balance to equal

$$v_e = B\eta_p eE = bE \qquad (3.55)$$

B is mechanical or thermal mobility defined by Equation (3.22) where $B = C/(3\pi\mu_g d)$, η_p is the number of electron charges per particle, e is the charge on one electron (1.603×10^{-19} coulomb), and E is field strength in

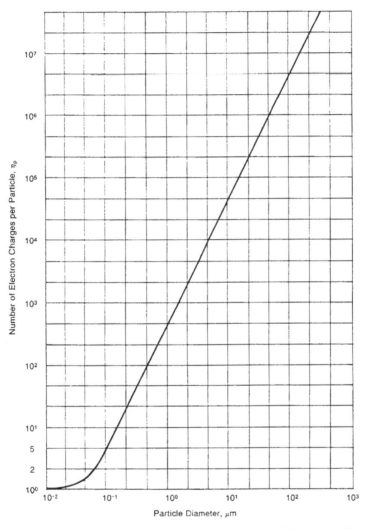

Figure 3.10 Calculated and measured maximum electron charge on single particles in dry air [15].

volt/cm (i.e., electrical potential divided by distance between discharge and collector electrodes). This is the quasistationary migrational velocity.

The term, b, in Equation (3.55) is particle electrical mobility. Rearranging the equation gives the definition $b = \eta_p eB$, which for a Slip Flow Regime size particle is

$$b = \frac{\eta_p eC}{3\pi\mu_g d} \tag{3.56}$$

Note that B has the units sec/g and b has units of $cm^2/(volt\ sec)$. These can be equated using the identity one coulomb equals 10^7 g $cm^2/(sec^2\ volt)$. For naturally charged particles, i.e., $\eta_p \cong 1$, electrical mobility, b, increases as particle size decreases. Charged particles in an electric field have a charge rate that increases about 10 times more rapidly than particle diameter as shown in Figure 3.10. Therefore, for particles in an electrically charged field, electrical mobility should increase as diameter increases.

The nonsteady-state electrical migrational velocity, v_e', of a particle initially moving toward the collector electrode with a zero velocity can be found from

$$v_e' = v_e[1 - \exp - (t/\tau)] \tag{3.57}$$

where τ is the relaxation time given by Equation (3.27).

Electrical migration velocities, v_e, are typically 0.15–0.30 m/sec in electrostatic precipitators (ESPs). Gas flow is turbulent and about 0.9–1.5 m/sec in ESPs. A fully charged 1-μm particle in an ESP at a 4- to 5-kV/cm field strength is theoretically collected with a force 3,000 times that of gravity. Actual particle migration velocity differs from theoretical due to factors such as (1) the "rapid" gas flow in ESPs which results in some particles not being fully charged, (2) collected particles may be reentrained back into the gas phase, and (3) field strength variations caused by dust layer buildups, rectified AC current oscillations, and power-source variations.

Theoretical fractional collection efficiencies for particles in a charged field are estimated using the Deutsch-Anderson equation as given by White [15]

$$\epsilon_i = 1 - \exp - \left(\frac{N_b A_c v_e}{\overline{N} Q F}\right) \tag{3.58}$$

where

N_b = particle concentration in laminar boundary region

$\overline{N}$ = average particle concentration
A_c = area of collector
Q = volumetric gas flowrate
F = gas flow unevenness factor: ranges from 1 for uniform flow to 2 for extremely uneven flow

3.6.5 PHORETIC FORCES

Phoretic forces are the result of an interaction between the particles and the gas molecules adjacent to the particles. Phoretic forces include diffusiophoresis, Stephan flow, thermophoresis, and photophoresis. They are weak forces in comparison to those discussed previously but become very critical as particle collection efficiency requirements increase. Phoretic forces must be recognized and used in a positive rather than in a negative sense. These forces become more significant as particle size decreases. They may act individually or combinations of several can be present simultaneously.

3.6.5.1 Diffusiophoresis

Diffusiophoresis is the net particle motion resulting from nonuniformities in the gas molecules of the suspending gas. It includes Stephan flow plus movement in the direction of the heavier or more concentrated gas molecule movement. It is caused by the differences in molecular impacts on opposite sides of the particles. The Stephan flow portion of diffusiophoresis results from the flow of gas molecules toward or away from the surface of a volatile liquid in the gas medium as a result of condensation or evaporation, respectively. That is, Stephan flow results in gas molecule, and hence particle movement, toward a liquid during condensation and away during evaporation.

This suggests that positive diffusiophoresis exists when particles to be collected in the presence of liquid droplets exist in a saturated and cooled gaseous medium. For example, spraying humidifies and cools a hot gas to improve particle collection. Hot dry gases develop a negative phoretic force by causing the liquid collection droplets to evaporate. This pushes small particles away from the collector.

Diffusiophoretic velocity, v_d, of 0.005- to 0.05-μm particles from air in water vapor systems is reported by Goldsmith and May [16] to be

$$v_d = -1.9 \times 10^{-4} \left(\frac{\Delta p}{\Delta x_d} \right) \qquad (3.59)$$

where the vapor pressure gradient, $\Delta p/\Delta x_d$, is in mb/cm and v_d is in cm/sec. Positive values of v_d indicate particle movement toward the liquid surface. It can be assumed that the vapor pressure gradient is linear and exists only within a specified boundary layer, Δx_d. Boundary layers are discussed in Section 3.6.6.

3.6.5.2 Thermophoresis

Gas molecules have a thermal velocity and therefore a kinetic energy which varies directly with temperature. Gas molecules on the hot side of a particle exert more force on the particles than do the gas molecules on the cold side. The resulting thermophoresis causes particles to move toward the colder region.

Thermophoretic velocity, v_T, of Free Molecule Regime spherical particles in an ideal polyatomic gas are given by Waldmann and Schmitt [17] as

$$v_T \cong -\frac{6\mu_g}{(8 + \pi)T\varrho_g}\frac{\Delta T}{\Delta x_T} \tag{3.60}$$

where

T = temperature of particle surface
ΔT = temperature difference over Δx_T, i.e., colder temperature minus warmer temperature
Δx_T = effective boundary layer over which thermophoresis can occur

Values for Δx_T can be approximated for spherical collectors using the information in Section 3.6.6. Positive values of v_T indicate velocities toward the colder region.

For larger particles in the Transition, Slip Flow, and Continuum Regimes and for particles with high thermal conductivities (e.g., metallic particles), thermophoretic velocities can be approximated using Brock's procedure [18] with Wachmann's thermal transfer values [19].

$$v_T \cong -\frac{(6.6)\lambda_g\mu_g}{\varrho_g dT}\frac{\Delta T}{\Delta x_T} \tag{3.61}$$

where λ_g is gas mean free path [see Equations (1.19), (1.20)].

3.6.5.3 Photophoresis

This is usually not present in air pollution control equipment because they operate enclosed and darkened. In the atmosphere, or where light is present, photophoresis can exist. Particle motion in the direction of light, i.e., away from the source, is considered by convention to be positive. Actual motion depends not only on where the light source is located and the type of light, but also on whether the particle reflects, absorbs, or transmits the light. Particle transparency, color, size, shape, and refractive index are all significant. Absorbed light can warm the side of the particle toward the light resulting in motion away from the light. Transmitted light can warm the side of the particle away from the light and cause the particle to move toward the light.

3.6.6 BOUNDARY LAYERS

External forces on a particle due to diffusiophoresis, thermophoresis, and Brownian (mechanical or thermal) diffusion can result because of gradients in vapor molecule concentration (e.g., water vapor), gas temperature, and particle concentration. It is usually assumed that these gradients are linear and exist only within a specific boundary layer. In the case of a spherical liquid droplet collector, the gradients are assumed to be linear and symmetrical. Johnstone and Roberts [20] have presented procedures for estimating effective boundary layer thicknesses.

Effective boundary layer thickness over which Brownian diffusion occurs is the smallest and is

$$\Delta x_B = \frac{D_c}{2 + 0.557Re_D^{0.5}Sc_p^{0.375}} \tag{3.62}$$

where

D_c = collector droplet diameter
Re_D = collector droplet particle Reynolds number, i.e., use D_c in numerator of Re_p, dimensionless
Sc_p = Schmidt number of particle = $\mu_g/(\varrho_p D_{PM})$, dimensionless

Effective boundary layer over which gas molecule mass transfer and, therefore, diffusiophoresis occurs near a water droplet is

$$\Delta x_d = \frac{D_c}{2 + 0.557Re_D^{0.5}Sc_w^{0.375}} \tag{3.63}$$

where Sc_w = Schmidt number of water vapor = $\mu_g/(\varrho_p D_{AB})$, dimensionless. D_{AB} is diffusivity of water vapor in air and is given by Equations (4.1) and (4.2).

Thermophoresis occurs over an effective boundary layer through which heat is transferred, which is

$$\Delta x_T = \frac{D_c}{2 + 0.557 Re_D^{0.5} Pr^{0.375}} \qquad (3.64)$$

where

Pr = Prandtl number = $C_p \mu_g / k_g$, dimensionless
C_p = gas specific heat at constant pressure
k_g = gas thermal conductivity

In wet scrubbers using free-fall water droplets as collectors, the magnitude of Δx_d and $\Delta x_T \cong D_c^{0.7}$.

The relative extent of these effective boundary layers over which the respective forces exist are shown in Figure 3.11. The value of Δx_B varies with collector diameter and particle collected size (this includes D_{PM}). For example, in saturated air at SC the effective Brownian diffusion boundary layer, Δx_B, for a 100-μm water droplet would be about

0.37 μm for 10-μm particles
0.91 μm for 1-μm particles
3.1 μm for 0.1-μm particles
13.6 μm for 0.01-μm particles

Under the same conditions, both Δx_d and Δx_T for a 100-μm droplet would be about 37 μm.

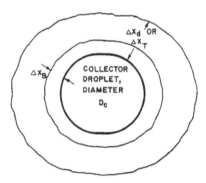

Figure 3.11 Relative magnitude of effective boundary layers around a spherical collector droplet.

3.7 PARTICLE COLLECTION

Most of the defining terms and basic particle behavior predicting theories have been presented in the preceding sections. This section shows how some of these ideas are related to particle collection in general. Specific collection devices and systems are discussed in Chapters 5 and 6, respectively. The reason that these subjects are not introduced until after Chapter 4 (Chapter 4 discusses the gas control mechanisms) is that many of these systems can simultaneously control *both* particles and gases. Diffusiophoresis and Brownian diffusion are not treated as separate sections but are included respectively in Section 3.7.3 on Impactor Systems and Section 3.7.5 on Filtration. These sections also show how to estimate total efficiency when several collection mechanisms occur simultaneously.

3.7.1 GRAVITATIONAL SETTLING

Gravitational force attempts to remove particles as discussed in Section 3.3.1 by causing them to fall at a terminal settling velocity. A simple vectorial analysis to resolve vertical and horizontal movements will result in a crude estimation of amount of particulate matter that settles from the gas medium. Combination of Equation (3.10) and linear horizontal velocity of the gas stream, v_x, with dimensions of the settling chamber gives

$$d = \sqrt{\frac{18\mu_g v_x h}{g(\varrho_p - \varrho_g)L}} \tag{3.65}$$

where h is chamber height and L is chamber length. Gas velocity in the vertical direction is often considered to be zero and quasistationary motion and spherical particles in the Continuum Regime are assumed. The particle diameter, d, represents the minimum-size particle completely removed by settling from the most adverse inlet position. Chambers designed to have low heights, h, are more effective. Equation (3.65) must be modified as discussed in Sections 3.3.1 and 3.3.2 to account for behavior of larger, smaller, and nonspherical particles.

Fractional removal efficiency for a specific-size particle by settling is ideally the ratio of v_s/h to v_x/L. However, experience shows that one-half this value is more practical so fractional removal efficiency is

$$\epsilon_i = \frac{0.5 v_s L}{v_x h} \tag{3.66}$$

where quasistationary motion is assumed and appropriate values of v_s are used.

For settling in the atmosphere, it is necessary to account for natural dispersion and eddy diffusivity in addition to gravitational settling. This is accomplished using Equation (1.35).

3.7.2 CENTRIFUGATION

Centrifugal separation of particles from gas streams is so important it should at least be mentioned by this name. It is the single most effective mechanism for removing particles from a gas stream. Inertial impaction and interception are centrifugal separation processes, but because they are unique to collection on targets they are treated as a separate subject. Furthermore, inertial impaction on fiber targets is also included under filtration. Movement of gases and particles within centrifugal devices was presented in Section 3.4 and a discussion of cyclones as control devices is given in Chapter 5.

3.7.2.1 Cell Sheets

A unique method for centrifugal collection of particles seems to occur in a scrubber that uses countercurrent rising gas and falling liquid with catenary grids. Catenary grids are open (e.g., 1/2 inch) with mesh pads shaped like a concave downward catenary [21]. The rising gas (air) causes the liquid (water) to form small circulation cells about 5 cm (2 inches) in height by 4 cm (1.4 inches) in width throughout a contacting zone about 46 cm (18 inches) deep above the grids. Stop-action, magnified pictures of these cells shows them to consist of many discrete sheets of liquid about 25 μm (0.025 mm) thick with a spacing between adjacent sheets of about 200–250 μm [22].

It is believed that these cells, containing many adjacent sheets of liquid, act like many small cyclones. Each pair of sheets behaves like a cyclone inlet at the top of each liquid cell. At typical gas velocities of 5.5 m/sec (18 ft/sec), centrifugal force could result in the collection of cell particles >0.65 μm with a specific gravity of 1.0.

3.7.3 IMPACTOR SYSTEMS

3.7.3.1 Impaction

Particle collection by impaction on small dry or wet targets is covered in this section. Systems that operate in this manner are called impactors to distinguish them from impingers. The distinction used here is that impinger targets are large surfaces. Inertial impaction is the major collection mechanism for both impactors and impingers. Commercial particle sizing devices called "impactors" operate like impingers as defined here because of the large collection surface. They do have small gas streams, however. These units are not part of this section.

Particle control by impaction is most often accomplished using small wet or dry targets. The wet targets are usually produced by atomizing a scrubbing liquid which then can move either cocurrent or countercurrent to the gas flow depending on how the device is constructed. The most important factor after design is the impaction parameter as given by Equations (3.6), (3.49), and (3.50). This is repeated in its most useful form

$$St = K_I = \frac{d^2 \varrho_p \Delta v C}{9 \mu_g D_c} \tag{3.67}$$

where Δv is the difference in velocity between the particle to be collected and the collector droplet. Always use D_c as the larger size object, whether it is the particle or the droplet. This equation is essentially Equation (3.50) for Slip Flow Regime size spherical particles assuming the gas density is negligible compared to the particle density. Any deviations should be appropriately considered.

Inertial impaction is a result of centrifugal force and could be determined theoretically. However, the problems involved are complex, so empirical relationships are used for relating impaction efficiency to impaction parameter as given in Figure 3.12 for various shape targets. The curve for cylindrical targets is expressed by the following two identical mathematical expressions

$$\epsilon_i = \frac{(\tfrac{1}{2} St)^2}{(\tfrac{1}{2} St + 0.35)^2} = \left(\frac{St}{St + 0.7}\right)^2 \tag{3.68}$$

for values of $K_I > 0.2$. Note that this equation would also closely represent the spherical collector efficiency curve. A slightly different curve presented by May and Clifford for cylinders is frequently used. It is not shown on Figure 3.12 but can be expressed as

$$\epsilon_i = \frac{(\tfrac{1}{2} St)^2}{(\tfrac{1}{2} St)^2 + 0.64} \tag{3.69}$$

The special case of sweeping impaction is discussed in Section 5.11.

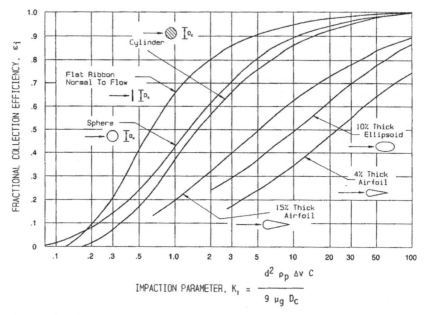

Figure 3.12 Estimated impaction efficiency as a function of inertial parameter for various targets.

3.7.3.2 Interception

Interception of particles on the collector is depicted in Figure 3.5 by particles 4 and 5 and usually occurs on the sides as shown just before reaching the top or bottom. Direct interception also includes particles that are deposited on the target because they move into position by Brownian diffusion. In an impactor, diffusion is usually not significant because gas velocities are high.

Collection of particles by interception in impactors results in an increase in overall collection efficiency. Small particles with a direct interception ratio, $d/D_c \ll 1$, follow the gas streamlines and result in an increase in individual fractional collection efficiency, $\Delta \epsilon_i$, of

$$\Delta \epsilon_i \cong \frac{(d/D_c)^2}{2.002 - \ln Re_D} \quad \text{(for } Re_D < 1) \tag{3.70}$$

when the targets are cylinders and assuming viscous flow theory. Viscous flow theory considers the fluid to have a finite viscosity and, therefore, the particle velocity is smaller as it approaches the target. Re_D in Equation (3.70) is collector, Re_p (i.e., use D_c for determining the value of Re_p).

Collection of small particles on spheres by interception using the same assumptions including $Re_D < 1$ gives

$$\Delta\epsilon_i \cong 3/2 \left(\frac{d}{D_c}\right)^2 \tag{3.71}$$

Larger particles, i.e., larger Re_D, move in straighter lines and collection efficiency increases due to interception on cylinders is

$$\Delta\epsilon_i = d/D_c \tag{3.72}$$

and on spheres is

$$\Delta\epsilon_i = \frac{2d}{D_c} \tag{3.73}$$

3.7.3.3 Multiple Collection Mechanisms

Impactor systems as described in this section achieve most of their particle removal capability by inertial impaction plus a small additional amount of removal due to interception. Particle removal by gravitational settling and Brownian diffusion is usually negligible because of the high velocities and low holdup times. Electrostatic precipitation can be significant if the system is so equipped and removal as discussed in Section 3.7.6 could be included. Eddy diffusion of turbulent gas in these systems can also increase amount of particle deposition, but it is difficult to obtain a value for this and it is only significant for particles <0.3 μm in size. Phoretic forces, specifically Stephan flow, should always be considered in impactor systems using liquid droplet collectors. This is necessary to optimize removal of the fine particles.

Total fractional collection efficiency for a given size particle, $\Delta\epsilon_{i,\text{Total}}$, by the various mechanisms in an impactor would include impaction, $\epsilon_{i,\text{Imp}}$ [see Figure 3.12 or equations such as Equation (3.68)], plus the additional collection due to interception, $\Delta\epsilon_{i,\text{Int}}$ [see Equations (3.70) to (3.73)].

$$\epsilon_{i,\text{Total}} = \epsilon_{i,\text{Imp}} + \Delta\epsilon_{i,\text{Int}} \tag{3.74}$$

If either positive or negative diffusiophoresis is present, the total efficiency would be influenced by

$$\epsilon_{i,\text{Total}} = (\epsilon_{i,\text{Imp}} + \Delta\epsilon_{i,\text{Int}})[1 + C_5 v_d \exp - (C_6 d^{C_7})] \tag{3.75}$$

where v_d is diffusiophoresis velocity with appropriate sign for positive or negative direction and C_5, C_6, and C_7 are constants that must be determined empirically.

3.7.3.4 Collision Impaction

Calvert, Inc. has developed a commercial scrubbing system that splits the gas stream containing particles then directs them back at each other head-on. The unit has the trademark name "Collision Scrubber." Each of the two gas streams enter a venturi scrubber throat where liquid (water) is introduced and pneumatically atomized as primary collection droplets. These droplets collect particles from the gas stream by inertial impaction as in a conventional venturi scrubber. The droplets are then accelerated by the gas stream as they continue to pass down their respective throats and into the collision zone where each of the streams enters from opposite sides. The droplets in the streams which are directed towards each other have an effective impaction velocity which is twice the velocity of a single gas stream. This doubled impaction velocity (Δv) increases the value of the impaction parameter and collection efficiency according to Equation (3.67) and Figure 3.12, respectively. It is possible that some of the collection droplets are also shredded in the collision zone interactions. This would increase the magnitude of the impaction parameter even more and further enhance impaction collection.

3.7.4 SPRAY SYSTEMS

Spray systems could be considered as special impactor systems whereby the collecting targets are droplets that fall downward through a rising gas which contains the particles. This is called scavenging or sweeping impaction. Examples of this are given in Section 5.11. Inertial impaction would dominate for particles >1 μm and for large droplets, but for smaller droplets the flowrates are sufficiently low and diffusion could be more significant. In this case, the fractional collection efficiency for <2-mm droplets in free-fall is

$$\epsilon_i \cong 1.68 Pe^{-2/3} \tag{3.76}$$

The Peclet number is defined for this purpose as

$$Pe = \frac{\Delta v D_c}{D_{PM}}$$

where Δv is difference in velocity between falling droplet and rising particle.

3.7.5 FILTRATION

Filtration is the collection of particles on a bed consisting of layers of a porous media. This includes fibers, permeable solids, and beds of small particles. The beds may be in place before filtration or may consist of deposited particulate matter. The filter bed can be operated wet or dry and may be stationary or in motion. Cleaning is usually achieved by mechanically disturbing the bed, by using a reverse flush fluid, or by a combination of these. Permeability of a filter is an indication of how much gas will flow through a clean new bed. The American Society for Testing and Materials (ASTM) definition of permeability for a cloth filter is the volumetric flowrate of air that will pass through 0.093 m² (1 ft²) of a clean new cloth at a pressure drop of 1.27 cm (0.5 in.) water.

Fabric filters are most common and are used as the filtration example. Initial collection of particles by filtration on dry cylindrical fibers occurs by the mechanisms of impaction, interception, sedimentation, diffusion, and electrostatic precipitation. Actual fabric filters have hairs on the fibers which also serve as initial collectors. The pores in the fibers become full of particles as the filter cake is formed and a portion of this cake remains each time the filter is cleaned. The cake actually does the effective filtering, but the collection mechanisms are similar.

Initial collection efficiency can be estimated by considering each mechanism. Inertial impaction efficiency alone is found by calculating K_I by Equation (3.67) and using Figure 3.12 for $\epsilon_{i,\text{Imp}}$. Collection improvement due to direct interception on the cylindrical fibers for $d/Dc \ll 1$ is obtained by Equation (3.70). Diffusional fractional efficiency, $\epsilon_{i,D}$, on the cylindrical fibers is obtained as noted by Friedlander [23] using

$$\epsilon_{i,D} = \frac{3.68 Pe^{-2/3}}{[2(2.002 - \ln Re_D]^{1/3}} \qquad (3.77)$$

where Pe is defined with Equation (3.76) but where Δv is the velocity difference between the particle and the collector. Re_D is collector, Re_p, and must be <1. In a filter, Re is typically from 10^{-4} to 10^{-1} at 0.5- to 5-cm/sec superficial gas velocities (1–10 ft/min). Electrostatic collection efficiency, $\epsilon_{i,E}$, would be estimated using procedures from Section 3.7.6 and sedimentation collection efficiency, $\epsilon_{i,S}$, could be approximated as the ratio of terminal settling velocity to local horizontal gas velocity.

Combined collection efficiency is influenced by the removal by each mechanism and the total initial fractional collection efficiency due to all mechanisms, $\epsilon_{i,\text{Total}}$, would be

$$\epsilon_{i,\text{Total}} = 1 - [1 - (\epsilon_{i,\text{Imp}} + \Delta\epsilon_{i,\text{Int}})](1 - \epsilon_{i,D})(1 - \epsilon_{i,E})(1 - \epsilon_{i,S}) \quad (3.78)$$

The projected area of one filter layer does not cover the entire cross-section area, so gas and particles not in line with a fiber have no chance of being captured. The total efficiency per single layer of fibers is then

$$\epsilon_{i,\text{Layer}} = (F)(\epsilon_{i,\text{Total}}) \quad (3.79)$$

where F is the fraction of cross-section area as fibers perpendicular to gas flow.

Total initial efficiency for a filter media of N layers would then be

$$\epsilon_{i,\text{Filter}} = 1 - (1 - \epsilon_{i,\text{Layer}})^N \quad (3.80)$$

Examination of the effectiveness of the various collection mechanisms shows that impaction and interception are most effective for larger particles but decrease as particle size decreases and are negligible for particles $< 1 \ \mu$m. Diffusion is only significant as particle size decreases below $0.01 \ \mu$m. Particles within about 0.01 to 1.0 μm in size are most difficult to collect. Friedlander [23] has developed an empirical relationship to show the combined effects of deposition by diffusion and interception of particles on glass fiber mats. This is

$$\epsilon_{i,D+\text{Int}} = 1.3Pe^{-2/3} + 0.7(d/D_c)^2 \quad (3.81)$$

Differentiation of this expression with respect to particle diameter and settling equal to zero gives an expression for determining particle size for least efficient collection. The following expression is averaged for particles from 0.01 to 1 μm in air at SC

$$Pe^{2/3}(d/D_c)^2 \cong 1 \quad (3.82)$$

This can be solved directly, but trial and error solution is probably quicker as the Cunningham correction factor, which is a function of d, and d appear in the Peclet number as particle diffusivity.

3.7.6 ELECTROSTATIC PRECIPITATION

In an ESP, charged particles are deposited by electrostatic force on the walls of a grounded collector. The electron charges on the particles must flow through this dust layer to complete the electrical circuit. The collected dust may be as thick as 1.5 cm before cleaning and 0.2 cm thick after cleaning. Electrical resistivity of the dust is an important factor as part of the voltage is lost across the dust layer. This causes the operating field strength to be reduced. If the applied voltage is raised too high in attempting to overcome this, excessive sparking occurs. During a spark and for a short time thereafter, the field strength may drop to near zero and little electrostatic precipitation occurs. If the applied voltage is not increased to overcome the voltage drop across the dust, the field strength may not be adequate to provide good electrostatic precipitation force. A certain amount of sparking is usually desirable to be sure voltage is maintained near the maximum and the required collection efficiency is sustained.

Particle electrical resistivity over about 2×10^{10} ohm cm is considered high and under these conditions large surface area is required for efficient particle collection. This occurs with low-sulfur coal fly ash at flue gas temperatures of about 180°C. Particles with a resistivity less than about 10^4 ohm cm have low resistivity and do not hold well to the collector surface. The desired particle resistivity for electrostatic precipitation is from 10^4 to 2×10^{10} ohm cm.

Particulates can be treated to vary resistivity. Particle resistivity also changes with temperature. At low temperature, surface electrical conduction dominates and total resistivity increases with temperature. At high temperature, volume resistivity dominates and total resistance decreases with temperature. Low-sulfur coal fly ash resistivity can be brought into the desirable range by operating the precipitator hot, i.e., about 380°C. If this is undesirable, a greater collection surface area must be used to achieve the necessary overall efficiency.

It is possible to modify the particle charging procedures to vary the electrostatic collection efficiency. Particles are charged by ions produced by the corona discharge. The corona is initiated on a wire at points of imperfection. Use of discharge electrode wires with spikes instead of a smooth wire may increase the corona discharge and improve collection efficiency.

The theoretical collection efficiency equation for electrostatic precipitation is given as Equation (3.58). Specific values for many of the terms in this equation are not practical to estimate accurately and the preferred procedure is to obtain pilot data from tests using exactly the same input particulate and gas composition and concentrations. Under specific operating conditions such as these, the test data give an overall efficiency as a function of the collector area, A_c, and volumetric gas flowrate, Q. That is, Equation (3.58) can be simplified for a specific system and operating conditions to show overall collection efficiency as

$$\epsilon_o = 1 - \exp - \left(\frac{A_c C_8}{Q} \right) \qquad (3.83)$$

where C_8 is an appropriate constant obtained for each test at a given ratio of collector area-to-gas flowrate. This ratio of collector area-to-gas flowrate is called specific collection area (SCA), and for high-efficiency units may range upward from about 30 m²/1,000 Am³ per hr (500 ft²/1,000 acfm).

Scale-up from the test data must duplicate all conditions such as types of discharge wires and collectors, configuration, gas and particle flow and distribution, and dust cake type and amount. Other equipment-related factors are presented in Chapter 5. Gravitational settling alone typically produces a 30–40% collection efficiency in an ESP when all power is off.

3.7.7 COALESCENCE

Coalescence is the growing and/or merging together of particulate matter and includes the actions of agglomeration, coagulation, and condensation. These actions produce larger particulate material and result in a decreased number concentration of particulates and a reduction of smaller particulates. Condensation could increase the particle number concentration. Evaporation could be included as a negative condensation and, in this case, particle size is reduced.

The kinetics of particles merging together due to Brownian motion can be described by a second-order rate equation

$$-\frac{dN}{dt} = KN^2 \qquad (3.84)$$

where N is the particle number concentration per cm³, t is time in seconds, and K is a rate constant in cm³/sec. K is approximately equal to $4/3kTC/\mu_g$, but empirical values range from 0.51×10^{-10} for larger to 3.7×10^{-10} cm³/sec for smaller particles.

Particles may be made to merge together more rapidly by increasing system turbulence, saturating the system with liquid vapor, charging the particles, and by the coupling of sonic energy. Saturation would only be effective if the particle has an affinity for the vapor such as hygroscopic liquid particles and deliquescent solids.

Coalescence is of greater significance when airborne particulates are in contact with each other for long periods of time as in the atmosphere or in sampling containers. In contrast, holdup time in collection devices is usually relatively short. However, this can still be an effective procedure for reducing particulate concentration, and process changes to incorporate coagulation may be beneficial.

3.7.8 CUT DIAMETER

Cut diameter, d_c, is defined in Section 1.2.1 as the diameter for which collection efficiency (and penetration) equals 50%. In the collection of particles, this becomes a convenient particle size to specify in indicating the particle-removal efficiency of a collector for a specific type of dust. Cut diameter is more applicable to systems such as scrubbers that operate mainly by a single collection mechanism. In this case, a continuous collection efficiency vs particle diameter trend is experienced. A dry filter, which operates with one mechanism for larger particles and another for smaller, as discussed in Section 3.7.5, may not be categorized as accurately in the cut diameter-efficiency relationship.

The 50% cut diameter expressed as a 50% aerodynamic cut diameter, d_{50}, is used to compare relative effectiveness of variations of essentially the same collection mechanism. This is shown in Figure 3.13 by Calvert [24]. All these data are for inertial impaction collection in wet-scrubbing devices. Curve 1 represents impingement collection of particles on the large wet surface of a 2.5-cm ring- or saddle-type packing. Packing depth is not too significant. Curve 2 is impingement in a liquid obtained by bubbling the gas through the liquid on sieve plates. The sieve plate holes are 0.5 cm in diameter and the liquid was bubbled into a foam of 0.4-g/cm³ density. The number of plates does not appear to be significant. Curves 3, 4, and 5 represent impaction of particles on cylindrical wet fibers 300, 100, and 40 μm in diameter, respectively. Curve 6 is the impaction of particles on pneumatically atomized droplets. The droplets are essentially spherical and energy to produce the droplets is accounted for in the gas pressure drop. The abscissa of Figure 3.13 shows the total system resistance to gas flow, including any atomization energy required from the gas stream, as gas phase pressure drop. Water power requirements are not included. The estimated gas blower power requirements are not included. The estimated gas blower power requirements were obtained using

$$kW = 6.1 \times 10^{-5}(\Delta P)(Q) \qquad (3.85)$$

where ΔP is pressure drop across the gas stream in cm H_2O and Q is flowrate in m³/hr. This actual power assumes 60% blower efficiency. Power in horsepower can be obtained from Equation (5.25).

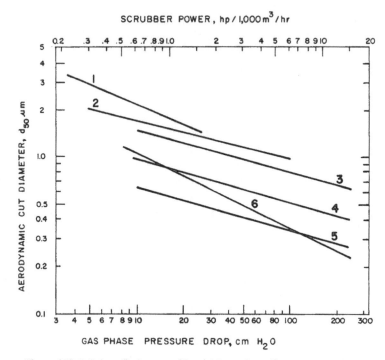

Figure 3.13 Relative effectiveness of inertial impaction collector arrangements.

Relationships such as those given in Figure 3.13 are useful first approximations of comparative particle-collection efficiency. The 90% aerodynamic cut diameters, d_{90}, in place of d_{50} may be more useful for evaluating high-efficiency collectors. It also appears that these curves for the various collection systems are more parallel. The d_{90} indicates size particle captured at 10% efficiency, i.e., penetration is 90%.

3.7.9 HIGH TEMPERATURE AND PRESSURE EFFECTS

Particle behavior is influenced by the gaseous medium containing the particles and both the particles and the gas are affected by changes in temperature and pressure. Effects on the gas are usually more significant. Particle and gas properties that change as temperature and pressure vary are summarized in Table 3.4. The gas-transport properties of viscosity, thermal conductivity, and diffusivity are only slightly affected by pressures below 20 atmospheres at standard temperature and even less at higher temperatures. For example, gas viscosity at 1,000°C is nearly independent of pressure up to 300 atmospheres. It is always necessary to account for the changes in volumetric flowrate in a confined system with changes in temperature and pressure. Gas viscosity increase with temperature rise is a highly significant effect and is covered in Section 1.3.2.

3.7.9.1 Cunningham Correction Factor

Change in the Cunningham correction factor is one of the more significant effects of high temperatures and pressures. This factor attempts to match the predicted behavior of Slip Flow and Transition Regime particles to experimentally observed behavior.

The constants in Equation (3.4) are functions of momentum and thermal energy accommodation of gas molecules impinging on the particle surface. Data available indicate the accommodation coefficients decrease with increasing temperature but are changed little with pressure changes. This makes the gas molecule resistance to particle motion decrease more rapidly than the simplified equations such as Equations (3.4) and especially (3.5) predict. For the smaller particles subject to use of the Cunningham correction factor, C, it may be necessary to use values up to 10 times the standard value for C obtained from Equation (3.4) as shown in Figure 3.14.

TABLE 3.4. Effect on Properties by Changes in Temperature and Pressure.

	Changes that Occur with Increased	
Property	Temperature	Pressure
Particle		
Volume	negligible	negligible
Density	negligible	negligible
Thermal conductivity	negligible	negligible
Dielectric constant	negligible	negligible
Diffusivity	increases	decreases
Electric mobility	decreases	negligible
Resistivity	increases then decreases	negligible
Viscosity	decreases	negligible
Gas		
Volume	increases	decreases
Density	decreases	increases
Molecular mean free path	increases	decreases
Viscosity	increases	increases
Thermal conductivity	increases	negligible
Diffusivity	increases	decreases

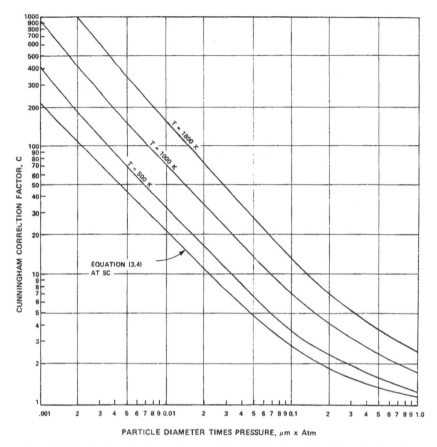

Figure 3.14 Cunningham Correction Factor predicted by Equation (3.4) at SC and by Rao [2] at elevated temperatures and pressures.

3.7.9.2 Impaction

The impaction parameter, K_I, is a direct function of relaxation time, τ, which in turn is affected by both C and μ_g [see Equations (3.27) and (3.67)]. Collection efficiency is directly related to the impaction parameter as shown in Figure 3.12. The effects of pressures up to about 15 atmospheres are only significant for particles less than 1 μm in air, but impaction of these small particles is not significant. High temperature results in a decrease in the value of K_I calculated by not correcting for temperature. For example, corrected values of K_I at 1,100°C are nearly 40% of the uncorrected values for any given particle size over 1 μm.

3.7.9.3 Interception

When interception follows the relationship given by Equation (3.70), the increased fractional efficiency, $\Delta\epsilon_i$, is proportional to $[2.002 - \ln Re_D]^{-1}$. Simplified, this says that $\Delta\epsilon_i$ increases with Re_D, i.e., $\Delta\epsilon_i \propto Re_D$. Re_D is the collector particle Reynolds number which equals $D_c\Delta v\varrho_g/\mu_g$. As pressure increases, ϱ_g increases and collection by interception should increase. As temperature increases, the μ_g increase dominates and collection should increase.

3.7.9.4 Particle Diffusivity

Particle diffusivity, D_{PM}, as shown by Equation (3.19) is a direct function of C and T and is indirectly related to μ_g. Collection efficiency in turn is directly related to both D_{PM} and Re_D as noted from Equation (3.77); Re is defined with Equation (3.76) and the previous discussion. Diffusivity is increased by increased temperature and decreased by increased pressure which is opposed to the effects on Re_D. Diffusivity is usually only significant for the <0.1-μm particles. The net effect on diffusional deposition is mainly due to temperature with increased deposition as temperature increases. It would decrease deposition with increased pressure to a lesser extent. Kornberg [25] predicts that the diffusional deposition efficiency on cylindrical fibers is proportional approximately to $T^{0.815}$.

3.7.9.5 Electrical Mobility

Electrical mobility, b, given by Equation (3.56) is directly related to the number of electron charges, η_p, and to C and is indirectly related to μ_g. The dominant effect is η_p. Ion diffusion charging of smaller particles increases significantly with temperature, increasing the value of η_p and hence increasing electrical mobility. This, in turn, increases migrational velocity [Equation (3.55)] and electrostatic collection efficiency [Equation (3.58)]. If the temperature is high enough to excite the gas molecules, ions are produced which discharge the particles and reduce collection efficiency. More data are needed in order to be more specific in this effect.

3.7.9.6 Separation Forces

Gravitational and centrifugal separation of particles from a gas are related to terminal settling velocity, Equation (3.15), and terminal centrifugal velocity, Equation (3.23). Where gas velocity and densities are negligible, these velocities depend directly on C and indirectly on μ_g. Therefore it would be expected that separation efficiency would decrease with increased temperature and not change significantly with pressure.

3.7.9.7 Phoretic Forces

Calvert and Parker [26] show that both thermophoresis and diffusiophoresis would be expected to vary with temperature and pressure. They both increase with temperature and decrease with pressure. The indirect effect of temperature and pressure on the liquid present and the condition of the gas could be more significant in diffusiophoresis collection.

3.7.9.8 Coalescence

The same reference [26] indicates that normal agglomeration of particles increases with temperature and decreases with pressure. However, if the system is in turbulent flow, eddy agglomeration decreases with temperature and increases with pressure.

3.7.9.9 Summary

Data available at this time indicate in general that particles larger than about 0.2 μm are collected with decreased efficiency as temperature and pressure increase and particles less than about 0.1 μm are collected with increased efficiency at higher temperatures but lower efficiency at higher pressures. The pressure effects are relatively small compared to the temperature effects except for low pressures less than 1 atmosphere.

3.8 CHAPTER PROBLEMS

3.8.1 TERMINAL SETTLING

a. Determine the terminal settling velocity and Re_p of a 5-μm spherical particle in still air at standard conditions. The particle density is 1.5 g/cc.
b. Repeat for a 0.5-μm particle.
c. Repeat for a 50-μm particle.
d. Repeat for a 500-μm particle.
e. Repeat for a particle described as a 0.3-μm diameter cylinder 1 μm long.

3.8.2 PARTICLE BEHAVIOR

A rectangular particle 0.7 μm on a side and with a density of 0.85 g/cm³ is free-falling in still air at SC. Determine:

a. Relaxation time
b. Particle velocity at 0.6τ seconds after start of fall
c. Distance particle has traveled in this time due to force of gravity
d. Particle thermal velocity
e. Distance particle travels in 6τ due to diffusional displacement
f. Particle mechanical mobility
g. Aerodynamic diameter

3.8.3 CHARGES ON PARTICLES

Consider three spherical particles in dry air at SC. The smallest is 0.2 μm in diameter. The next is 0.7 μm in diameter and the largest is 1.2 μm in diameter. Calculate for each:

a. Mechanical mobility
b. Electrical mobility for natural particles
c. Electrical mobility for fully-charged particles
d. The electrical migrational velocity for each type if they are between parallel plate electrodes spaced 15 cm apart with a voltage potential of 30 KV
e. The theoretical fractional collection efficiency if the quantity ($N_b A_c / \overline{N} Q F$) equals 0.50

3.8.4 GRAVITY SETTLING CHAMBERS

A sinter plant plans to use a simple settling chamber to remove the large particles of quartz ($\varrho_P = 2.6$) and iron oxide ($\varrho_P = 4.5$). Volumetric flow rate is 1/4 million cubic meters of gas per hour at 20°C. The chamber design is shown in Figure 3.15.

a. Determine the minimum size particle completely removed.
b. Obtain data and plot percent removed vs particle diameter for 100- and 200-μm particles (include point obtained from part a. above and note that this device is about 2–4% effective for the removal of particles <40 μm).

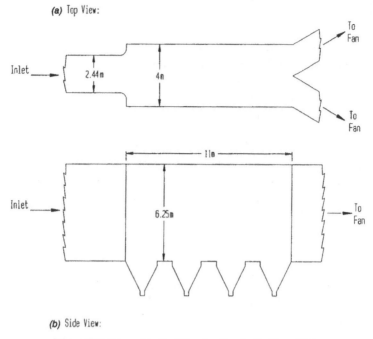

(a) Top View:

(b) Side View:

Figure 3.15 Schematic of settling chamber for Problem 3.8.4.

3.8.5 FILTRATION

Gas containing spherical unit density particles at SC is passing through a glass fiber filter media at a superficial gas velocity of 1.3 cm/sec. The filter media consists of 3 layers of clean 20 μm cylindrical fibers. Each layer cross section area is 35% fibers and 65% open pore space. The media is vertical so no sedimentation occurs and there is no electrostatic charge.

a. Determine the collection efficiency for 0.1-, 1.0-, and 10-μm particles.

b. Estimate the size of particle that results in the lowest collection efficiency.

3.8.6 COALESCENCE AND ELEVATED TEMPERATURE

a. Determine the change in concentration due to coalescence of "large" and "small" particles in 10 minutes. Do this when initial concentrations are 3.5 × 10⁵ and 3500 particles/cm³.

b. Assume that the filter in Problem 3.8.5 is made of 20-μm diameter steel wires and operated at 480°C with all other conditions as noted previously. Determine the collection efficiency on 0.1-μm particles and compare with normal temperature results (check all values even though some are negligible).

REFERENCES

1 Davies, C. N. 1943. *Proc. Phys. Soc.* (London), 15(6).

2 Rao, A. K., M. P. Schrag and L. J. Shannon. 1975. "Particle Removal from Gas Streams at High Temperature/High Pressure," U.S. Environmental Protection Agency Report EPA-600/2-75-020.

3 Calvert, S. et al. 1972. *Scrubber Handbook.* APT, Inc.

4 Oseen, C. 1927. *Neure Methoden und Ergebnisse in der Hydrodynamik.* Leipzig: Akadmische Verlag.

5 Klyachko, L. 1934. *Otopl. i. Ventril*, No. 4.

6 Strauss, W. 1966. *Industrial Gas Cleaning.* Elmsford, NY: Pergamon Press, Inc.

7 Fuchs, N. A. 1964. *The Mechanics of Aerosols.* Rev. ed. Elmsford, NY: Pergamon Press, Inc.

8 Thomas, J. W. "Gravity Settling of Particles in a Horizontal Tube," *J. Air Poll. Control Assoc.*, 8(1):32–34 (1958).

9 Slinn, W. G. N. 1976. "Theory of Diffusive Deposition of Particles in a Sphere and in a Cylinder at Small Fourier Numbers," *Atmos. Environ.*, 10(9):789 (1976).

10 Davies, C. N. 1946. *Proc. Royal Soc.*, B 133:298.

11 Gormley, P. and M. Kennedy. 1949. *Proc. Roy. Irish Acad.*, 52A:162.

12 Ranz, W. E. and J. B. Wong. 1952. "Impaction of Dust and Smoke Particles," *Ind. Eng. Chem.*, 44:1371.

13 Lundgren, D. A. 1969. "A Sampling Instrument for Determination of Particulate Composition, Concentration and Size Distribution Changes with Time," NAPCA Symposium in Advances in Instrumentation, Cincinnati.

14 Whitby, K. T. and B. Y. H. Liu. 1966. "The Electrical Behavior of Aerosols," in *Aerosol Science.* C. N. Davies, ed. New York: Academic Press, Inc.

15 White, H. J. 1963. *Industrial Electrostatic Precipitation.* Reading, MA: Addison Wesley Publishing Co.

16 Goldsmith, P. and F. G. May. 1966. "Diffusion and Thermophoresis in Water Vapour Systems," in *Aerosol Science.* C. N. Davis, ed. New York: Academic Press, Inc.

17 Waldmann, L. and K. H. Schmitt. 1966. "Thermophoresis and Diffusiophoresis of Aerosols," in *Aerosol Science.* C. N. Davies, ed. New York: Academic Press, Inc.

18 Brock, J. R. 1962. *Coll. Sci.*, 17:768.

19 Wachmann, H. 1962. *ARS J.*, 32:2.

20 Johnstone, H. F. and W. H. Roberts. 1949. "Deposition of Aerosol Particles from Moving Gas Streams," *Ind. Eng. Chem.*, 41:2417–2423.

21 1984. U.S. Patent No. 4,432,914, "Mass Transfer Contact Apparatus," U.S. Dept. Commerce Patent and Trademark Office, Vol. 1039, No. 3.

22 Hesketh, H. E. 1987. "The Fluidized Bed Wet Scrubber for Control of Gases and Particulates," *Proceedings of the Pollution Control and Clean Air, Asian-Pacific Conference*, Singapore.

23 Friedlander, S. K. 1977. *Smoke, Dust and Haze/Fundamentals of Aerosol Behavior*. New York: John Wiley & Sons, Inc.

24 Clavert, S. 1977. "How to Choose a Particulate Scrubber," *Chem. Eng.*, 84(18): 54–68.

25 Kornberg, J. P. 1973. "High Temperature Filtration of Particles by Diffusion," Ph.D. Thesis, Harvard School of Public Health, Boston, MA.

26 Calvert, S. and R. Parker. 1977. "Effects of Temperature and Pressure on Particle Collection Mechanisms: Theoretical Review," EPA-600/7-77-002.

CHAPTER 4

Gas Control Mechanisms

THIS chapter deals with the removal of gaseous pollutants from the carrier gas, which is usually air. The control mechanisms are chemical engineering unit operations which relate to the physical changes that take place. These include fluid flow, heat and mass transfer, condensation, absorption, and adsorption. Chemical reactions may take place at the same time and can be considered in addition to the unit operations. Sometimes a general concept such as mass transfer is used to include simultaneously several individual unit operations such as condensation, absorption, and adsorption. However, in most cases the individual concepts are presented, as they are more easily visualized this way.

Gaseous pollutant streams are like chemical process streams. The major difference is that the gaseous pollutant concentrations are usually lower than chemical process stream concentrations. These low concentrations make it possible to consider gaseous pollutants as ideal gases, even though some may consist of large molecules which in more concentrated form are not ideal gases. The carrier gas, air, is also an ideal gas in the operating range of most air pollution control equipment. Ideal gas relationships are given in Section 1.3. If these principles have not been reviewed it would be advisable to do so at this time. High temperatures and pressures cause air to deviate from ideal gas behavior. This is discussed at the end of this chapter and in Sections 1.3.2 and 3.7.9.

Several important dimensionless relationships related to gas behavior have already been presented. Mach number, Ma, as a basic definition is presented in Chapter 1. Air is considered an incompressible fluid when $Ma \ll 1$. Defining equations to relate gas properties of density, viscosity, molecular weight, mean free path, and acoustic velocity are also given in Chapter 1. Flow Reynolds number, Re_f, as an expression of system turbulence is presented in Chapter 3 along with particle Reynolds number, Re_p.

4.1 DIFFUSIVITIES

Movement of gas molecules through a gaseous fluid is necessary to initiate mass transfer such as condensation, absorption, and adsorption. This gaseous diffusion is not to be confused with atmospheric eddy diffusion created by temperature and pressure gradients and the earth drag forces. Atmospheric diffusion relates to the movement of parcels of air.

In some operations, such as absorption, it is important that the absorbed gas molecule moves through the liquid and away from the surface of the absorbing liquid. This can be accomplished by diffusion of gas through the liquid fluid. This occurs at a different rate than diffusion of a gas molecule in the gas fluid.

4.1.1 GAS DIFFUSIVITY IN GASES

Movement of gaseous pollutants, A, through the air carrier gas, B, can be given as the diffusivity of A through B, D_{AB}. This equals diffusivity of gas B through gas A, D_{BA}. In units of cm²/sec this is given by the modified Gilliland [1] equation as

$$D_{AB} = 1.8 \times 10^{-4} \frac{(T)^{1/2}}{[(\tilde{V}_A)^{1/2} + (\tilde{V}_B)^{1/2}]^2} \frac{M_A}{\varrho_A} \left[\frac{1}{M_A} + \frac{1}{M_B} \right]^{1/2} \qquad (4.1)$$

where

   T = absolute temperature, K
   M = molecular weight
   $\tilde{V}$ = molecular volume of gas when a liquid at the normal boiling point, cm³/g mole (see Table 4.1)

The ratio M_A/ϱ_A in Equation (4.1) can be replaced by RT/P by use of the ideal gas law where R is the universal gas constant, 82.05 atmospheres cm³/(g mole K), and P is pressure in atmospheres.

Gas diffusivities of pollutant A through gas B can also be obtained experimentally using Stephan's procedure [2] of placing a small tube nearly filled with liquid pollutant A in a stream of flowing air as shown in Figure 4.1. The tube is kept in a constant temperature bath at the air temperature during the tests and rate of diffusion is determined by measuring the rate

TABLE 4.1. Molecular Volume of Some Gases in Liquid State at the Normal Boiling Point.

Substance	Molecular Volume (cm³/g mole)
Air	29.9
H_2	14.3
CO_2	34.0
CO	30.7
O_2	14.8
N_2	36.0
NO	23.9
NO_2	32.2
SO_2	40.4
SO_3	47.8
C_2H_6	51.8
C_6H_5COCl	136.1
H_2O	18.8

of fall of the diffusing liquid level by

$$D_{AB} = \frac{RT}{P \ln (p_{B_1}/p_{B_2})} \frac{\rho_{Al}}{M_A} \frac{L_2^2 - L_1^2}{2t} \tag{4.2}$$

where

ρ_{Al} = density of liquid A, g/cm³
p_{B_1} = partial pressure of air at L_1
p_{B_2} = partial pressure of air at L_2
L_1 = initial liquid level, cm
L_2 = final liquid level, cm
t = time for liquid level to change, sec

Locations of L_1 and L_2 are shown in Figure 4.1. At L_1, concentration of the pollutant is very low and assumed to be zero, so by use of Dalton's law the value of p_{B_1} at this point is essentially total pressure. The partial pressure of air directly over the liquid, p_{B_2}, is obtained from vapor-liquid-equilibrium (VLE) data. VLE is discussed in Section 4.3.5. These diffusivity equations are applicable for gases near standard conditions. See Section 4.7 for corrections at other conditions.

Example Problem 4.1

Determine the diffusivity of SO_2 in air at SC.

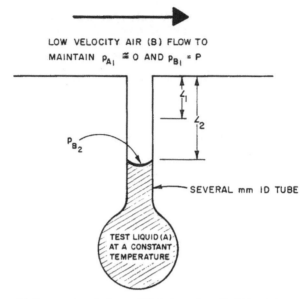

Figure 4.1 Test setup to obtain gas diffusion rate from liquid-level change rate.

Solution

Equation (4.1) is used. For SO_2

$$M_A = 64 \text{ g/g mole}$$
$$\widetilde{V}_A = 40.4 \text{ cm}^3/\text{g mole}$$

For air, use

$$M_B = 29 \text{ g/g mole}$$
$$\widetilde{V}_B = 29.9 \text{ cm}^3/\text{g mole}$$

$$\varrho_A = \frac{M_A P}{RT} = \frac{(64)(1 \text{ atm})}{(82.05)(293)} = 2.66 \times 10^{-3} \text{ g/cm}^3$$

$$D_{AB} = (1.8 \times 10^{-4}) \frac{(293)^{1/2}}{[(40.4)^{1/2} + (29.9)^{1/2}]^2}$$

$$\times \frac{64}{2.66 \times 10^{-3}} \left[\frac{1}{64} + \frac{1}{29}\right]^{1/2}$$

$$D_{AB} = 0.119 \text{ cm}^2/\text{sec}$$

Example Problem 4.2

Find the diffusivity of methyl ethyl ketone (MEK) in air at 25°C from liquid MEK level-change data. The liquid level changes from 4.05 to 5.75 cm at a system pressure of 751 mm Hg (0.988 atm) in 55.61 hr. The liquid density is 0.8 g/cm³.

Solution

Solve using Equation (4.2)

Values for MEK

$$M_A = 72 \text{ g/g mole}$$
$$\varrho_A = 0.8 \text{ g/cm}^3$$
$$L_1 = 4.05 \text{ cm}$$
$$L_2 = 5.75 \text{ cm}$$
$$p_{A_1} \cong 0 \text{ mm Hg}$$
$$p_{A_2} = 90.6 \text{ mm Hg (vapor pressure of pure MEK at 25°C) [3]}$$

Values for air, using Dalton's law Equation (1.12),

$$p_{B_1} \cong 751 - 0 = 751 \text{ mm Hg}$$
$$p_{B_2} = 751 - 90.6 = 660.4 \text{ mm Hg}$$

$$D_{AB} = \frac{(82.05)(273 + 25)}{(0.988 \text{ atm}) \ln (751/660.4)} \frac{0.8}{72} \frac{5.75^2 - 4.05^2}{(2)[(3,600)(55.61)]}$$

$$D_{AB} = 0.089 \text{ cm}^2/\text{sec}$$

4.1.2 GAS DIFFUSIVITY IN LIQUIDS

Once absorbed, it is necessary for the dissolved gas molecules to move away from the surface before more gas molecules can be absorbed. This diffusion of gas through a liquid can be estimated using gas A in liquid B diffusivity, D'_{AB}, in cm²/sec obtained from the Stokes-Einstein equation as modified by Wilke [4]

$$D'_{AB} = 7.4 \times 10^{-10} \frac{(C_9 M_B)^{1/2} T}{\mu_B \widetilde{V}_A^{0.6}} \tag{4.3}$$

where C_9 is a constant for the liquid solvent. It is equal to about 2.6 for water, 1.9 for methanol, 1.5 for ethanol, and 1.0 for benzene, ether, heptane, and similar solvents. This equation is good for dilute solutions. A diffusivity of 1.61×10^{-5} cm²/sec for SO_2 in water at standard conditions, $\mu_B \cong 0.01$ g/(cm sec), can be calculated. Compared with Example Problem 4.1, this shows as one would expect that the diffusivity of SO_2 is much greater in air than in water (about 7,400 times greater).

4.2 MASS TRANSFER

In order to remove a gaseous pollutant from the carrier gas, it is necessary to transfer a mass of pollutant from the gas phase to another phase. The other phase is usually a liquid, as in absorption, but may also be a solid (or a liquid), as in adsorption. Gas molecules must be transferred through a boundary region which can be considered as two "films" — one film each for the liquid and for the gas phases. These films represent the resistance to the passage of molecules which may be passing through the film in either direction. At steady-state conditions, the number of molecules moving in each direction is constant but not necessarily equal. At equilibrium, the number of molecules moving in each direction is equal. The net transfer rate is influenced by the driving forces of temperature, pressure, concentration, and affinity of the solvent for the gas molecules (solubility).

Concentration of molecules on the gas side is affected by both eddy and molecular diffusion. Normally, eddy diffusion is created to maintain a uniform concentration of molecules in the bulk of the gas phase. Molecular gaseous diffusion provides the means by which the gas molecules move to the film. On the liquid side, a similar situation exists where eddy diffusion is created to maintain essentially a constant concentration in the bulk of the liquid and liquid diffusion moves molecules to and from the film. If the molecules of gas that pass through the film and into the liquid react chemically or are otherwise physically removed to keep them from passing back through the film, the net rate of transfer from the gas to the liquid is increased.

The net overall resistance is considered to be due to the combined gas and liquid films and is equal to the reciprocal of the sum of the reciprocal of the individual resistances. Usually, the resistance of one of the films is significantly larger and it is convenient to attribute the total resistance to the passage of molecules through the film to the film with the largest resistance. For example, if the molecules can pass rapidly through the gas side of the film, then the net resistance to the passage of molecules is attributed to the liquid film and this is called "liquid-phase controlling." If most of the resistance is in the gas film, then it is called "gas-phase controlling" and the resistance due to the liquid side is neglected. Gas molecules usually pass more rapidly through gas films, so "gas-phase controlling" absorptions are more effective for pollutant removal.

The rate at which mass is transferred from one phase to another is proportional to the area and the driving force. The proportionality constant is called the mass transfer coefficient (rate constant). Separate equations can be written for each film. The equation for the gas film is

$$N_A = k_G A (p_{AG} - p_{Ai}) \tag{4.4}$$

For the liquid film, the equation is

$$N_A = k_L A (c_{Ai} - c_{AL}) \tag{4.5}$$

where

k_G = gas film mass transfer coefficient, g mole/(hr cm^2 atm)
k_L = liquid film mass transfer coefficient, g mole/(hr cm^2 g mole/cm^3)
N_A = rate of transfer of A through the film, g mole/hr
A = surface area of mass transfer, cm^2
p_{AG} = partial pressure of A in gas phase, atm
p_{Ai} = partial pressure of A at vapor–liquid interface, atm
c_{Ai} = concentration of A at vapor–liquid interface, g mole/cm^3
c_{AL} = concentration of A in liquid phase, g mole/cm^3

The driving force for mass transfer is the difference in partial pressures and the difference in concentrations in Equations (4.4) and (4.5), respectively. The rate of transfer of molecules of A from the gas phase to the liquid phase must be the same through both films even though one film is controlling because there is no accumulation of molecules between the films.

4.3 GAS ABSORPTION

Gas absorption is the taking in of a gas or vapor by a liquid and includes both physical and chemical absorption. In absorption, the material taken up would ultimately be distributed throughout the entire absorbent (liquid) phase. Absorption is distinguished from adsorption in that the latter is a surface phenomenon with the material taken up being distributed over the surface of the adsorbing material.

In order to be absorbed, pollutant molecules must diffuse to the gas film, pass through both the gas and the liquid films, and finally diffuse into the liquid. As the net transfer of molecules through either film is the same, we can set Equations (4.4) and (4.5) equal to each other.

$$N_A = k_G A(p_{AG} - p_{Ai}) = k_L A(c_{Ai} - c_{AL}) \qquad (4.6a)$$

It is not common to assign values of pressures and concentrations at the interface, therefore overall mass transfer coefficients are used with pressure and concentration values that reflect the overall or complete gradient across both films. The overall mass transfer equations using overall absorption coefficients can be written for each film. Combining these in the same manner as Equation (4.6a),

$$N_A = K_G A(p_{AG} - p_{AL}) = K_L A(c_{AG} - c_{AL}) \qquad (4.6b)$$

where

N_A = rate of absorption of A, g mole/hr
A = surface area for absorption, cm^2
K_G = overall gas absorption coefficient, g mole/(hr cm^2 atm)
K_L = overall liquid absorption coefficient, g mole/(hr cm^2 g mole/cm^3)
p_{AL} = partial pressure of A if it were in equilibrium with a liquid solution having that concentration of A, atm

The p_{AL} and the c_{AG} terms are pseudoquantities. The term p_{AL} is not a partial pressure of A in the liquid; however, it refers to a vapor condition that could exist above the liquid. Note that no defining statement is made for c_{AG} which could be thought of as being the concentration of A in the gas in units of g mole/cm^3. Equation (4.6b) can also be utilized for extraction processes where it is desirable to know the rate of desorption (extraction) of a dissolved gas from a liquid. In this case, the signs would be reversed for all terms in parentheses in both parts of the right hand of Equation (4.6b).

Values for partial pressures and concentrations in the gas phase can be obtained by the use of Dalton's law expressed as Equation (1.14) and the other gas laws in Section 1.3 assuming ideal gases. At high temperatures and pressures, the gases may no longer be ideal and other equations of state must be used. Values for quantities related to the solutions can be found by the procedures in the following section.

4.3.1 SOLUTION LAWS

4.3.1.1 Raoult's Law

The behavior of solutions can be predicted by Raoult's and Henry's laws.

Raoult's law is for *concentrated* ideal solutions where the components do not interact. At equilibrium conditions, this equation can be written as

$$p_{AL} = x_A P_A^o \tag{4.7}$$

where

x_A = mole fraction of A in solution

$$= \frac{\text{moles } A}{\text{moles total liquid}}$$

P_A^o = vapor pressure of pure A at the same temperature and pressure as the solution

Vapor pressures of pure substances at various temperatures and pressures are available in many of the numerous chemistry, physics, and engineering handbooks (values for water can also be easily obtained from the steam tables; see also Section 4.3.5).

Example Problem 4.3

Determine the equilibrium vapor pressure of water over a solution of 30% glycerine in water at 100°C and 1 atm.

Solution

This solution is concentrated in water so Raoult's law can be used. The molecular weight of water is 18 and of glycerine is 92, so the mole fraction of water is

$$x_A = \frac{70/18}{70/18 + 30/92} = 0.923$$

The vapor pressure of pure water at 100°C (the boiling point) is 760 mm Hg. The resulting vapor pressure of water in equilibrium with this solution is

$$p_{AL} = (0.923)(760) = 701.5 \text{ mm Hg}$$

The vapor pressure of glycerine at this temperature is essentially zero, so the total pressure is 701.5 mm Hg. This solution will not boil at 100°C. The boiling point can be estimated by determining the temperature that will cause the water to have a vapor pressure of $760/0.923 = 823.4$ mm Hg. From the steam tables, this is about 103°C.

4.3.1.2 Henry's Law

Henry's law is for *dilute* solutions and can be written at equilibrium conditions as

$$p_{AL} = c_{AL}H \qquad (4.8)$$

where

p_{AL} = equilibrium partial pressure of A over solution, atm
c_{AL} = concentration of A in liquid phase, g mole/cm³

H = Henry's law constant, $\dfrac{\text{atm cm}^3}{\text{g mole } A}$

Henry's law constants are determined empirically. This law is a special form of Raoult's law and for consistency is written as such. The significant difference between the two laws is that Henry's law is for dilute solutions and Raoult's law is for concentrated solutions. The reader is cautioned that Henry's law is written in many different ways, two of the more frequent being $c_{AL} = p_{AL}H'$ and $p_{AL} = x_A H''$. Units of H should always be examined to determine the form of equation for which it is intended to be used.

Values of the Henry's law constant are given in Figure 4.2 for some gases in water over the temperature range from 0 to 35°C. H for gases in organic liquids can be estimated using

$$H = \frac{22,400}{L \widetilde{V}_L N_T} \qquad (4.9)$$

where

$\widetilde{V}$ = molecular volume of liquid at 20°C, cm³/g mole
N_T = total number g mole per cm³ liquid
L = Oswald coefficient, cm³ gas/cm³ liquid

$$\cong \exp\left[\frac{-\Upsilon}{40} + 0.1\right]$$

Υ = surface tension, dyne/cm

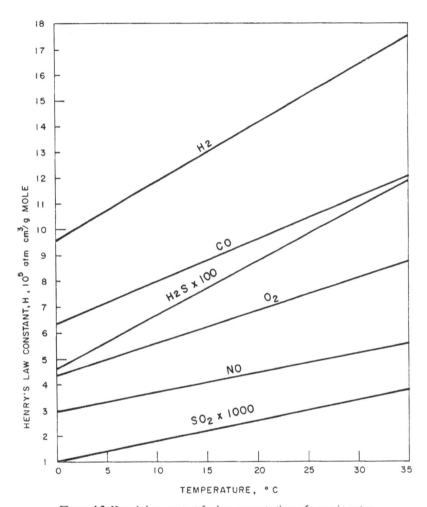

Figure 4.2 Henry's law constant for low concentrations of gases in water.

Figure 4.2 values of H are more accurate at low concentrations of solute and accuracy decreases as concentration increases. For example, for SO_2 at $c_{AL} \cong 0.001$, values of H shown are good; at $c_{AL} \cong 0.01$, values of H are about twice those shown; and for $c_{AL} \cong 0.1$, values of H are about 2.5 times those shown.

Here is an example as to how one must interpret a graph such as Figure 4.2. At 30°C, the value of $H_{SO_2} \times 1000 \cong 3.3 \times 10^5$. Therefore, $H_{SO_2} = 3.3 \times 10^2$ atm cm³/g mole.

A comparison of Equation (4.8) and Figure 4.2 shows that, in general, the solubility of a gas decreases as temperature increases. That is, H increases with temperature, so at a given p_{AL} the value of c_{AL} must decrease.

Example Problem 4.4

A gas stream at 20°C and 700 mm Hg (0.921 atm) contains 1.5% H_2S. Determine the equilibrium concentration of H_2S in water at these conditions and determine what could be done with pressure and temperature to increase this.

Solution

This would be a dilute solution, so Henry's law, Equation (4.8), should be used. At equilibrium, using Equation (1.14)

$$p_{AL} = p_{AG} = y_A P_T$$

combining Equations (1.14) and (4.8) and rearranging

$$c_{AL} = \frac{y_A P_T}{H}$$

For ideal gases, the volume fraction = mole fraction. Therefore, $y_A = 0.015$. From Figure 4.2, $H \cong 8.8 \times 10^3$ atm cm³/g mole. Solving for equilibrium concentration of H_2S in water under a gas containing 1.5% H_2S

$$c_{AL} = \frac{(0.015)(0.921)}{8.8 \times 10^3} = 1.57 \times 10^{-6} \; \frac{\text{g mole } H_2S}{\text{cm}^3 \text{ solution}}$$

Or expressed as mole fraction in this dilute solution

$$x_A = (1.57 \times 10^{-6}) \frac{18 \text{ g/g mole}}{1 \text{ g/cm}^3} = 2.83 \times 10^{-5} \text{ mole fraction}$$

From the combined equations, it can be seen that increasing pressure increases solubility of H_2S and decreasing temperature decreases H and increases solubility.

4.3.2 MASS TRANSFER COEFFICIENTS AND HENRY'S LAW CONSTANT

The mass transfer coefficients can be related to each other at equilibrium conditions. When Henry's law applies, appropriate quantities in Equations (4.6a) and (4.6b) can be substituted for, yielding the relationships

$$K_L = K_G H \tag{4.10}$$

$$\frac{1}{K_G} = \frac{1}{k_G} + \frac{1}{k_L} \tag{4.11}$$

$$\frac{1}{K_L} = \frac{1}{k_L} + \frac{1}{Hk_G} \tag{4.12}$$

The fractional total resistance to mass transfer due to the gas film then becomes

$$\frac{\dfrac{1}{k_G}}{\dfrac{1}{K_G}} = \frac{K_L}{Hk_G} \tag{4.13}$$

and the corresponding equation for the fractional total resistance to mass transfer due to the liquid film is

$$\frac{\dfrac{1}{k_L}}{\dfrac{1}{K_L}} = \frac{HK_G}{k_L} \tag{4.14}$$

4.3.3 INTERFACIAL AREA

The contact surface area between two phases in an absorber could be an essentially impossible quantity to evaluate. In the case of a simple wetted-wall column operating with a smooth continuous film, the contact area may be estimated, but even here surface irregularities exist and exact contact area is not known. For cases such as spray columns, packed beds, or even wetted-wall columns with irregular liquid surface, calculation of the area would only be a guess and probably not very accurate. For this reason, an interfacial area, which is a pseudoquantity, is used. It is

$$a = \frac{A}{V} \qquad (4.15)$$

where

a = interfacial area per unit of bulk volume
A = actual contact area
V = total volume of *empty* absorber

The interfacial area term is then combined with the overall transfer coefficient and the product of the coefficient times a is empirically determined. An example of empirically obtained $K_G a$ data for a specific system [5] is presented as Figure 4.3. The system represents a countercurrent absorption of 1% CO_2 in air by 4% NaOH solution at SC in a 76-cm diameter column. The gas rate was maintained at 15 g mole/(hr cm²) up through 3 m of 7.6-cm ceramic "Intalox saddles." Note here that $K_G a$ increases to a maximum as liquid rate increases. Changing any of the fixed parameters (gas rate, packing size, concentrations, etc.) will result in new values of $K_G a$.

Empirically determined interfacial area-gas transfer coefficient values

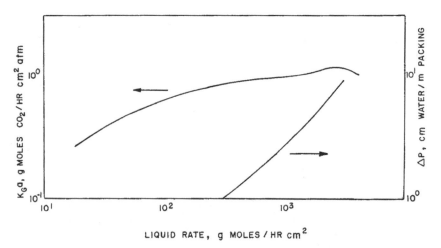

LIQUID RATE, g MOLES / HR cm²

Figure 4.3 $K_G a$ versus liquid rate for CO_2 absorption in 4% NaOH [5].

are useful in obtaining related data. For example, Equation (4.10) can be rewritten

$$K_L a = K_G a H \qquad (4.16)$$

and Equation (4.6b) can be rewritten using the interfacial area

$$N_A = K_G a V (p_{AG} - p_{AL}) \qquad (4.17)$$

4.3.4 LOG MEAN PRESSURE DIFFERENCE

In countercurrent absorbers, the component being absorbed from the gas phase has a decreasing partial pressure in the gas phase and a decreasing equilibrium partial pressure from the liquid phase in contact with the gas as the gas passes through the system and is absorbed. (Cocurrent absorbers have the opposite relationship.) The equations that have been presented up through Equation (4.17) require point values, so an average partial pressure difference for the system is necessary. For cases where significant changes occur, use a log mean pressure difference as shown by the following modification of Equation (4.17)

$$\Delta N_A = K_G a V (p_{AG} - p_{AL})_{\ln} \qquad (4.18)$$

where

ΔN_A = g moles of absorbable material entering the tower minus g moles absorbable material leaving the tower per hour, all in the gas phase (or equivalent in terms of liquid phase which equals gas phase values when no holdup changes occur)

$$(p_{AG} - p_{AL})_{\ln} = \frac{(p_{AG} - p_{AL})_1 - (p_{AG} - p_{AL})_2}{\ln \dfrac{(p_{AG} - p_{AL})_1}{(p_{AG} - p_{AL})_2}} \qquad (4.19)$$

The subscript "ln" signifies log mean difference and is the natural log. The subscripts 1 and 2 refer to the pressures at the bottom and pressures at the top, respectively. When the log mean pressure difference is in atmospheres and the volume is in m³, units of $K_G a$ are g mole/(m³ atm).

4.3.5 VAPOR LIQUID EQUILIBRIUM

Equilibrium between a gas and a liquid is obtained when the rate of absorption of the gaseous component from the gas phase is equal to the rate of release of the absorbed gas from the liquid phase. Theoretically, equilibrium is only possible when the two phases have been in contact for an infinite period of time, but it is possible to approach nearly equilibrium conditions in a reasonable period of time for many substances. The vapor-liquid equilibrium constant, K_i, is expressed

$$K_i = \left(\frac{y_A}{x_A}\right)_i \tag{4.20}$$

where

i = each individual specific concentration ratio at a given temperature and pressure

x_A = mole fraction A in solution

y_A = mole fraction A in gas

The value of K_i changes with concentration and also changes with temperature and pressure. Equilibrium data can be used to construct equilibrium curves such as shown in Figure 4.4. The slope of this curve at any point is K_i.

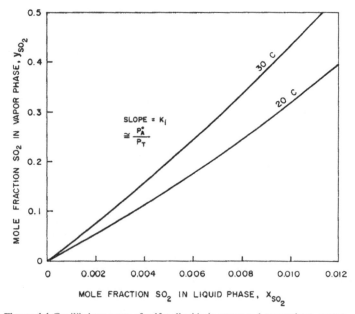

Figure 4.4 Equilibrium curve of sulfur dioxide in water at 1 atmosphere pressure.

Equilibrium curves can also be approximated using the generalized equations presented already. Dalton's law for an ideal gas, Equation (1.14), and Raoult's law for an ideal solution, Equation (4.7), are equal at equilibrium conditions

$$y_A P_T = x_A P_A^o$$

and can be rearranged to give the equation of a line passing through the origin

$$y_A = \frac{P_A^o}{P_T} x_A \qquad (4.21)$$

The ratio of the vapor pressure of the pure component at that temperature and pressure over the total pressure is sometimes equal to the slope of the equilibrium curve. However, this is only approximate and was derived for ideal gas and concentrated ideal solution conditions.

For the special case where Henry's law applies, Equation (4.8), for dilute ideal solutions, we can obtain another approximate expression for the slope of the equilibrium curve

$$y_A = \frac{H}{P_T} c_A \qquad (4.22)$$

For a dilute binary solution, the mole fraction of A can be expressed

$$x_A = \frac{c_A}{c_A + c_B} \cong \frac{c_A}{c_B}$$

because $c_A \ll 1$ and $c_B \cong 1$. As a result, $x_A \cong c_A$ and, therefore, for the special case of a very dilute ideal binary system

$$y_A \cong \frac{H}{P_T} x_A \qquad (4.23)$$

In addition to the physical absorption of a gas into a liquid, there is also the possibility that a chemical reaction can occur. This is called chemisorption. If there is sufficient chemical reaction, it is no longer possible to consider the solutions ideal because of the interacting of molecules. For example, when SO_2 is absorbed by water, this is mainly a physical absorption mechanism with little chemical reaction. However, when SO_2 is absorbed in sodium hydroxide, there is an extensive chemical reaction (evidenced in this case by the release of heat of reaction). When chemisorption occurs, the component absorbed is no longer present to exert a partial pressure and the above equations cannot be used. It is quite useful at times to utilize both chemical reactions and diffusion to improve the collection of gaseous pollutants.

In air pollution control, concentration of gaseous pollutants is often very low. If the concentration range is small enough, the vapor-liquid equilibrium curve may be considered to be a straight line. For example, at low concentrations it is convenient to use $y = 0.77x$ for the ammonia-water equilibrium curve and $y = 33x$ for the SO_2-water equilibrium curve.

It may not always be possible to obtain directly from literature values of the vapor pressure of a pure substance, P_A^o, at the desired temperature for use in Equation (4.7). Where the heat of vaporization of the liquid, ΔH_v, can be assumed constant, the Clapeyron equation (which is based on one of the Maxwell relations) may be useful. Units of ΔH_v are cal/g mole. The Clapeyron equation is

$$\frac{dp}{dT} = \frac{\Delta H_v}{T\Delta V} \qquad (4.24)$$

where

p = equilibrium vapor pressure of liquid
ΔV = change in volume
T = absolute temperature

Based on ideal gases and neglecting liquid-volume change compared to gas volume, the Clauses-Clapeyron equation is derived

$$\frac{d(\ln p)}{dT} = \frac{\Delta H_v}{RT^2} \qquad (4.25)$$

R is the universal gas constant equal to 1.987 cal/(g mole K). Assuming constant ΔH_v over small temperature changes at constant pressure this integrates to

$$\ln \frac{p_2}{p_1} = \frac{\Delta H_v}{R} \frac{T_2 - T_1}{T_1 T_2} \qquad (4.26)$$

where subscripts 1 and 2 refer to initial and final conditions, respectively.

Without limits this is

$$\ln p = -\frac{\Delta H_v}{R}\frac{1}{T} + C \qquad (4.27)$$

C is a constant of integration.

Note that Equation (4.27) is in the form of a straight line on semilog coordinates. Figure 4.5 shows vapor pressure data for several pure liquids and or 5% ammonia. There is some slope to these lines indicating ΔH_v is not a true constant.

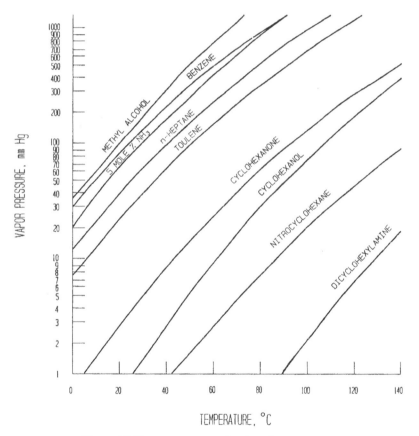

Figure 4.5 Vapor pressures as a function of temperature.

Example Problem 4.5

Determine the vapor pressure of methyl ethyl ketone at 60°C. From various handbooks and index literature, ΔH_v is 8,149.5 cal/g mole and vapor pressure is 90.6 mm Hg at 25°C.

Solution

Use Equation (4.26)

$$\ln \frac{p_2}{90.6} = \frac{8,149.5}{1.987} \left[\frac{60 - 25}{(298)(273 + 60)} \right]$$

$\therefore p_2 = 90.6 \exp 1.4466 = 384.9$ mm Hg

4.3.6 ABSORPTION DRIVING FORCE

Mass transfer is proportional to driving force as shown by Equations (4.4) and (4.5) for point conditions and Equations (4.18) and (4.19) for systems such as absorbers. Countercurrent, cocurrent, and crosscurrent absorbers are all common in air pollution control. Although there are many different variations of each type, Figure 4.6 shows how spray absorbers can be used in all three ways. Both gas and liquid must be uniformly distributed across the entire bed. Not shown for the cocurrent unit is a liquid separator which is usually centrifugal. All units require mist eliminators to remove the liquid droplets from the exit gas streams.

A countercurrent absorber is used as the example to show driving force for absorption. In these devices, the scrubbing liquid for absorption enters the top with a mole fraction concentration of dissolved material, x_2. After being evenly distributed, it falls through the rising gas by gravity. The gas stream containing pollutants to be absorbed at a mole fraction concentration, y_1, enters at the bottom where it is distributed and forced up through the falling spray droplets. Absorption of the pollutant gas occurs as the rising gas stream contacts the liquid droplets. Amount of absorption depends on the mass transfer coefficients, droplet area, and driving force. Concentrations at the bottom are x_1, y_1 and at the top are x_2, y_2. The line joining these points is called the operating line and it may be straight or curved. If the inlet liquid contains no solute, the concentration of x_2 would be equal to 0 and the operating line would extend to the y axis. Figure 4.7 is an example showing points x_1, y_1 and x_2, y_2, a straight operating line, and a vapor-liquid equilibrium curve. The vertical distance from either of x_2, y_2 or x_1, y_1 to the equilibrium curve represents the equilibrium driving force

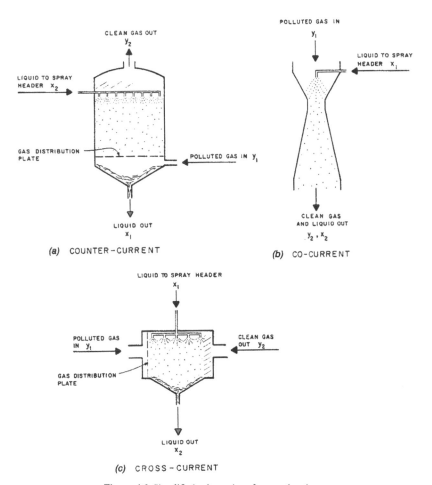

Figure 4.6 Simplified schematics of spray absorbers.

at the top and bottom, respectively, of this countercurrent absorber. The greater the distance, theoretically, the greater the driving force attempting to produce absorption. Concentrations of the absorbable material in the liquid and gas phases at various positions within the absorber are represented by the operating line (this is discussed further in the next section). Figure 4.7 shows the average driving force in the absorber as represented by the distance from the midpoint of the operating line to the equilibrium curve. If there is sufficient solubility of the gas pollutant being absorbed in the liquid, and if there is sufficient liquid to absorb the gas, and if adequate contact time is available, it should be possible to reduce the concentration

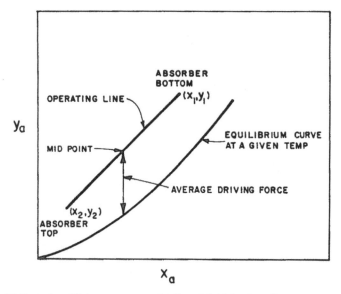

Figure 4.7 Typical equilibrium curve, operating line, and driving force for a countercurrent absorber.

of the pollutant in the gas stream to nearly the equilibrium concentration. This is not usually practicable. If such were the case, the lower end of the operating line would approach the equilibrium curve and very low pollutant emissions would occur.

4.3.7 ABSORBER OPERATING LINES

The operating line for a countercurrent absorber may be a straight line as depicted in Figure 4.7. It is frequently possible to use this linear relationship to describe actual gas and liquid concentrations inside the gas scrubber, but this approximation is not always accurate. In order to understand how the operating line is derived, visualize a countercurrent absorber as shown in Figure 4.8. L represents the total moles in the liquid phase in g mole/(cm^2 hr) and G is the total moles in the gas phase in g mole/(cm^2 hr). The subscripts 1 and 2 indicate the bottom and top, respectively, of the absorber as has been discussed.

Assuming no accumulation and no chemical reaction, the total mass balance on the absorber at steady state is the sum of materials in equals the sum of materials out

$$G_1 + L_2 = G_2 + L_1 \tag{4.28}$$

The overall balance on the component A which is absorbed, can be written as

$$G_1 y_1 + L_2 x_2 = G_2 y_2 + L_1 x_1 \qquad (4.29)$$

where x and y equal mole fraction A in liquid and gas, respectively.

If the amount of material absorbed from the gas phase is small, there may be no appreciable change in the total number of moles in the gas phase and G_1 may be assumed to equal G_2. Also, if only a small amount of material is absorbed and if no liquid volatilizes or condenses, then L_1 almost equals L_2. For these conditions, Equation (4.29) can be rewritten to express the slope of the operating line as

$$\text{Slope of operating line} = \frac{L}{G} = \frac{y_1 - y_2}{x_1 - x_2} = \frac{\Delta y}{\Delta x} \qquad (4.30)$$

Countercurrent absorbing can be likened to passing gas and liquid phases countercurrently through a series of separate contacting stages as shown in Figure 4.9. The subscripts in this example refer to the stage from which each component comes. In each of the n theoretical stages, there is

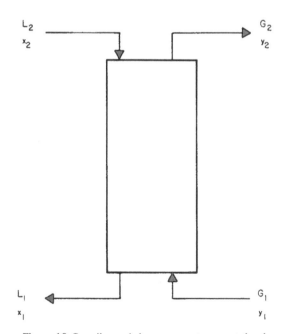

Figure 4.8 Overall mass balance on countercurrent absorber.

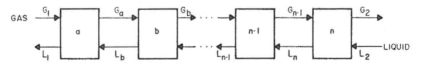

Figure 4.9 Diagrammatic sketch of countercurrent contact stages.

good mixing and sufficient time to permit equilibrium between the vapor and the liquid to be obtained. Instead of having separation facilities and connecting piping as shown, put all of these stages together in one continuous device, such as the countercurrent tower in Figure 4.8 which could be a vertical cylinder shell containing a number of horizontally spaced plates. These plates, for example, can be perforated to permit the gas to flow upward through them, yet contain liquid on them for contacting the gas. The top section of such a column is shown in Figure 4.10. If equilibrium is reached on each plate, then these are theoretical stages or plates. In actual practice, equilibrium is not attained because the stage efficiencies are not 100% on each stage. Therefore, there are more actual plates than theoretical plates.

A mass balance on the component absorbed can be made around the top (n^{th}) plate shown in Figure 4.10. This produces the equation

$$G_{n-1}y_{n-1} + L_2x_2 = G_2y_2 + L_nx_n \qquad (4.31)$$

When this is rearranged, we obtain the equation of a line

$$y_{n-1} = \frac{L_n}{G_{n-1}} x_n + \frac{G_2y_2 - L_2x_2}{G_{n-1}} \qquad (4.32)$$

This is the equation of the operating line at this plate. In order to be a straight line from x_2,y_2 to x_1,y_1 with the slope L/G, it is necessary that: (a) the concentration of absorbable component be low so there would be no appreciable change in L or G due to transfer of this component; (b) the solvent be relatively nonvolatile so there would be essentially no loss of the liquid phase; (c) the carrier gas, which comprises the bulk of the gas phase, should not be soluble in the liquid solvent; and (d) no condensation occurs.

Example Problem 4.6

Assume the gas in Example Problem 4.4, which contains 1.5% H_2S at

20°C, is to be scrubbed in a countercurrent absorber to reduce the outlet concentration to 0.5% by volume H_2S. The liquid-to-gas, L/G, ratio to be used is 1.4 times the minimum liquid flowrate and the inlet gases are already saturated with water vapor, so L and G are essentially constant. These data are shown in Figure 4.11.

Solution

The minimum liquid rate (or minimum L/G) occurs when the slope of the operating line is a minimum. This is represented schematically in Figure 4.12, where inlet water is noted as containing no H_2S. This figure depicts vapor-liquid equilibrium existing at the bottom of the tower between inlet gas and leaving liquid. If so, x_1 at the bottom of the column equals 2.83×10^{-5} as calculated in Example Problem 4.4

$$\text{Minimum } L/G = \frac{\Delta y}{\Delta x} = \frac{0.015 - 0.005}{2.83 \times 10^{-5}} = 353 \text{ moles } H_2O/\text{mole gas}$$

Desired L/G then is 1.4 times this or 495 moles H_2O/mole gas. This is a very large liquid rate and, therefore, not a good absorption system. Chemisorption is much better for this.

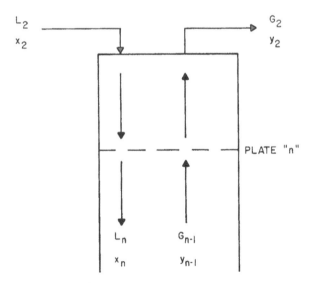

Figure 4.10 Mass balance on top plate of countercurrent absorber.

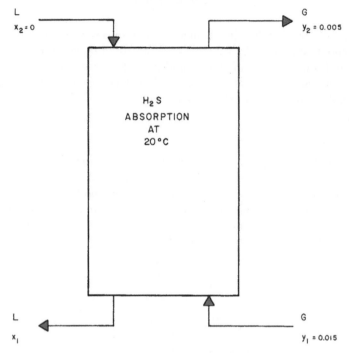

Figure 4.11 Absorber schematic for Example Problem 4.6.

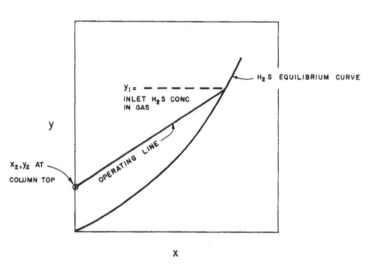

Figure 4.12 Schematic of H_2S vapor-liquid equilibrium curve and operating line at minimum L/G for Example Problem 4.6.

4.3.8 CONTACT STAGES AND STAGE EFFICIENCY

The existence of a difference between theoretical (ideal) contact stages and actual contact stages was noted in Section 4.3.7. Attempts are made to design absorbers for maximum efficiency by trying to optimize physical contact mechanisms such as good mixing and high mass transfer rate as well as by designing for maximum contact time. Even so, it is unusual to allow equilibrium to be achieved in actual practice because this could require an "infinite" (relatively long) contact time. The number of theoretical contact stages can be estimated by "stepping off" the contact points on an equilibrium diagram. This is done as shown in Figure 4.13 by drawing horizontal and vertical lines connecting the operating and equilibrium lines. Figure 4.13 shows a countercurrent absorber with inlet liquid containing no solute. The position on the equilibrium curve indicated as x' is the theoretical equilibrium concentration of liquid *leaving* the top plate. At this same point, the composition of the vapor, y', is the theoretical equilibrium concentration of vapor which is rising from, that is, *leaving*, the top plate. The intersection of each horizontal and vertical line on the equilibrium curve indicates a similar situation for the successive plates. In the example given in Figure 4.13, there are approximately 2.25 theoretical plates (ideal contact stages).

Assuming that the equilibrium curve remains the same, the number of ideal contact stages can be varied by changing the slope of the operating

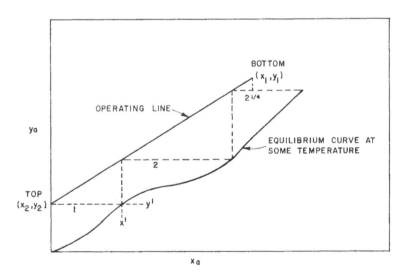

Figure 4.13 Ideal contact stages for countercurrent absorber.

line, L/G. Remember that when either L or G are varied, this not only changes the slope of the operating line, but also changes the locations of points x_1,y_1 and x_2,y_2 because of resulting concentration changes. For example, in a countercurrent absorber, decreasing L will decrease the slope, increase the concentration of x_1, increase the concentration of y_2, and it may or may not have a significant effect on contact time (decreasing G will obviously increase the contact time).

Actual absorber average stage efficiencies can be established by dividing the number of theoretical contact stages estimated by the stepping-off procedure by the number of contact stages actually present in the absorber.

Keep in mind that absorption efficiency and stage efficiency are not the same. For example, an absorber may reduce a gaseous pollutant concentration from 2,000 to 50 ppm. If the total number of moles of gas has not changed significantly during passage through the absorber, absorption efficiency is $[(2,000 - 50)/2,000](100) = 97.5\%$. Actual stage efficiencies would typically be closer to about 85% for good gas-phase controlling absorption systems.

4.3.9 MASS TRANSFER COEFFICIENTS AND STAGE EFFICIENCY

Absorption systems can be either gas-phase or liquid-phase controlling depending on which film offers more resistance to the molecules being absorbed. Gas-phase controlling means that the mass transfer through the gas film is slow compared to the liquid-film transfer. Gas-phase controlling is desirable in that it usually signifies a faster overall transfer rate than when liquid phase is controlling because the absorbing molecules can usually pass through the gas film faster.

Calvert [6] reports the following rules: gas phase controls when the value of k_G approaches the value of K_G (the values of k_G are larger than values of K_G). Gas phase controls when $k_G \leq 1.1\ K_G$. A low value of k_G means there is poor transfer through the gas phase—this occurs when values of H are $\leq 2 \times 10^2$ atm cm³/g mole. Liquid phase controls when $k_G > 10\ K_G$. Large values of k_G mean good transfer through the gas film. This usually occurs when $H > 2 \times 10^5$.

The various mass transfer coefficients for absorption can be calculated using diffusivities. The individual gas film coefficient in g mole/(hr cm² atm) is calculated by

$$k_G = \frac{7,200}{RT} \left(\frac{D_{AB}}{\pi t_s}\right)^{1/2} \tag{4.33}$$

and the individual liquid-film coefficient in units of g mole/(hr cm² g mole/

cm³) is estimated by

$$k_L = 7,200 \left(\frac{D'_{AB}}{\pi t_l}\right)^{1/2}$$ (4.34)

The contact times t_g and t_l for the gas and liquid phases, respectively, are reported to range from 0.01 to 0.5 sec but are a very doubtful value. The gas constant, R, equals 82.05 atm cm³/(g mole K). Diffusivities for gas in air at low concentrations are about 0.1 cm²/sec and for gas in liquid are about 1.5×10^{-5} cm²/sec at low concentrations. It might be more appropriate to note that rapid, easy absorptions are gas-phase controlling and slow absorptions are usually liquid-phase controlling.

Predicted absorber contact-stage efficiencies for gas-phase controlling systems can be estimated using Figure 4.14 which plots contacting efficiency against a term called "overall gas transfer units" for both cocurrent and countercurrent systems. Numerical values of the overall gas transfer units can be estimated by dividing the numerical value of $y_1 - y_2$ by the y magnitude of the average driving force. The magnitude of this driving force is obtained by subtracting the y value at the midpoint of the operating line from the y value obtained when the driving force line intersects the

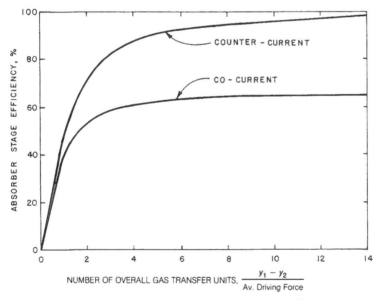

Figure 4.14 Approximate contact stage efficiencies for gas-phase controlling absorption at optimum operating conditions.

equilibrium curve. This is depicted on Figure 4.7. It will require fewer actual contact stages to provide a specific amount of absorption for a given system when the number of overall gas transfer units is smaller, even though absorption efficiency for each stage is lower. This is due to the increasing of the liquid rate and the operating line relative to the equilibrium curve. The contact-stage efficiencies shown in Figure 4.14 are optimum values which can only be obtained by proper design and operation of an absorber.

Addition of complexing agents to the absorbing liquid to tie up the absorbed pollutants by chemical reaction increases the driving force available which increases the absorption rate. This effectively reduces the back pressure due to concentration and results in a lower number of transfer units required. This is chemical absorption (chemisorption) and can increase the value of k_L by up to 12 times the value experienced when only physical absorption exists.

4.4 MOISTURE CONTENT OF GASES

Wet scrubbing systems subject the gases, which are usually air or air-like gases, to water or water solutions. The carrier gases usually contain some water vapor before scrubbing and may even be saturated. If they are not saturated, adiabatic saturation occurs in the scrubber. This happens whether gaseous pollutants are being controlled by an absorption mass transfer process or whether particulate matter is being captured. One of the assumptions for a straight operating line in an absorber is that neither the liquid nor the gas phases experience significant exchange in molar flowrate. This is not true when hot unsaturated gases are scrubbed.

Flue gases contain water vapor because of the water formed from the combustion of the hydrogen in the fuel plus that free water which may be present in the fuel. The moisture content of coal combustion gases is typically about 6% water, which is far from saturation water content at flue-gas temperatures. Stack gases leaving the final economizers may be about 150°C. At any temperature over 100°C, saturation would consist of essentially 100% water vapor at atmospheric pressure. Cool water sprayed into hot unsaturated gases attempts to saturate the gases. Evaporation of the water cools the gases until adiabatic saturation is reached. Flue gases at 150°C reach adiabatic saturation at about 55°C, at which time they contain about 17% water vapor. This amounts to a considerable exchange of water from the liquid to the gas phase and increases the total mass of exhausted gases. Volume of the final saturated and cooled exhaust gas depends on *both* temperature and mass. Under these conditions it would be only about 88% of the original hot gas volume. Adiabatic saturation in a scrubber occurs very rapidly as long as adequate water is available for saturating the gas.

It is important to be able to determine the moisture content of gas systems so several procedures are given in the following sections. It is assumed that the main gas in these systems is air near standard pressure. Instrumental humidity methods are not discussed but could be used.

Chemical Equation Method. Chapter 2 gives several examples of flue gases and how combustion equations can be used to show composition of any gas. Water concentration would equal moles of water divided by total moles of gaseous products times 100. This gives percent by volume water directly. If water vapor is added to this gas and the final water content is known, the final gas volume can be established by use of the original moles of dry flue gas and the ideal gas relations.

Wet/Dry Bulb Method. One of the easiest methods for obtaining water content of a gas is by measuring the wet and dry bulb temperatures. This can be accurate if the temperatures are not too high, the gas is mainly air, pressures are near normal, and if adequate moisture is present on the wick during the entire reading period.

The humidity, H, in mass of water per mass of dry air is obtained from a psychrometric chart (or table) as shown in Figure 4.15. The wet bulb temperature, t_w, is read at the saturation line, then the adiabatic saturation line is followed until the dry bulb temperature, t_d, is reached. This gives

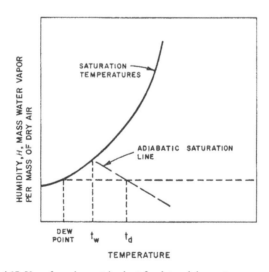

Figure 4.15 Use of psychrometric chart for determining water vapor content.

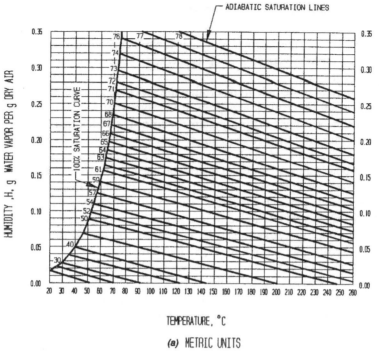

(a) METRIC UNITS

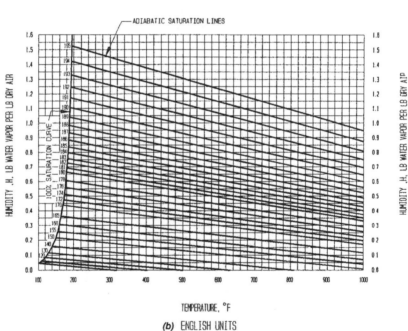

(b) ENGLISH UNITS

Figure 4.16 High-temperature psychrometric chart for air-water systems near standard pressure.

the humidity. Water content of the gas (air) as percent moisture by volume is

$$\% \ H_2O = \frac{161H}{1 + 1.61H} \tag{4.35}$$

Figure 4.16 is a high-temperature psychrometric chart constructed for air water systems near normal pressure.

For systems at other than standard pressure (29.92" Hg), it is necessary to use corrected data such as given in Figure 4.17 [7]. This nomograph, developed by Entropy Environmentalists Inc., is a simple correction chart to account for system pressures that vary by 3–5" Hg from standard. This chart can also be used for high or low atmospheric pressure corrections. The procedure is to use the absolute pressure and the wet bulb temperature to fix a line that intersects the "Mixture Content at Saturation" scale. If the wet and dry bulb temperatures are equal, i.e., the gas is saturated, this would be the moisture content. If it is not saturated, use this saturated mixture content point and the dry bulb-wet bulb temperature difference value to locate a second line. The percent moisture can then be read where this second line intersects the curved "Moisture Content" line.

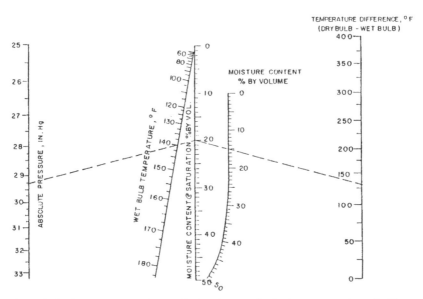

Figure 4.17 Air-water vapor psychrometric chart (corrects for absolute pressure changes). (Courtesy Walter Smith, President, Entropy Environmentalists, Inc.)

Vapor Pressure Method. The vapor pressure method is useful when the wet bulb temperature is high and when pressure is not near standard atmospheric pressure. Figure 4.16 does not show saturation lines above 75°C, so this procedure would be used in that range. This method incorporates the Carrier equation. It is accurate up to about 200°C and can be used for flue gases containing up to 15% CO_2. The equation includes factors to account for transfer of heat by conduction and radiation and accounts for diffusion and vaporization. The actual water vapor pressure, p_a, in the gas in cm Hg is

$$p_a = p_w - \left[\frac{(P_D - p_w)(t_d - t_w)}{1,547 - 1.44t_w} \right] \qquad (4.36)$$

where

p_w = saturated vapor pressure of water at duct wet bulb temperature from steam tables, cm Hg
P_D = absolute pressure in duct, cm Hg
t_d = dry bulb temperature in duct, °C
t_w = wet bulb temperature in duct, °C

The water content is then

$$\% \ H_2O = \frac{p_a}{P_D} 100 \qquad (4.37)$$

Condensation Method. Gas-sampling trains utilize impingers in an ice bath to condense moisture. The saturated gases at 0°C leaving the impingers should contain no liquid droplets. The total volume of water in the sampled gas equals the volume of condensed water in gaseous form, V_c, plus the volume of saturated vapor leaving the meter, V_s. V_s is usually negligible, but it can be found by

$$V_s = \frac{p_w}{P_I} V_M \qquad (4.38)$$

where

p_w = saturation vapor pressure of water at the impinger temperature from steam tables, cm Hg (at 0°C this would be 0.458 cm Hg)
P_I = absolute pressure in impinger (i.e., barometric pressure ± impinger static pressure), cm Hg

V_M = corrected volume of metered gas at meter conditions, m^3

The value of V_c in m^3 can be obtained for the same conditions by

$$V_c = M_c\left(\frac{2.24 \times 10^{-2}\ m^3/g\ mole}{18\ g/g\ mole}\right)\left(\frac{T_M}{273}\right)\left(\frac{76}{p_M}\right) = \frac{3.46 \times 10^{-4}\ M_c T_M}{p_M}$$

(4.39)

where

M_c = mass of condensate, g
T_M = meter temperature, K
p_M = absolute pressure at meter (i.e., barometric pressure ± meter static pressure), cm Hg

The water content in the gas is then

$$\%\ H_2O = \left(\frac{V_c + V_s}{V_M + V_c + V_s}\right)100$$

(4.40)

Drying Tube Method. This procedure is ideal for low concentrations of water vapor where exact water content is required. Usually three or more tubes in series are used. If no change in weight occurs in the last drying tube, the sum of weight gains in the other tubes equals the water in the gas. Calculate water content using a procedure similar to the condensate method with Equations (4.39) and (4.40). The tubes contain a desiccant chemical.

Example Problem 4.7

Sulfur dioxide is absorbed from a flue gas in a small scrubber operating at 16.5-cm H_2O pressure drop with a 2.5-liter scrubbant liquid per m^3 of inlet gas. Barometric pressure is 75.77 cm Hg. Other data are summarized in Table 4.2. Determine whether the data seem consistent and calculate the SO_2 removal efficiency.

Solution

A mass balance in = out based on Nm^3/hr dry gas would show whether the volumetric flowrates are consistent because there are both temperature and mass changes due to adiabatic saturation.
Use Boyle's and Charles' ideal gas relations to make the pressure and

TABLE 4.2. SO_2 Scrubbing System Test Data.

	Absorber	
	Inlet	Exit
Gas Flow, actual m³/hr	2,680	1,808
Static Pressure, cm H_2O	−4.6	−21.6
Dry bulb temperature, °C	261	52
Wet bulb temperature, °C	58	52
Orsat, % CO_2	8.3	
O_2	10.8	
CO	0.1	
SO_2, ppm	1,400	100

temperature corrections (recognizing that the specific gravity of mercury is 13.6 times greater than water) for flowrates at SC

$$\text{Volumetric flow in} = (2,680)\left(\frac{293}{273 + 261}\right)\left(\frac{75.77 - \frac{4.6}{13.6}}{76}\right)$$

$$= 1,460 \text{ Nm}^3/\text{hr}$$

$$\text{Volumetric flow out} = (1,808)\left(\frac{293}{273 + 52}\right)\left(\frac{75.77 - \frac{21.6}{13.6}}{76}\right)$$

$$= 1,591 \text{ Nm}^3/\text{hr}$$

This puts the volumetric flowrates on a more comparable basis, but water of saturation was added in the scrubber so the flowrates should be compared on a dry basis.

From Figure 4.16, inlet and exit humidities are about 0.043 and 0.096 g H_2O/g dry air, respectively. Using Equation (4.35), the water contents are 6.5% in and 13.4% out. Dry gas flowrates are

$$\text{in} = (1,460)\left(\frac{100 - 6.50}{100}\right) = 1,365 \text{ Nm}^3/\text{hr dry}$$

$$\text{out} = (1,591)\left(\frac{100 - 13.4}{100}\right) = 1,378 \text{ Nm}^3/\text{hr dry}$$

This suggests that more ($\sim 1\%$) gas may come out than goes in. This should be expected for a system operating under negative pressure.

SO_2 removal efficiency neglecting gas volume change due to added water vapor could be calculated by using the volumetric concentrations in and out

$$\left(\frac{1,400 - 100}{1,400}\right)100 = 92.86\%$$

The additional water vapor in the exit gas serves to dilute exit SO_2 concentration. Correcting this to have the same inlet and outlet quantity of gas gives an SO_2 outlet concentration of

$$\left(\frac{100 \text{ m}^3 \text{ } SO_2}{10^6 \text{ m}^3 \text{ gas out}}\right)\left(\frac{1,591 \text{ m}^3/\text{hr gas out}}{1,460 \text{ m}^3/\text{hr gas in}}\right)$$

$$= \frac{109 \text{ m}^3 \text{ } SO_2}{10^6 \text{ m}^3 \text{ gas in}} \text{ or } 109 \text{ ppm } SO_2 \text{ on inlet gas basis}$$

SO_2 removal efficiency is now 92.21%.

4.5 GAS ADSORPTION

Adsorption is the taking up of a gas (or liquid or dissolved substance) on the surface of a solid or a liquid. (Liquid surface adsorption is not greatly significant and is not discussed further in this section.) The surface of any solid contains some adsorbed material. Certain finely divided solids such as activated carbon and silica gel adsorb large quantities of materials because of the great amounts of surface area available and because of the surface properties. Both physical attraction and chemical reaction can take place during adsorption, giving rise to the terms "chemical" and "physical" adsorption. Physical adsorption consists of attracting gas molecules, usually by electrostatic or van der Waals forces, which results from gas molecule polarity and strongly positive or negative ions on the surface of the adsorbing solid. Chemical adsorption (chemisorption) usually consists of physical adsorption accompanied by a chemical reaction. Physical adsorption is most important for gas separation work.

In adsorption, the term *adsorbent* means the adsorbing solid and *adsorbate* is the adsorbed material. Adsorption processes may consist of contacting the solid with a gas mixture or vice versa to remove any or all odor, taste, moisture, solvents, or other pollutants from the gas. The adsorbed species (and the adsorbent) may or may not be recovered by regeneration of the adsorbent.

4.5.1 ADSORPTION AND DESORPTION

The advantage of gas adsorption lies in the fact that it is possible to purify gases containing only small amounts of pollutants. The captured gases can be recovered and/or the adsorbent can be reused by desorption if one or more of the following recovery cycles are utilized: (a) temperature gradients, (b) pressure gradients, and (c) concentration gradients. These cycles or "swings" require semibatch-type operating systems. The polluted gas stream is introduced into the adsorption system under one or more of the favorable conditions of low temperature, high pressure, or high concentration. The purified gases leaving the adsorber can be released into the atomsphere until the adsorbent becomes nearly saturated. When the polluted gas stream "breaks through," it is shut off and a clean air or other gas stream is introduced to strip off the adsorbed pollutants at one or more of the following conditions of high temperature, low pressure, or low concentration. The recovery effluent gas stream usually contains a high concentration of the pollutant gas making it possible to collect and use them by some method such as absorption or condensation. The entire process can be carried out in either fixed or movable bed systems.

The gaseous molecules, when captured by an adsorbent, give off a heat of adsorption, making the process exothermic. The heat released varies depending on the magnitude of the electrostatic force of the physical attraction (which depends on the polarity of both the adsorbent and the adsorbate) as well as on the chemical reactions which may take place. Highly exothermic adsorption processes, e.g., 10–100 kcal/g mole gas adsorbed, indicate that chemical adsorption has occurred. In these cases, it may not be possible to desorb the gases for recovery of either the gas or the adsorbent.

Factors which affect adsorption are

(1) Operating conditions: improved adsorption is obtained by low temperature, high pressure, and high adsorbate gas concentration
(2) Gas being adsorbed: molecule size, boiling point, molecular weight, and polarity
(3) Adsorbent: surface polarity, pore size, and pore spacing
(4) Adsorber design:
   • Adequate surface area and holdup time are needed and these are functions of unit size.
   • Provide good gas distribution so that all the area is available.

- pretreatment of inlet gases to remove adsorbent contaminants and to remove gases that can be removed by more economically effective processes to prevent unnecessary overburdening of the system
- adequate regeneration and cool down
- replacement of unusable adsorbent

4.5.2 HEAT OF ADSORPTION

Significant amounts of heat can be released when a gas molecule becomes bonded to the surface of an adsorbent. The amount of adsorption energy released depends upon the type and extent of chemical adsorption that occurs for the particular vapor-adsorbent combination. This heat can be exessive if allowed to build up and could ignite the system. Consider the consequences if the system consists of a volatile vapor and a volatile adsorbent. This heat buildup can be measured directly or it can be estimated for a particular system using adsorption isotherms and the Clausis-Clapeyron equation modified for heat of adsorption, ΔH_a. Rearranged Equation (4.26) gives

$$\Delta H_a = R \ln \frac{p_2}{p_1} \left(\frac{T_1 T_2}{T_2 - T_1} \right) \qquad (4.41)$$

Total system heat buildup would equal the number of moles of vapor adsorbed times ΔH_a. Units of ΔH_a are cal/g mole.

Example Problem 4.8

A nonregenerable activated carbon adsorber is to be used to control gasoline evaporative emissions. Test data were obtained to give the adsorption isotherms shown in Figure 4.18. Estimate the heat of adsorption for this particular carbon-gasoline adsorption system.

Solution

Obtain from Figure 4.18 two pairs of data points for a region in which the system will be expected to operate based on anticipated gasoline-vapor pressure to adsorber. For example, if the adsorbent capacity is 50 cm³ gasoline per g carbon, then Figure 4.18 gives

$$p_2 = 60 \text{ mm Hg @ } t_2 = 40°C \text{ (313 K)}$$

$$p_1 = 20 \text{ mm Hg @ } t_1 = 10°C \text{ (283 K)}$$

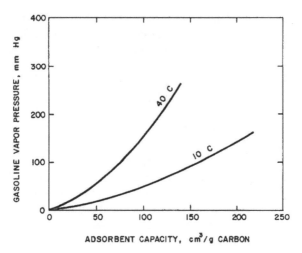

Figure 4.18 Adsorption isotherms for the carbon-gasoline adsorber of Example Problem 4.8.

Equation (4.41) gives heat of adsorption

$$\Delta H_a = 1.99 \text{ cal/(g mole K) } \ln \frac{60}{20} \frac{(313)(283)}{30} = 6{,}455 \text{ cal/g mole}$$

If the capacity were 100 cm³/g, ΔH_a would be ~ 7,750 cal/g mole.

4.5.3 ADSORBENTS

Adsorption, a surface phenomena pertaining to the taking up of gaseous molecules on the surface, requires adsorbents that have a large amount of surface area for a given bulk volume or weight of material. Surface areas in the range of 4×10^5 cm²/g or more are desired. Little of the surface area is obtained by simply breaking up the material into smaller and smaller sizes. The surfaces of adsorbents when magnified are highly irregular, which greatly increases the total effective surface available for adsorption. Adsorbents typically have sizes ranging from 4 to 20 U.S. Sieve Series mesh (4.76-mm to 841-µm sieve opening). Particles smaller than 20-mesh size are usually called powders while material from 20- to 4-mesh size are frequently described as pellets or beads. In addition to the high surface area required, this material should have a high oxidation temperature to prevent oxidation and adequate strength so as to be structurally stable.

Common types of adsorbents are the carbons, nonmetallic oxides, metallic oxides, and combination oxides. Carbon is one of the older adsorbents and has enjoyed a high degree of popularity because of its versatility, economy, and availability. Adsorbent carbon is also known as activated carbon, active carbon, and activated charcoal. It can be made from soft coal, fruit pits, hardwood, and other organic material of direct vegetative origin, or it can be synthesized commercially from the reduction of other organic chemicals. Typical properties of some adsorbents are shown in Table 4.3.

4.5.3.1 Activated Carbon

Activated carbon is produced by heating organic solids to about 900°C in a reducing atmosphere. This produces the porous particle structure desired with a density range of from 0.08 to 0.5 g/cm³. Selective sizing of this material to obtain the small 70- to 80-mesh material would result in particles with an effective mean diameter of about 250 microns. Assuming a spherical configuration, a quick calculation would indicate that the surface area is approximately 1,850 cm²/g. The estimated internal surface area of active carbon (additional surface area due to the irregularity of the surface) is in the vicinity of 1.4×10^7 cm²/g. This shows that *total* surface area of an adsorbent consists of mainly the internal surface area plus the small amount of normal external surface.

Activated charcoal is used to remove odors from living and working areas. For example, only about one kg of activated charcoal will purify the air for one year in a 125-m³ normal living area, a 50-m³ commercial area, a 19-m³ kitchen, toilet, sewage, and some manufacturing areas or a 6-m³ highly odorous area. There are some reactive gases which do not have an affinity for straight activated charcoal. In these cases, use of charcoal impregnated with proprietary substances may be helpful. This is true with gases such as ammonia, formaldehyde, hydrogen chloride, nitrogen dioxide, and sulfur dioxide. Activated charcoal capacity for various substances is divided into four broad categories. The amount of material adsorbed in the two more reactive groups averages 0.333 and 0.167 mass of adsorbate per mass of activated carbon. Nearly all organic compounds with a molecular weight over 45 and a boiling point over 0°C are effectively retained by activated carbon and they can usually be desorbed by steam.

4.5.3.2 Molecular Sieves

Molecular sieves are adsorbents that have been commercially available

TABLE 4.3. Typical Properties of Adsorbents.

Adsorbent	Form	Pore Volume (cm³/g)	Reactivation Temperature (°C)	Max. Gas Flow [cm³/(min g)][a]	Specific Heat, C_p [Cal/(g K)]	Typical Adsorbates
Activated carbon	pellets	0.62–.81	100–500	—	0.25	HC larger than C_4H_{10}, organic compounds, iodine
	beads (G)[b]		100–500	—	0.25	
Silica gel	beads (G)	0.44	120–230	75	0.22	CH_4 through C_4H_{10}, C_2H_4 through C_4H_8, H_2O, SO_2, H_2S
	beads (S)[c]		150–230	75	0.25	
Activated alumina	beads (G)	0.37	175–300	50	0.22	H_2O, H_2S, oil vapors
	beads (S)		175–300	50	0.25	
Molecular sieves	pellets	—	150–300	50	0.23	See Table 4.4
	beads (G)	—	150–300	50	0.23	
	beads (S)	—	150–300	50	0.23	

[a]Increases as adsorbent size increases and may depend on maximum pressure drop allowable.
[b]G = Granules.
[c]S = Spheroids.

260

since about 1954. These materials can be tailor-fitted to adsorb particular size and type of gas molecules. They can be natural crystalline zeolites or synthetic metal hydrates. The synthetic metal alumino-silica hydrates can be precision-made with pore diameters that vary by no more than one angstrom ($Å = 10^{-8}$ cm). The metal ions may be calcium, sodium, magnesium, potassium, or any combination of these.

Molecular sieve pores are custom-sized to adsorb only desired molecules. Typically, molecules up to 0.5 Å larger than the pore size will be adsorbed into that particular sieve because of molecule elasticity. Larger molecules are reflected. Molecular sieves contain metallic positive ions. Polar gas molecules are attracted and retained better in the sieves. A partial listing of molecular sieves is given in Table 4.4 and a listing of gas molecule sizes is given in Table 4.5. A new molecular sieve material called silicalite has been developed from a polymorph of silica. This sieve will adsorb small nonpolar organic molecules the size of benzene and smaller, but will not adsorb the polar water molecules.

Molecular sieve regeneration can consist of heating (e.g., steam), decreasing pressure, or stripping with a nonadsorbent purge gas. The adsorbed material in molecular sieves can also be displaced with a purge of adsorbable liquid, frequently water, then dried and reused.

4.5.3.3 Other Amorphous Adsorbents

Activated carbon, activated alumina, and silica gel are examples of amorphous (noncrystalline) adsorbents. Metallic oxides have the advantage of being able to withstand higher temperatures. However, these materials are not used alone as adsorbents because they are not electrophilic enough (i.e, they do not "like" electrons). They are insulators and, therefore, must be modified by the addition of ionic binders. Activated alumina is a highly porous and granular form of aluminum oxide with a specific gravity of 0.8. It is popularly used for the adsorption of moisture from gases. This material can be reactivated by heating to 175–300°C to drive off the moisture.

Many of the nonmetallic oxides can be readily obtained because they exist almost naturally. For example, diatomaceous earths, which are also known by the names kieselghur and the trade name Sil-O-Cel, are obtained from the calcification of diatomite, which is a soft earthy rock composed of siliceous shells of small aquatic plants. Although the true specific gravity of this material is approximately 2, the apparent specific gravity of the finely divided calcined form is less than 0.5 Fuller's earth is another nonmetallic oxide. It is a claylike powder which can be used after drying directly in its natural state. Silica gels are amorphous silica which are pro-

TABLE 4.4. Molecular Sieves (Linde Type).

Sieve			
Type	Nominal Pore Diameter (Å)	Cation	Molecules Adsorbed[a]	Remarks
3A	3	Potassium	<3-Å effective diameter (e.g., H_2O and NH_3)	Used for drying and dehydration
4A	4	Sodium	<4-Å diameter (e.g., ethanol, H_2S, CO_2, C_2H_4, and C_2H_6)	Scavenge water from solvents and saturated hydrocarbons (HC)
5A	5	Calcium	<5-Å diameter (e.g., n-C_4H_9OH, n-C_4H_{10}, C_3H_8 to $C_{22}H_{16}$)	Separate N-paraffin from branched and cyclic HC
13x	10	Sodium	<10-Å diameter	Drying, H_2S, and mercaptan removal (gas sweetening)

[a]Each type adsorbs listed molecules plus those of preceding types.

TABLE 4.5. Molecular Diameters.

Gas Molecule	Diameter (Å)
Hydrogen	2.4
Oxygen	2.8
Carbon monoxide	2.8
Carbon dioxide	2.8
Nitrogen	3.0
Water	3.15
Ammonia	3.8
Methane	4.0
Ethane	4.2
Ethylene	4.25
Methanol	4.4
Ethanol	4.4
Propane	4.9
n-Butane to n-$C_{22}H_{46}$	4.9
Cyclohexane	6.1
Toluene	6.7
Benzene	6.8
Carbon tetrachloride	6.9
Chloroform	6.9

duced commercially by the reaction of sodium silicate and sulfuric acid. These adsorbents have pore sizes ranging from 20 to 10,000 Å. Commercial silica gels, like carbon and the other nonmetallic oxides, are not capable of withstanding high temperatures. Silica gels will break down at temperatures of 250°C and above.

4.5.3.4 Modified Adsorbents

Use of impregnated activated carbon to increase the adsorptive affinity for certain gases has been noted. The addition of other chemicals to an adsorbent can be used to promote a chemical reaction that can tie up the adsorbate with the additive. This induced chemical adsorption can greatly increase the rate of adsorption as well as increase the capacity of the system, but if not properly chosen, can lead to degradation of the adsorbent causing it to be structurally weaker and to wear away quicker due to attrition from the gas flow. The additive can also react chemically with the adsorbent, especially at higher temperature, which could weaken and ultimately destroy the effectiveness of the adsorbent.

Occasionally, it is possible to impregnate an adsorbent with a catalyst which can increase the rate of chemical reaction between the various pollutants that are adsorbed. Catalytically stimulated oxidation reactions with air are examples. The modified adsorbents are often not capable of being regenerated. For example, lead acetate impregnated on carbon causes chemical adsorption of hydrogen sulfide from a gas stream. The resulting lead sulfide deposit cannot be recovered without destroying the carbon, so the carbon adsorbent must be discarded when saturated.

4.5.4 CAPACITY

Adsorption saturation capacity and adsorber working capacity are considerably different. The capacities for carbon adsorption noted in the previous section assume completely fresh carbon. In regenerable adsorption systems, the adsorbent is recycled and contains some adsorbate which cannot be economically recovered. This makes the capacity of recycled adsorbent less than that of new adsorbent. In addition, emission limits require the shutdown of an adsorber long before the total saturation capacity is reached. This is called the breakthrough point. It is these combinations of factors that set the working capacity for a specific system and operation. Usually this is established by obtaining empirical data which is plotted as adsorption isotherms.

As an example, assume some hypothetical grade of activated carbon is used to adsorb a specific organic vapor. Data show that this type and size carbon should hold 0.3 g of this vapor per g carbon. The system economics specify that 4 g steam per g vapor adsorbed will be permitted for regeneration. Operation of the adsorber of the designed configuration shows that breakthrough on the first cycle occurs when one-third of the saturation capacity is reached. From this, it is known that the carbon bed only contains an average of 0.1 g vapor per g carbon when the concentration of vapor in the emissions reaches the maximum allowable limit. After several adsorption-desorption cycles are run, the system may show that only about 8% of the saturation capacity is adsorbed each cycle due to the added constraint of incomplete regeneration. In this case, the adsorber can be expected to remove only 0.024 g vapor per g carbon each cycle. Therefore, working capacity is 0.024 g vapor/g carbon and saturation capacity is 0.3 g/g carbon in this hypothetical example.

4.5.5 APPLICATION

Adsorbers are ideally suited for the removal of low-concentration gaseous impurities. Particulates and/or condensation of liquids should be minimized as these could coat the adsorption surface and render the system useless. High concentrations of adsorbable gases, including water vapor, would rapidly load up an adsorber and require frequent regeneration. Condensation of water vapors before entering the adsorber may be more economical for high water vapor concentrations. Adsorbers are effective for solvent recovery, odor removal, and dehumidification. Odor removal by adsorption is usually 90% complete in about 0.04-sec contact time. Oxidation by combustion and absorption of odors require a 10 times longer contact time to achieve the same results. Desorbed solvents are usually quite pure and frequently provide a high investment return when sold or reused.

4.6 CHEMICAL REMOVAL PROCESSES

4.6.1 GAS TO PARTICULATE CONVERSION

If a gaseous pollutant can be converted to a particle, the pollutant can be removed by particle control mechanisms. There are numerous examples of gas to particle conversion procedures. This occurs naturally in the atmosphere, for example, when sulfates and other substances are formed from the presence of SO_2, nitric oxides, and hydrocarbons. Particle formation, growth, diffusion, coagulation, and sedimentation affect the time rate of change and ultimately air quality. Such topics are part of aerosol science studies and are presented in depth by Friedlander [8].

Several gas to particle conversion systems for control of air pollution emissions have been studied for SO_2 removal. Gaseous ammonia and SO_2 react to form a sulfate precipitate. The particle formation is highly efficient, but removal of the fine particulate formed and of unreacted ammonia gas can be a problem. Particulate alkali oxides and SO_2 also form sulfate particulates in gas-phase reactions. These reactions are slower and not as effective. Reactive alkali material, such as sodium ores, react rapidly with SO_2 to form particulate sulfates. Efficiencies of over 70% SO_2 removal have been obtained by placing these ores on a filter as a precoat then passing the gases containing SO_2 through this filter cake.

4.6.2 OTHER CHEMICAL PROCESSES

Oxidation and reduction processes can be used to control gaseous pollutants. These are discussed in Chapter 2. Other systems, such as dilution and masking, have been used but these are not control mechanisms and are therefore not presented here. Process modifications, in both chemical and physical aspects, should always be considered as potential methods for reducing emissions and for converting them to a type more easily controlled.

For example, use of water-base instead of oil-base paint can eliminate solvent emission problems.

4.7 HIGH TEMPERATURE AND PRESSURE EFFECTS

The effects of high temperature and pressure on a gas as related to particle behavior are presented in Section 3.7.9 and the significant gas properties affected are summarized in Table 3.4. Table 3.4 notes that a temperature increase causes an increase in gas volume, molecular mean free path, viscosity, thermal conductivity, and diffusivity. Only density, which is inversely proportional to volume, decreases. The gas transport properties of viscosity, thermal conductivity, and diffusivity are only slightly affected by pressure increases below 20 atmospheres.

Section 1.3 shows how changes in temperature and pressure affect gas volume, partial pressure, acoustic velocity, density, molecular mean free path, and viscosity. Appropriate corrections must be made for these gas properties to not only account for gas behavior but to predict the behavior of particles in the gas. This is especially true for the properties of viscosity and volume. The significant gas properties that have not been discussed as a function of temperature and pressure are diffusivity and specific heat.

4.7.1 GAS DIFFUSIVITY

The ratio M_A/ϱ_A in Equation (4.1) can be replaced by the ideal gas law equivalent of RT/P. This shows that diffusivity of a gas in a gaseous medium is

$$D_{AB} \propto \frac{T^{3/2}}{P} \tag{4.42}$$

Gases do deviate from ideality and this theoretical relationship is only approximate. Empirical data as provided, for example, by Reid and Sherwood [9], show that D_{AB} often increases more rapidly than predicted by the 3/2 exponent.

The empirical Loschmidt-von Obermayer equation [10] shows that final diffusivity, D_{AB_f}, is related to initial diffusivity and to final temperature, and pressure, P_f, by

$$D_{AB_f} = D_{AB} \left(\frac{T_f}{T}\right)^{m_1}\left(\frac{P}{P_f}\right) \tag{4.43}$$

where m_1 averages about 2.6 and ranges from about 1.5 to 3.5 for many common organic air pollutants.

4.7.2 THERMAL CONDUCTIVITY

Gas thermal conductivity, k_g, increases with temperature according to Gambill [11] as a function of gas viscosity, μ_g, specific heat at constant pressure, C_p, and molecular weight, M

$$k_g = \mu_g \left(C_p + \frac{2.48}{M} \right) \tag{4.44}$$

where

$k_g = \text{cal/(cm sec K)}$
$\mu_g = \text{g/(cm sec)}$
$C_p = \text{cal/(g K)}$

Gas viscosity can be found by Equation (1.21). Specific heat of air in cal/(g K) from the data of Liley [12] at 1 atmosphere pressure for temperatures from 100 to 1,000°C is

$$C_{p,\text{air}} = 3.227 \times 10^{-5} t + 0.2455 \tag{4.45}$$

where t is in degrees centigrade.

Specific heat of other gases could be found from tables of data or by using the thermodynamic formula

$$C_p = T \left(\frac{\partial S}{\partial T} \right)_P \tag{4.46}$$

where S is entropy. Changes in C_p with pressure may be estimated by thermodynamic Maxwell relations [13]

$$\Delta C_p = \int_{P_1}^{P_2} T \left(\frac{\partial^2 V}{\partial T^2} \right)_P dP \tag{4.47}$$

From van der Waals equation of state

$$\left(\frac{\partial^2 V}{\partial T^2} \right)_P = -R \left[\frac{2aV^{-3} - 6abV^{-4}}{(P - aV^{-2} + 2abV^{-3})^2} \right] \left(\frac{\partial V}{\partial T} \right)_P \tag{4.48}$$

and

$$\left(\frac{\partial V}{\partial T}\right)_p = \frac{R}{P - (a/V^2) + (2ab/V^3)} \qquad (4.49)$$

where R is the universal gas constant and a and b are van der Waals constants which are characteristic of the gas.

4.8 CHAPTER PROBLEMS

4.8.1 DIFFUSIVITY OF GASES IN AIR

a. Benzene is a hazardous pollutant. The diffusivity is to be measured experimentally. Air is passed through a tube at 30°C and 780 mm Hg. The benzene bulb is kept at 30°C. Initial liquid level is 3.5 cm and final level 46.28 hr later is 5.5 cm. Density of benzene is about 0.876 at these conditions and vapor pressure can be obtained from Figure 4.5. Calculate the diffusion coefficient of benzene under these conditions.
b. Estimate the diffusivity of SO_2 in air at 400°C and 1 atm and compare with the value at SC.
c. Estimate the diffusivity of SO_2 in water at 90°C and 1 atm and compare with the value at SC. Viscosity of water at 90°C is 0.3 centipoises.

4.8.2 FLUE GAS HUMIDITY

Determine the percent water in:
a. A 200°C flue gas with a wet bulb temperature of 50°C
b. Same, but wet bulb is 58°C
c. A flue gas from which 350 ml of water is condensed in the impingers while 2 Nm³ of gas is sampled. The impinger pressure is minus 10 cm Hg, barometric pressure 750 mm Hg, and meter temperature is 100°C. Assume meter is at barometric pressure.
d. The same gas as part b above. Use duct pressure of 1 atm with vp method.

4.8.3 AMMONIA ABSORPTION

Ammonia is to be removed in a countercurrent absorption system by scrubbing with pure water. Assume this is gas-phase controlling absorption, the equilibrium line is $y = 0.77x$ and the slope of the operating line is 1.0. Gas enters at 400 Nm³/hr and contains 8.5% ammonia. Concentration of ammonia leaving should not exceed 3.5%. Assume SC and consider the transfer of mass between phases to be negligible. The vapor pressure of pure ammonia at 20°C is 867.12 kPa.

a. Draw a sketch showing y_1, x_2, y_2, and all other data.
b. Calculate the number of moles of ammonia transferred to the water.
c. Calculate the concentration of ammonia in the exit water.
d. Determine the partial pressure of ammonia in all inlet and outlet streams.
e. Find $K_G aV$ for this system.
f. Determine the number of theoretical stages in the tower.
g. Estimate the number of overall gas transfer units.
h. Find the number of actual stages in the tower.

4.8.4 BUTANE ADSORPTION

A specific type of activated carbon is used to adsorb butane (C_4H_{10}) from an air stream at 40°C. With a given system configuration, the saturation capacity is 10 g butane/100 cc carbon. When the purge rate is 10 bed volume/min for 10 minutes, the working capacity is 6 g butane/100 cc carbon at an adsorption-desorption temperature differential of −15°C. The adsorbent pore volume is 0.7 cm³/g and the density is 0.3 g/cm³. Inlet butane concentration is 1.5%, and outlet concentration is *not to exceed* 200 ppm. Gas flowrate is 600 m³/hr, and the adsorbing portion of the cycle is 2 1/2 hrs.

a. Sketch a curve of outlet concentration versus time for a 320-minute period of time. Neglect any time required for cooling and assume the cycle repeats as soon as possible.
b. Specify how many grams of adsorbent are required under these conditions.
c. What is the purge temperature and what might be used for the purge gas?
d. List four ways to increase the system capacity without changing any quantities, times, concentrations, or components.

REFERENCES

1 Gilliland, E. R. 1934. *Ind. Eng. Chem.*, 26:68.
2 Sherwood, T. K. and R. L. Pigford. 1952. *Absorption and Extraction*. New York: McGraw-Hill Book Company, pp. 16, 56, 131, 287.

3 Norman, W. S. 1961. *Absorption Distillation and Cooling Towers*. New York: McGraw-Hill Book Company, p. 21.

4 Wilke, C. R. 1949. *Chem. Eng. Prog.*, 45:218–224.

5 1967. "Design Information for Packed Towers," Norton Bulletin DC-10, U.S. Stoneware.

6 Calvert, S. 1968. "Source Control by Liquid Scrubbing," in *Air Pollution*. A. Stern, ed. New York: Academic Press, Inc., Chapter 46.

7 Smith, W. S. 1985. "Air Water Vapor Psychrometric Chart," *The Entropy Quarterly*, 6(4).

8 Friedlander, S. K. 1977. *Smoke, Dust and Haze/Fundamentals of Aerosol Behavior*. New York: John Wiley & Sons, Inc.

9 Reid, R. C. and T. K. Sherwood. 1958. *The Properties of Gases and Liquids*. New York: McGraw-Hill Book Company.

10 Quinn, E. C. and C. L. Jones. 1936. *Carbon Dioxide*. Stanford, CT: Reinhold Publishing Corp., p. 45.

11 Gambill, W. R. 1963. "Prediction and Correlation of Physical Properties," in *Chemical Engineers' Handbook*. R. H. Perry, ed. New York: McGraw-Hill Book Company, Section 3.

12 Liley, P. E. 1963. "Physical and Chemical Data," in *Chemical Engineer's Handbook*. R. H. Perry, ed. New York: McGraw-Hill Book Company, Section 3.

13 Hougan, O. A., K. M. Watson and R. A. Ragatz. 1965. *Chemical Process Principles, Part II*. 2nd ed. New York: John Wiley & Sons, Inc.

CHAPTER 5

Control Devices

THE previous chapters have included background material related to the need for control and the basic control theory mechanisms, as well as to the production, dispersion, and disposal of pollutants. When all of this information is compiled, it can be used to determine what can be accomplished and what basic mechanisms should be considered. This chapter presents information on specific control devices and shows how particulates and gaseous pollutants may be controlled.

Almost any control device has the ability to remove both particulate and gaseous pollutants if variations in the process or control system are considered. For example, variations in process operating conditions can change type and/or quantity of emissions; additions of chemicals can convert gases to particles; any temperature, pressure, or concentration variations can change particles to gases; liquid irrigation of a dry particle collector can enable it to remove gases also, and particle agglomeration and/or electrostatic enhancement can improve the efficiency of low-energy particle collectors. Although some of these factors are discussed in this chapter, most are not detailed specifically. However, the reader should keep these alternatives in mind when actually considering control devices and control systems.

Techniques for reducing emissions to the atmosphere and techniques for managing toxic and hazardous wastes that are included here can be likened to several of the familiar forces of nature. Specifically these are

wind—representing cyclones and filters
lightning—representing electrostatic precipitation
precipitation—representing scrubbers
fire—representing incinerators and combustion systems

271

Incineration and combustion have been discussed in great detail as potential energy conversion devices in Chapter 2 and are not included in this chapter. However, control devices discussed in this chapter are often used with incineration and combustion systems.

The first part of this chapter presents devices that are predominantly particle collectors. Then, devices that are or can be both particle and gaseous collectors are presented, followed by gas-removal devices. Respectively, the section titles are mechanical collectors, electrostatic precipitators, filters, scrubbers (wet and dry), absorbers, and adsorbers. Hybrid devices, gas-conditioning, and scale-up are covered at the end of the chapter. This chapter deals with control of pollutants as far as specific type, size, and configuration of control devices relate to the effectiveness of control. The practical aspects of elementary system designs and costs are covered in Chapter 6.

5.1 MECHANICAL COLLECTORS

Mechanical collectors are simple devices that serve extremely well as precleaners preceding high-efficiency control devices. They use gravitational and/or centrifugal force to remove particulate matter. This includes those designed with no moving parts, such as settling chambers and cyclones where the collection energy comes from the inlet gas stream, as well as devices that utilize a mechanically-driven blade or disc to provide part or all of the collection energy. Gravitational settling chambers are presented in Section 3.6 as the basic example for particulate settling. Most blade or disc devices are basically wet scrubbers as they have a liquid spray to provide much of the collection and to wash the device clean of collected particles. For this reason, these collectors are covered in this chapter as wet scrubbers. Cyclone separators are presented as the example of mechanical collectors.

5.1.1 CYCLONE SEPARATORS

Cyclones obtain their name by similarity with natural atmospheric storms. These devices are among the oldest types of particulate-control equipment. They can be highly efficient, but usually are medium-to-low efficiency, low pressure-drop (and therefore low operating-cost) precleaning devices. There are numerous variations of design ranging from complex arrangements of multiple small cyclones in a common hopper to single large units of various configurations. Certain design factors are very important to the successful operation of the unit.

In a dry cyclone, dusty inlet gases are forced to swirl around inside the cylindrical body producing a centrifugal force that moves the particles to the wall. This action is depicted in Figure 5.1. The particles, whether liquid or solid, move out of the small opening at the end of the cone and the cleaned gases leave at the other end. The inlet can be either at the top or the bottom as shown in Figure 5.1. A newly proposed high-performance cyclone is depicted in Figure 5.1(c). This design uses a split gas flow, part of which enters the cyclone through a tangential inlet and part of which enters through a radial inlet. A flow control valve is used to do the gas splitting. This produces a smoother, stronger vortex flow which improves particle collection. It also helps reduce the traditional problems with variable and low flow rates. Liquid sprays could be introduced into any of these designs. The following sections discuss only standard, top-inlet, tangential-entry, dry cyclones.

5.1.2 CYCLONE DIMENSIONS

There are several basic design dimensions that have been standardized by those who study cyclone collection efficiency. These are given in Table 5.1 as dimensionless ratios based on a cyclone body diameter, D_c, of 1.0. These high-efficiency design configurations are designated by Stairmand [1] and Swift [2] and the conventional designs are by Lapple [3] and Swift. Figure 5.2 locates the symbols and depicts a high-efficiency unit. Low-efficiency units are often called "rock collectors" and would appear fat compared to the slim lines of high-efficiency units. Volumetric flowrates are roughly approximated by $5{,}480D_c^2$, $6{,}850D_c^2$, and $15{,}350D_c^2$ m³/hr for high-efficiency, conventional, and low-efficiency designs, respectively, with D_c in meters. In English units with D_c in feet, these would be $300D_c^2$, $375D_c^2$, and $840D_c^2$ ft³/min, respectively. For example, a high-efficiency cyclone volumetric flowrate in acfm is $300D_c^2$, where D_c is in feet. Inlet velocities should be from 50–90 ft/sec for high-efficiency units.

Cyclone design constraints are:

(1) Avoid short circuitry: keep $L_i < L_o$

(2) Avoid sudden contraction: keep

$$W_i < (1/2)[D_c - D_o]$$

(3) Keep vortex inside cyclone:

$$L_o + 1 \leq L_{cy} + L_{co}$$

$$L_o < L_{co}$$

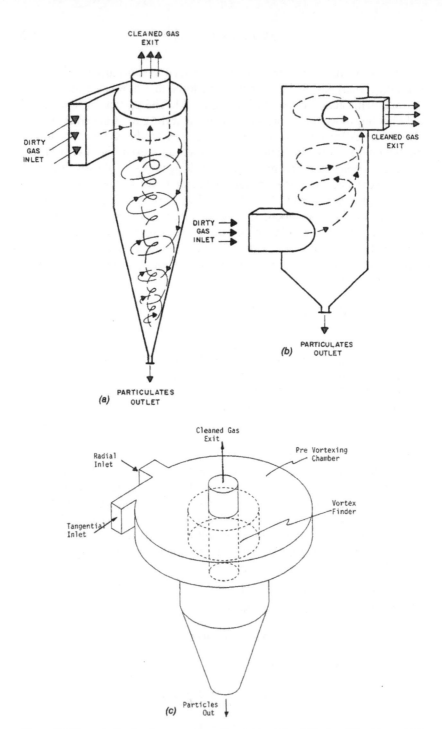

Figure 5.1 Conceptualized cyclone separator action: (a) top inlet; (b) bottom inlet; (c) new high performance cyclone.

TABLE 5.1. Dimensionless Design Ratios for Tangential Entry Cyclones.

		High-Efficiency		Conventional	
Symbol[a]	Nomenclature	Stairmand [1]	Swift [2]	Lapple [3]	Swift [2]
D_c	Body diameter	1.0	1.0	1.0	1.0
L_i	Inlet length	0.5	0.44	0.5	0.5
W_i	Inlet width	0.2	0.21	0.25	0.25
L_o	Outlet length	0.5	0.5	0.625	0.6
D_o	Outlet diameter	0.5	0.4	0.5	0.5
L_{cy}	Cylinder length	1.5	1.4	2.0	1.75
L_{co}	Cone length	2.5	2.5	2.0	2.0
H_o	Overall height	4.0	3.9	4.0	3.75
D_B	Bottom diameter	0.375	0.4	0.25	0.4
ℓ	Natural length	2.48	2.04	2.30	2.30
G	Configuration factor[b]	551.3	699.2	402.9	381.8

[a]See Figure 5.2.
[b]Dimensionless constants, not ratios.

(4) Avoid excessive operating costs: keep

$$\Delta P < 10'' \text{ water gauge}$$

Usual industrial capacities range from 120–50,000 cfm (200–85,000 m³/hr) per cyclone. Inlet velocities range from 50–90 ft/sec (1,500–2,700 cm/sec) for high-efficiency units, as noted above, and lower for others. A quick calculation shows that *industrial* high-efficiency units are usually about 20 cm and larger in diameter. Pressure drop in cyclones is proportional to the square of the inlet velocity. Low-efficiency units have the lowest velocities and pressure drops. Even so, pressure drops of only 15 cm H_2O maximum are typical for high-efficiency units and pressure drops of about 5 cm H_2O are normal.

The last two entries in Table 5.1, natural length and configuration factor, are not defined in Figure 5.2. Natural length, ℓ, of a cyclone is considered to be the length required for a complete gas vortex configuration. This is given by Leith and Licht [4] as

$$\ell = 2.3 D_o (D_c^2 / L_i W_i)^{1/3} \qquad (5.1)$$

and should be less than the overall cyclone height minus the gas outlet length, L_o, to maintain the full gas vortex. Configuration factor, G, is a dimensionless ratio dependent upon design configuration. Values predicted by Leith and Licht are given in Table 5.1. The ratio $G/\Delta P$ should be maximized for optimum performance [5].

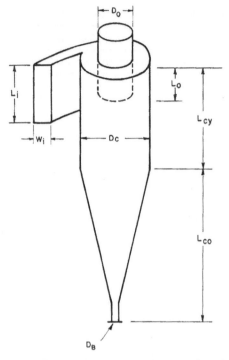

Figure 5.2 Tangential entry cyclone design configurations.

5.1.3 PRESSURE DROP

Cyclone pressure drop, ΔP, is dependent mainly on the velocity. It can be estimated by the Shepherd-Lapple method [6] as modified by Briggs [7] for dust loading

$$\Delta P = 8.19 \times 10^{-3} \varrho_g v_i^2 \left(\frac{L_i W_i}{D_o^2}\right)\left(\frac{1}{0.0057 C_i + 1}\right) \qquad (5.2)$$

where ΔP is in cm H_2O and

v_i = inlet velocity, cm/sec
C_i = inlet dust loading, g/m³

If a neutral vane is used inside the cyclone on the inlet between the cylindrical body shell and the outlet extension, multiply Equation (5.2) by 0.5. Vanes can become plugged by sticky ducts. Equation (5.2) shows that pressure drop decreases as dust loading increases.

5.1.4 CYCLONE EFFICIENCY

Industrial efficiencies can be calculated using particle cut diameter, d_c, which is estimated by

$$d_c = \sqrt{\frac{9\mu_g W_i}{2\pi N_e v_i(\varrho_p - \varrho_g)}}$$
(5.3)

where

μ_g = gas viscosity
W_i = cyclone inlet width
N_e = effective number of gas spirals (~ 8 for high efficiency and ~ 5 for conventional industrial cyclone)
v_i = inlet gas velocity

Units for the terms in Equation (5.3) must be consistent.

Theodore and DePaola [8] have curve fit the Lapple efficiency data [9] to show that cyclone individual particle size fractional collection efficiency, ϵ_i, is

$$\epsilon_i = \frac{1}{1 + (d_c/d)^2}$$
(5.4)

where d is diameter of individual particle in question, with units consistent with those for d_c.

Theodore also shows that overall fractional efficiency, ϵ_o, is

$$\epsilon_o = \frac{1}{1 + (d_c/\bar{d}_M)}$$
(5.5)

where $\bar{d}_m$ is the particulate mass mean diameter with the same units as d_c.

Cyclones have the capability of operating at elevated temperatures and pressures. The temperature limitation is relative to the material of construction. Fluidized bed combustor hot cyclones typically operate at up to 1600°F (870°C). Pressurized combustors would operate at elevated pressure also. Cyclone collection efficiencies at high temperatures and pressures do not follow conventional cyclone efficiency theories. Increased pressure tends to increase collection efficiency for a given particle diameter up to about 5 atmospheres pressure, after which the change with higher pressures becomes less noticeable. Temperature effects are the reverse in that cyclone efficiency decreases with increased temperature up to about 750°F (400°C), after which the changes become less significant.

Parker and co-workers reported data [10] to show that at temperatures from 930–1300°F (500–700°C) and at 5 atmospheres pressure the combined effects resulted in a decrease in particle collection efficiency as particle size increased up to about 4.5–6 μm, then an increase in efficiency as particle size increased up to about 15 μm. In the 5-cm diameter test cyclone, maximum collection efficiencies at 15 μm were about 85% at 500°C and about 25% at 700°C.

Parker et al. [10] presented empirical data for high temperature/pressure cyclone operation from which an equation is developed to predict aerodynamic cut diameter, d_{50}. For d_{50} in microns and for cyclones $\geq$4.5 cm in diameter, this is

$$d_{50} = 330x^{-0.58} \tag{5.6}$$

in which x, as defined below, is good from about 10^2 to 10^5:

$$x = Re_c St_M^{0.5} \tag{5.7}$$

In Equation (5.7), the Stokes Number for the mass mean particle diameter, St_M, is

$$St_M = \frac{(\overline{d}_M)^2 \varrho_p v_i C}{9\mu_g D_c'} \tag{5.8}$$

where

v_i = cyclone inlet velocity
C = dimensionless Cunningham factor
D_c' = the collector dimension expressed as the hydraulic diameter of the cyclone inlet, or for rectangular inlets

$$D_c' = \frac{2W_i L_i}{W_i + L_i} \tag{5.9}$$

where W_i and L_i are cyclone inlet width and length respectively. The Reynolds Number based on the cyclone collector, Re_c, is

$$Re_c = \frac{D_c' v_i \varrho_g}{\mu_g} \tag{5.10}$$

When the cyclone is <4.5 cm in diameter, d_{50} is found by

$$d_{50} = \frac{D_c}{4.5} (330x^{-0.58}) \tag{5.11}$$

where D_c is cyclone diameter in cm.

Normally, industrial cyclones are not smaller than 6 inches (15 cm) in diameter to prevent plugging. Yet, particle sizing using small cyclones is very important. Small cyclones operated in parallel can be used to provide a simple and accurate size and size distribution for particulate matter. Values for the aerodynamic cut diameter, d_{50}, are usually determined based on empirical data. For example, values of d_{50} in microns for these very small cyclones are

$$d_{50} = aQ^{-b} \qquad (5.12)$$

where Q is the gas volumetric flowrate in ℓpm and a and b are constants pertinent to the gas flowrate and the cyclone. For 10-mm diameter cyclones

$$\left.\begin{aligned} a &\cong 6.17 \\ b &\cong 9.75 \end{aligned}\right\} \text{ when } 1.25 \le Q \le 5.4 \text{ ℓpm}$$

and

$$\left.\begin{aligned} a &\cong 16.1 \\ b &\cong 1.25 \end{aligned}\right\} \text{ when } 5.5 \le Q \le 23.0 \text{ ℓpm}$$

For 25-mm cyclones,

$$\left.\begin{aligned} a &\cong 123.68 \\ b &\cong 0.83 \end{aligned}\right\} \text{ when } 65 \le Q \le 350 \text{ ℓpm}$$

Example Problem 5.1

(a) Estimate the overall collection efficiency of a 24″ (61-cm) diameter high-efficiency cyclone on coal dust with a mass mean diameter of 9 μm and a density of 1 g/cc. The gas is air at SC and the inlet velocity is 66 ft/sec (20 m/sec).

Solution

Consistent units are required, so cgs values are chosen. Use the Stairmand high-efficiency design data from Table 5.1. This gives

$$W_i = 0.2D_c = (0.2)(61) = 12.2 \text{ cm}$$

From Equation (5.3), cut diameter is

$$d_c = \sqrt{\frac{(9)[1.83 \times 10^{-4} \text{ g/(cm sec)}](12.2 \text{ cm})}{2\pi(8)(2{,}000 \text{ cm/sec})(1 \text{ g/cm}^3)}}$$

$$= 4.47 \times 10^{-4} \text{ cm or } 4.47 \ \mu\text{m}$$

Then overall efficiency using Equation (5.5) is

$$\epsilon_o = \frac{1}{1 + \left(\dfrac{4.47}{9}\right)} = 0.668 \text{ or } 66.8\%$$

(b) Estimate the individual collection efficiency of this system on 10 μm particles.

Solution

Use Equation (5.4) to obtain

$$\epsilon_i = \frac{1}{1 + \left(\dfrac{4.47}{10}\right)^2} = 0.833 \text{ or } 83.3\%$$

(c) Calculate the aerodynamic cut diameter on this coal dust for the same system if operation is at 1550°F and 5 atm with all other parameters remaining the same.

Solution

Use Figure 1.7 for air at 843°C (1550°F)—figure is more accurate than equation at this temperature:

$$\mu_g = 4.3 \times 10^{-4} \text{ g/(cm sec)}$$

$$\varrho_g = (1.20 \times 10^{-3} \text{ g/cm}^3 \text{ @ SC}) \frac{293}{273 + 843}\left(\frac{5}{1}\right)$$

$$= 1.58 \times 10^{-3} \text{ g/cm}^3$$

Use Table 5.1 again for

$$L_i = (0.5)D_c = 30.5 \text{ cm}$$

From Equation (5.9)

$$D_c' = \frac{(2)12.2(30.5)}{12.2 + 30.5} = 17.4 \text{ cm}$$

Then Equation (5.8) gives

$$St_M = \frac{(9 \times 10^{-4} \text{ cm})^2(1 \text{ g/cm}^3)(2,000 \text{ cm/sec})(1)}{(9)(4.3 \times 10^{-4} \text{ g/cm sec})(17.4 \text{ cm})}$$

$$= 0.0241$$

By Equation (5.10)

$$Re_c = \frac{(17.4 \text{ cm})(2,000 \text{ cm/sec})(1.58 \times 10^{-3} \text{ g/cm}^3)}{4.3 \times 10^{-4} \text{ g/(cm sec)}}$$

$$= 127,900$$

Then, by Equation (5.7),

$$x = (127,900)(0.0241)^{0.5} = 1.99 \times 10^4$$

Finally, from Equation (5.6),

$$d_{50} = 330(1.99 \times 10^4)^{-0.58} = 1.06 \ \mu\text{m}$$

(Note that aerodynamic cut diameter, d_{50}, is equal to d_c because particle density is 1.0.)

5.2 ELECTROSTATIC PRECIPITATORS

Electrostatic precipitators (ESP) have been used as effective particulate-control devices for many years. Equations (3.58) and (3.83) show that efficiency is directly related to dust concentration, collector area, and migrational velocity, and indirectly related to gas flowrate. Migrational velocity in turn is related to electric mobility (which increases as particle size decreases, assuming fully charged particles) and to electrical field strength. The field strength for a given space depends on electrical resistivity of the layer of deposited particulate dust and on the voltage applied. Field strength oscillates as dust builds up and is rapped off both the discharge and collector electrode, as the rectified AC voltage oscillates and as system sparking occurs. Efficiency is dependent on these and numerous other factors, some of which oscillate continuously as the ESP operates.

ESP designers have increased collection efficiencies to keep pace with emission limit regulations by improving system performance, increasing collecting area, and adjusting dust resistivity. In general, average U.S. fly ash ESP efficiency has increased from about 94% in 1950 to about 99% in the mid-1970s. This could represent a threefold increase in collection area for a specific dust if only collection area is increased. Other types of collection devices such as filters and scrubbers can be electrified to improve performance. These are discussed under hybrid devices (Section 5.7).

5.2.1 UNIT SPECIFICATIONS

5.2.1.1 General

ESPs are usually large and, because of this, require careful design. Improved performance is obtained by reducing gas flow unevenness, dust flow unevenness, dust reentrainment, and gas sneakage. Gas sneakage is the term used to account for particles that pass through nonelectrified regions of the precipitator. This can occur at the top of a unit where electrical distribution, plate support, and rapper systems are located and at the bottom where the dust hoppers are attached. These areas can be seen in the ESP cutaway of Figure 5.3. This figure shows a simple single-stage precipitator (where dust charging and collection occur in the same stage). One or more diffuser plates over the inlet to close off 40–60% of the opening area each and inlet straightening vanes help eliminate gas and dust unevennesses. A single plate is shown in Figure 5.3.

Rapping to free the dust deposits from the collecting and discharge electrodes results in most of the reentrainment of dry dusts if the system is properly designed. Use of several precipitation field sections in series reduces overall lossess due to entrainment because less dust enters each section. This obviously increases ESP size. Longer, less frequent rapping may also reduce reentrainment. Pilot-scale models are used to test gas flow patterns, dust entrainment, and gas flow unevenness.

Best design for "no" sneakage may still have about 0.01% dust sneakage. For very low reentrainment systems, it could be estimated that the collection efficiency is reduced by about 1% for every percent of dust sneakage. As reentrainment increases, the efficiency reduction becomes greater.

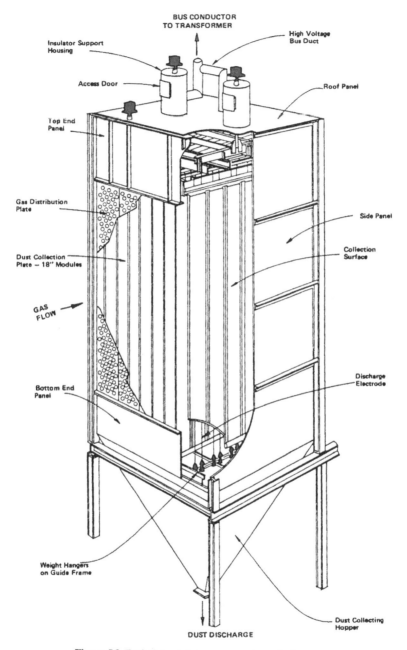

Figure 5.3 Typical precipitator cross section—single stage.

5.2.1.2 Spark-Over

The ESP collection efficiency is related through migration velocity to the electrical field strength which is in units of volts/cm. The voltage can be increased until sparking (spark-over) occurs at which time voltage drops momentarily. The greatest amount of power that could be delivered during ESP operation would be the area under a voltage-time curve where the voltage is just below spark-over voltage. Spark-over voltage continuously varies due to changes in gas and dust composition and temperature. In order to know if the system is operating near the spark-over voltage, it is desirable to have frequent sparking (e.g., 60–100/sec), but this reduces effective delivered voltage.

Microcomputers are being used to automatically respond to spark-overs and adjust voltages so as to maintain the highest power possible at all times without continuous operator adjustment. These systems also respond to internal upsets such as shorts and flooded hoppers.

5.2.1.3 Electrode—Type and Arrangement

There are two basic precipitator designs—European and American. The American discharge electrodes are usually free-hanging electrical wires or wire frames as shown in Figure 5.3. The wires are often 2.8-mm-diameter straight wire. Collecting electrodes are often plates, such as in Figure 5.3, spaced equidistant from the wire frames and about 20 to 30 cm apart from each other. It has been noted that the discharge corona starts at rough spots or imperfections on the discharge electrode. Therefore, rough surfaces on discharge electrodes can promote increased corona discharge when dust resistivity is higher than desired. Ten different types of electrodes available in European-design ESPs are shown in Figure 5.4. The spiked emitter types improve ESP operation when dust resistivity is high.

Needle electrodes consisting of sharp needles attached to both the upwind and downwind sides of positively charged plates are also available. The collection occurs on negative plates spaced between the positive plates as well as on the positive plates. This system operates at low voltage with close plate spacing to maintain the desired field strength. This is possible as no wires exist between the plates.

Collecting electrodes consist of relatively long metal panels about 40 to 50 cm wide by up to 14 m long. Six to eight panels are secured together as an electrically continuous and mechanically rigid plate. Often, these panels contain special folds for rigidity as shown in Figure 5.5. Some new precipitators are increasing plate spacing from the 20–30 cm noted to up to 1 m. Operating voltages on the wide-space precipitators are up to 200 kV, which is four times normal operating voltages.

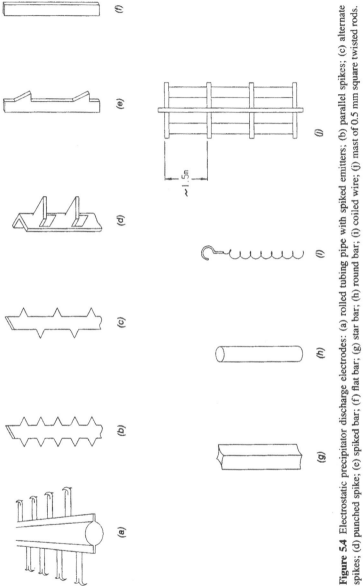

Figure 5.4 Electrostatic precipitator discharge electrodes: (a) rolled tubing pipe with spiked emitters; (b) parallel spikes; (c) alternate spikes; (d) punched spike; (e) spiked bar; (f) flat bar; (g) star bar; (h) round bar; (i) coiled wire; (j) mast of 0.5 mm square twisted rods.

285

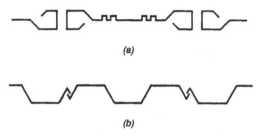

(a)

(b)

Figure 5.5 Top views of two types of electrostatic precipitator collection electrodes: (a) individual panel design; (b) semi-interlocked design.

5.2.1.4 Electrode Cleaning

Both collecting and discharge electrodes must be cleaned of dust to reduce the electrical resistance of the dust layers and permit continued operation. Dust buildup on wires can be difficult to remove and, occasionally, the dust "doughnuts" or even "grapefruit" must be removed by hand cleaning. The collection plates can be cleaned by striking the plates with a hammer or vibrator device. Figure 5.6 shows a hammer–anvil arrangement. In this system a rotating hammer strokes the anvil attached to the plate. The European-designed rappers strike at the bottom of the plate with a force of up to 1,000 g's. Dust layers which build up to thicknesses of 1.25 to 2 cm are usually cleaned to 0.2–0.3 cm. Some tall units are rapped at several locations along the height.

Discharge electrodes can be cleaned by similar hammer-rapping devices or by vibrators. Most systems rap or vibrate multiple rows of wires simultaneously as shown in Figure 5.7.

5.2.1.5 Dust Resistivity

Dust resistivity of 10^9 to 10^{10} ohm-cm is desired for ESPs. As resistivity

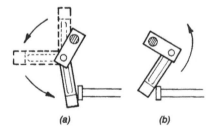

(a)   *(b)*

Figure 5.6 Collecting plate rapping systems: (a) late strike; (b) early strike.

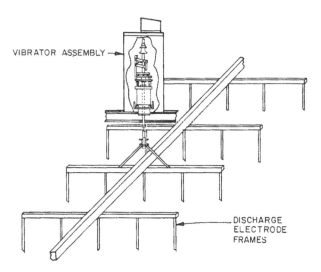

Figure 5.7 Discharge electrode vibrator.

increases above this, it becomes difficult to maintain adequate corona discharge and back ionization can occur. Back ionization occurs in dust pockets where air ionizes due to the high-voltage drop across the dust. The positive ions formed migrate away from the positive collection electrode, freeing collected particles and stopping precipitator operation due to electrostatic forces. Dust will still be removed, as a 30% or more particle collection effectiveness is achieved due to gravitational settling alone.

Particle resistivity varies with temperature and composition as shown in Figure 5.8 for fly ash. Particle volume resistivity dominates at low temperatures and surface resistivity dominates at high temperatures, yielding the series resistance values seen in the figure. Fly ash resistivity is highly dependent upon SO_3 which comes from the combustion of sulfur. A few ppm (5–40) of SO_3 can condition fly ash to produce the same favorable resistivities resulting from natural-acid conditioning due to the burning of 5% sulfur coal.

Other substances besides sulfur can serve as natural conditioning agents for fly ash. High sodium-to-silica ratio in low-sulfur western coal provides good fly ash conditioning because sodium oxide is a good electrical conductor. Next in order of importance are lithium, iron, and potassium oxides. Data from Frisch and Dorchak [11] show that western low-sulfur coal fly ash resistivity based on the ratio of sodium oxide to silicon dioxide can be estimated by

$$\Omega \cong 6.4 \times 10^{11} \, (Na_2O/SiO_2)^{-2}$$

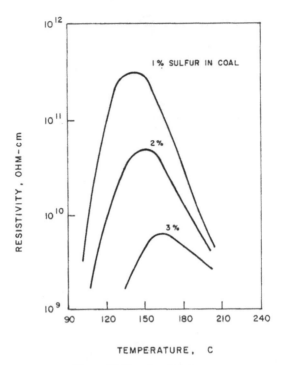

Figure 5.8 Fly ash resistivity.

where Ω is resistivity in ohm-cm and Na_2O and SiO_2 are percent by weight. Low-sulfur (0.25%) western coal with low sodium (0.3% Na_2O) in the ash would be expected to have a fly ash resistivity of about 10^{12} ohm-cm at 175°C. The same fly ash with about 2% Na_2O would have a resistivity of about 10^{10} ohm-cm and, at 5% Na_2O, about 3×10^9 ohm-cm.

If desired resistivity is not achieved by natural or added chemical conditioning, then fuel blending and/or hot precipitator operation could be considered. Hot precipitators operate at temperatures greater than 350°C. Normal "cold" operation is at 120–150°C. The added gas volume at the high temperatures increases ESP unit size.

5.2.2 COLLECTION EFFICIENCY

ESP collection efficiency is based on test data obtained for each specific dust and system. Equation (3.83) for overall ESP collection efficiency is rewritten here in a modified form to account for some of the actual operating variations

$$\epsilon_o = 1 - \exp\left[-\omega\left(\frac{36A}{Q}\right)^n\right]$$ (5.13a)

In this form, ω represents the corrected migration velocity in cm/sec and is a value obtained by test data. The ratio A/Q is collecting electrode area in m^2 over gas flowrate in m^3/hr and is known as a specific collecting area, SCA. The exponent n seems to vary from about 0.5 to 1 depending on the value of SCA. Units are dimensionalized by the value 36.

The quantities A, Q, and ϵ_o in Equation (5.13a) can be measured constants in a given test system. Corrected migration velocity, ω, can be determined by rearranging the equation as

$$\omega = -\frac{\ln (1 - \epsilon_o)}{(36A/Q)^n}$$ (5.13b)

This value of ω is a corrected average to account for dust size, size distribution, gas and dust unevenness within the particular system, charging effectiveness including sneakage, reentrainment, dust resistivity as accounted for by effective field strength, and any other peculiarities of the specific system. Values of n close to unity should be used for SCA less than about 25-m^2 collecting electrode area per 1,000 m^3/hr gas flow (about 450 $ft^2/1,000$ cfm) and close to 0.5 above this. This is shown in Figure 5.9 where the dashed portion of the lines is Equation (5.13) with $n = 1$ and the solid portion is for $n = 0.5$. The lines would extend straight from the dashed end if n equaled 1 for all values of SCA. Using Figure 5.9 or Equation (5.13), collection efficiencies at various SCA can be estimated. High-efficiency ESPs for fly ash usually require SCA of about 25 m^2 per 1,000 m^3/hr or greater, indicating that migration velocity is relatively low when fly ash is the dust. It can be seen from Figure 5.9 that *doubling* the SCA from 25 to 50 increases the efficiency by only 2 percentage points from 93.5% when ω is 3 cm/sec and by about 0.7 percentage point from 97.5% when ω is 4 cm/sec. Efficiency could be raised as much by keeping SCA constant and raising ω from about 3 to 4.5 and from 4 to 6, respectively.

Loss of particles due to reentrainment and saltation affect precipitator efficiency. Saltation is the loss of deposited material that occurs when a particle being precipitated strikes and frees already-deposited material. Reentrainment losses increase as gas velocity increases and particle adhesion-cohesion decreases. For these reasons, particles leaving ESPs may have about the same size and size distribution as inlet particles even

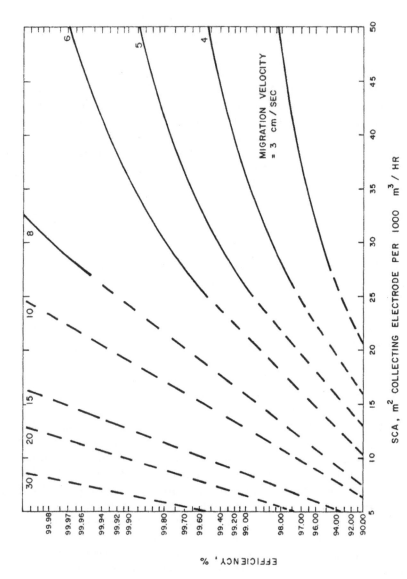

Figure 5.9 Electrostatic precipitator collection efficiency as a function of specific collecting area and migration velocity.

though removal efficiency is in the high ninety percentile. These factors should be accounted for in the pilot testing and determining of ω.

Section 5.2.1.5 notes that sulfur content of fossil fuel is critical in establishing a specific dust resistivity value. This is so reliable that Hesketh [12] has been able to develop an empirical procedure for estimating SCA for coal-fired boilers. In English units, this is

$$SCA = 433 S^{-0.59}(100 - E)^{-0.24} \qquad (5.14)$$

where

SCA = specific collecting area, ft^2 collecting area/1000 acfm gas flow
S = as-received coal sulfur, %
E = particulate collection efficiency, %

Multiply Equation (5.14) by 0.0546 to obtain SCA in units of m^2 per 1000 m^3/hr. This procedure assumes standard DC charging and collecting and does not include possible advantages of pulse-powering which could be accounted for by Equation (5.13). The procedure enables one to estimate SCA for a given coal and efficiency or to estimate efficiency given SCA and coal.

5.2.2.1 Voltage Control

In addition to the theoretical aspects relative to collection efficiency in ESPs, there are the practical aspects of voltage and unit design. Precipitators are more efficient as voltage is increased between the emitting and collecting electrodes. For a given system, increased voltage increases field strength, which enables better particle charging and collection. Precipitators should be powered at the greatest voltage practicable without excessive sparking to obtain the highest efficiency. Limited sparking is acceptable as it locates the point of maximum voltage for operation, but continued high rates of sparking can damage the emitting electrodes. Automatic voltage control and regulation are essential to achieve this condition of maximum voltage without excessive spark-over.

Automatic spark response is part of the system. Spark detection reduces transformer-rectifier voltage to zero when a spark occurs. Voltage is kept at zero for only a short time (e.g., < 16 milliseconds) then reapplied at a "fast ramp" to some predetermined fraction of the spark-over voltage. The voltage is then increased at a "slow ramp" rate until sparking occurs again. This allows the voltage to be maximized and yet protects the ESP.

5.2.3 DESIGN PARAMETERS

Specific values for ESP design vary with type of dust, size of unit, efficiency required, and vendor. Some typical values used for fly ash precipitators are summarized in Table 5.2. Note that there are about 520,000 Am³/hr of emissions at 175°C per 100 MW of electrical generating rate.

Precipitator gas velocities vary depending on the stickiness of the dust and the desired efficiency. For example, the approximate critical velocities at which dusts are blown off collectors are 50 cm/sec for carbon black, 200 cm/sec for fly ash, 300 cm/sec for cement, and 500 cm/sec for liquid droplets. Aspect ratio (AR) of an ESP is the effective precipitator length to effective height. Currently AR ranges from 1.5 to 2.0 which shows that ESPs are typically longer than they are high. The 30% removal due to settling results in most collection occurring in the lower half of a precipitator.

Corona current varies with migration velocity for a specific unit and dust. Actual current also includes shorts in the system due to broken wires and/or dust buildup and arcing during operation. Figure 5.10 shows how theoretical corona current increases with migration velocity for several average particle sizes. This is based on a SCA of 11 m² per 1,000 m³/hr.

The latest design modifications involve discharge wire size, plate spacing, and pulse energization. Historically, high corona currents were achieved by the use of discharge wires about 0.3 cm (1/8 inch) in diameter or by the use of electrodes with multiple discrete discharge points. However, the recent trend is to use larger, smooth-surface electrodes about 1 cm (3/8 inch) in diameter. These electrodes provide significant performance improvement, especially for high-resistivity dust operation, because of the high field strengths and high useful current densities possible.

ESP plate spacing has been studied by the U.S. utility industry for many

TABLE 5.2. Typical Design Parameters for Fly Ash Precipitators.

Parameter	Range
Gas velocity	75–200 cm/sec
Migration velocity	1.4–15 cm/sec
Specific collecting area	5–45 m² per 1,000 m³/hr
Corona power	30–300 W per 100,000 m³/hr
Corona current	5–75 nano amp/cm²
Plate area	500–7,500 m² per electrical set
Number of high-tension sections in gas flow direction	2–8
Degree of sectionalization	0.23–2.35 high-tension bus sections per 100,000 m³/hr

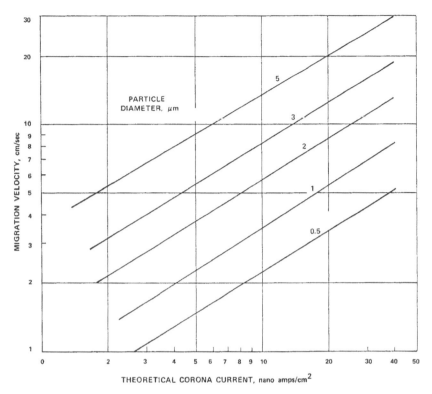

Figure 5.10 Electrostatic precipitator migration velocity as a function of theoretical corona current and particle size.

years. Data from these operations have demonstrated that units with 12″ (30.5-cm) plate spacing out-perform those with 9″ spacing (23 cm) on high-resistivity dusts when electrical sectionalization and corona emitters are equivalent for the two types of systems. Europe and Japan have been using 15″ (38-cm) plate spacing. Wider spacing is preferred for high resistivity material. However, economics may dictate design as units with wider plate spacing cost more because higher voltage electrical equipment is required.

ESP pulse powering is an old concept recently reintroduced. Pulse energization occurs when a high-frequency, high-voltage pulse is superimposed on the ESP base voltage. This generates a more intense, uniformly distributed corona on the discharge electrode which increases the collection efficiency of high-resistivity dusts. Back corona with the subsequent dust reentrainment is also suppressed by this procedure. This procedure is commercially available but would be best used for the retrofit of existing systems which are in noncompliance or for systems which have varying dust resistivity.

5.3 FILTERS

Filters are among the most effective particle collection devices. They can often obtain higher collection efficiencies than any other type of unit. However, the dry fabric filters (baghouses) are limited by temperature, adhesiveness of the dust and hydroscopicity. Like dry precipitators, they usually produce a dry product. There have been a multitude of basic filtering media and filtering arrangements developed recently that permit very high temperature and wet gas filtration. Wet filters a e considered wet scrubbers and charged filters are included with hybrid devices.

Filters consist of fibrous media, felted media, and porous media. For air pollution control, the fibrous a d felted media are usually contained in baghouses. The filter media stops the particles by any or all of impaction, interception, diffusion, settling, and electrical attraction. As the cake of deposited dust builds up, most of the filtering is achieved by the sieving action of the filter cake. For example, under similar filtering conditions one may find very fine particle filter collection efficiencies of

	Light Cloth	Heavy Cloth
New Cloth	2%	30%
Cleaned Cloth	13%	67%
Cloth plus Cake	65%	80%

5.3.1 BAGHOUSES

5.3.1.1 General

Baghouses are the most common air pollution control filtration systems. Fabric filter material, whether fibers or felt, is formed into cylindrical or envelope bags and suspended in the baghouse. The dirty gases pass through the fabric leaving the dirt on either the inside or outside of the bag depending on filtering arrangement. The fabric is then cleaned by knocking the cake free and allowing it to fall into the bottom hoppers.

The permeability of fabric filters varies with the tightness of the cloth. The American Society for Testing and Materials (ASTM) procedure for determining new cloth permeability consists of passing air through the cloth at a 1.27-cm (0.5-in.) water pressure drop. Permeability from 3 to 10 m³/min per m² cloth are typical. This can also be expressed as filter velocity or air-to-cloth ratio, A/C, which is 5 to 17 cm/sec or 5:1 to 17:1 (10–33.5 ft/min) for these clean-cloth permeabilities. During actual filtration through a cloth and cake, the A/C ratios are typically 0.5 to 5 cm/sec (1–10 ft/min).

All sections of a baghouse filter do not operate continuously. During the cleaning cycle, at least a portion of the filter is not operating. The entire filter unit is usually designed in such a manner that filtering can still take place in part of the baghouse while cleaning is taking place in another part. The exception to this is the reverse jet unit which *may* be operated and cleaned at the same time. The four basic baghouses are categorized by the type of cleaning methods. These are the reverse pulse, shaker, reverse flow, and reverse jet baghouses.

5.3.1.2 Reverse Pulse

The reverse pulse is also known as the pulse jet filter. These units were not developed until the late 1960s and have become so popular that they account for well over half of all filter sales. These units have a short but strong pulse of air of about 100 millisecond duration. The cleaning air pulses pass through the top of the bag and into the bag to expand it as shown in Figure 5.11. The filter cake which has built up on the outside of the bag falls into the hopper and is removed. Because filtering occurs on the outside, each bag is supported by an internal cage.

There are two types of pulse cleaning. The older type uses conventional jet cleaning action with 90–120 psi air pulsed through a vertical tube which in turn takes in six to seven times more air than the pressurized air. This results in a weakened pulse which quickly loses cleaning capability. The second type is a low-pressure pulse jet of 30–35 psi air which blows directly into the bags through a header. This technique results in less secondary air being taken in so the pulse is quicker and penetrates deeper into the bag. Both types of pulse cleaning result in more cake being removed from the bags than by other cleaning methods. Because of this, heavier bag fabric of needled felts is used.

Advantages of pulse jets are longer bag life and smaller size. Bag service is extended by the low bag movement during cleaning which causes little wear compared to other cleaning methods. Unit size can be smaller because of higher filtering velocities, closer bag placement, and no out-of-service requirement for cleaning. Pulse jets, with their heavier felt cloths, can operate at 3–4 times higher filtration velocities (e.g., at 4–5 cm/sec). Bag movement during cleaning is small so they can be placed closer together which reduces the size of the unit. The unit can continue to operate

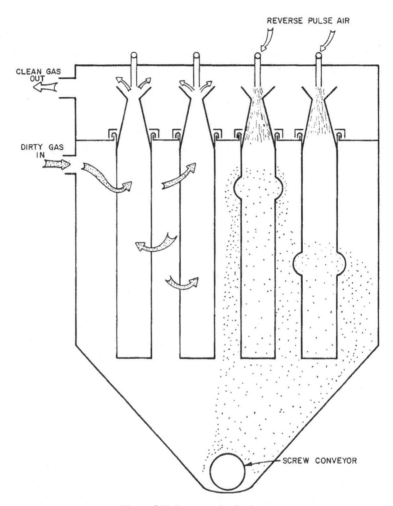

Figure 5.11 Reverse pulse baghouse.

during cleaning as the cleaning pulse is only momentary, so little or no additional filter capacity is required for a continuous operation. The disadvantage is that these filters are less efficient.

5.3.1.3 Shaker

Shaker baghouses are the old reliable method of bag cleaning. They are usually the most efficient type of filter. Dust cleaning is a function of number of shakes, shaking frequency, shaking amplitude, bag resonance frequency, bag tension, dust adhesion, and bag movement acceleration. Typical shaking amplitudes range up to about 5 cm. When the bags are properly tensioned at about 2–20 newtons of force (0.5–5 lb), bag accelerations of over 8 g's result in little additional dust removal. Shaking is periodic on a timed basis and consists of up to several hundred shakes per cleaning. Shaking more than 200–250 times per cleaning cycle does not normally result in significant additional removal of filter cake. This low-energy cleaning process leaves a significant amount of residual dust cake on the filter after cleaning so the filter media is always cake plus cloth. This makes the filtering more efficient and permits a light woven filter cloth to be used. Low air velocities of less than 3 cm/sec are used during the filtering. During shaking, air flow to the filter is stopped or the portion of the baghouse being shaken is isolated. Shaking durations of one minute are common.

5.3.1.4 Reverse Flow

Reverse flow (also known as reverse gas) baghouses are cleaned by forcing the air (or gas) backwards through the bags in a direction opposite that of normal filtering. Some of these systems are like shaker baghouses in that woven bags are used while others are like pulse jets in that felt fabric envelopes are used. Either type have filtering characteristics similar to the shaker filter in that the cleaning is at a lower energy level leaving deposits of cake to do the filtering.

Units that use bags filter on the inside of the woven bags. Reverse flow cleaning air collapses the bags inward. When the bag is brought back on-line it expands and the dislodged cake falls into the hopper through the open end at the bottom of the bag. By contrast, each felt envelope filter element is supported by an internal rigid wire mesh frame and separator and filtering occurs on the outside. This is shown in Figure 5.12. These units operate at 1–3 cm/sec velocities. Compartments are isolated periodically and cleaning air is blown into the open front end of each envelope. It requires about 0.1 to 1 m³ of air per m² of filter area for cleaning.

5.3.1.5 Reverse Jet

This is also known as the traveling ring filter. A jet manifold surrounds each bag and continuously moves up and down the bag. Jets of air from the manifold blow dust off the bag. These filters are similar in filtering characteristics to shaker and reverse flow filters.

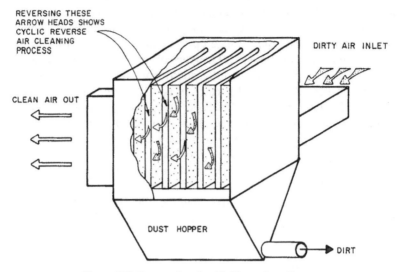

REVERSING THESE
ARROW HEADS SHOWS
CYCLIC REVERSE
AIR CLEANING
PROCESS

DIRTY AIR INLET

CLEAN AIR OUT

DUST HOPPER

DIRT

Figure 5.12 Reverse air unit with felt envelope filters.

5.3.2 PRESSURE DROP

Deposited particles on the fabric cause most of the filter pressure drop. The pressure drop of a new cloth can be noted at startup time for various flowrates. The ratio of terminal pressure drop of a filter before cleaning to new-cloth pressure drop is about 200:1. Pressure drop ratio for before cleaning to after cleaning is typically 100:1.

Total pressure drop equals residual pressure drop of the cleaned cloth plus cake pressure drop. Cake pressure drop buildup over the time interval, t, can be estimated by

$$\Delta P = 3.6 \times 10^3 K_1 v_g^2 C_o t \tag{5.15}$$

ΔP = cake-fabric pressure drop, cm H_2O
K_1 = cake-fabric resistance coefficient, cm $H_2O/[(g\ dust/cm^2)(cm/min)]$
v_g = gas superficial velocity, cm/sec
C_o = inlet particulate concentration, g/cm^3
t = time, min

Although K_1 can be estimated by the Kozeny-Carman procedure [13], it is better to measure values of all terms except K_1 in Equation (5.15) and then calculate an empirical K_1 for the specific dust and filter unit. Values of K_1

depend on the filtering substance. For cake filtration, K_1 depends on particle size according to one variation of the Kozeny-Carman equation

$$K_1 = 7.09 \times 10^{-5} \frac{\mu_g s^2}{\varrho_p} \frac{1 - \epsilon_p}{\epsilon_p^3} \qquad (5.16)$$

where

s = ratio of particle surface area to particle volume in cake layer, cm^{-1}
  ($\cong 6/d$ for spheres barely touching)
ϵ_p = porosity of void volume in cake layer, dimensionless fraction

Values for porosity can be estimated for spherical particles using

$$\epsilon_p = \frac{\ln d_p + 6.14}{15.36} \qquad (5.17)$$

where d_p is particle diameter in microns.

As particle size decreases, the resistance coefficient increases and, therefore, filter pressure drop increases. Experimentally measured values of K_1 obtained from Equation (5.15) might range from about 122 for 0.1-μm particles to 0.03 for 100-μm particles. Equation (5.16) will overpredict for the smaller particles and underpredict for the larger particles.

Cooper and Hampl [14] give a simplified form for finding cake resistance, K_1. This expression for fly ash filtering data gives

$$K_1 = 110.5 \overline{d}^{-1.06} \qquad (5.18)$$

where $\overline{d}$ is mass mean diameter in microns.

An even simpler, rule-of-thumb procedure for coal-fired combustion facilities using fiberglass bags is presented by Carr [15]. This shows that typical U.S. reverse air systems have a tubesheet pressure drop, in inches of water, of three times the air-to-cloth ratio (in units of acfm/ft^2, which is ft/min). This is accurate to $\pm 20\%$ for units operating with bags that are 1–2 years old. Pressure drop is less for bags with <1 years service and greater for >2 years service.

Recent data [16] suggest that for coal-fired filter systems, dust cohesivity is one of the most important ash characteristics—even more important than particle composition, shape, and size. The more cohesive ash produces filter cake with higher porosity and, therefore, lower dust cake pressure drop. Low cohesive ash consists of smoother particles which pack tightly and eventually bleed through the filter fabric.

Example Problem 5.2

Estimate the cycle time and the air-to-cloth ratio needed to prevent the pressure drop increase during the filtering cycle from exceeding 10.5 cm H_2O. Fly ash is being filtered in a reverse air plus shaker unit at 1.4 cm/sec filter velocity. Inlet dust concentration is 4.57×10^{-6} g/cm^3 (2 gr/ft^3) and the particle mass mean diameter is 8 μm. The bags are graphite/silicon-coated fiberglass, 20-cm diameter by 6.7 m long. The coal is low-sulfur bituminous.

Solution

Determine the filter-cake resistance using Equation (5.18)

$$K_1 = (110.5)(8)^{-1.06} = 12.19 \text{ cm } H_2O/[(\text{g dust/cm}^2)(\text{cm/min})]$$

[Equation (5.16) would predict a lower value of K_1 equal to 0.9 assuming unit density and spherical particles.]

The cycle time from Equation (5.15) is

$$t = \frac{(10.5)}{(3.6 \times 10^3)(12.19)(1.4)^2(4.57 \times 10^{-6})} = 26.71 \text{ min}$$

Many of these systems operate with cycle times of about 30 minutes. Determine A/C using Carr's rule-of-thumb

$$\Delta P' = 10.5 \text{ cm} = 4.1''$$

$$= 3 \times (A/C)$$

$$\therefore A/C = \frac{4.1}{3} = 1.4 \text{ ft/min}$$

or 0.7 cm/sec

5.3.3 COLLECTION EFFICIENCY

Filters are among the most efficient particle-collection devices, although they are not amenable to all types of dust cleaning requirements. Wet, sticky dusts can rapidly plug dry filters. Fabric filters in the utility industry have been consistently yielding 99.9% efficiency while ESPs are performing at 99 to 99.8% efficiency. Fuel sulfur and ash metallic oxides concentrations do not affect filter operation. Collection efficiencies up to 99.99% are common with the shaker filters which give higher efficiencies because more residual dust cake remains after cleaning. The woven fabrics used in shaker filters are not good filters by themselves, but the fabric-cake combination is excellent. Concentrations of dust released from shaker filters would be expected to be two to three times lower than from pulse jets under similar filtering conditions. Filter collection efficiencies appear to be relatively constant regardless of inlet concentration variations. Therefore, increases in inlet loadings increase emissions more so than in some other collectors, e.g., ESPs.

Filtration efficiency in any given filter unit depends on the collection mechanisms of gravitational settling, sieving, inertial impaction, interception, and diffusion as noted in Table 5.3. These relationships are for a felt fabric which is thicker and less easily penetrated than a woven filter fabric. Filters usually have a minimum efficiency for particles about 0.4 to 0.6 μm in diameter.

Pulse jet particulate emissions are unusual in that the particles released may be large in size. This is a result of agglomeration and the high cleaning energy, even though collection efficiencies are in the high 99.9– 99.99% range. Individual fractional particle size collection efficiency for a pulse jet has little practical significance because size and number concentration change radically over the filtration cycle due to the high cleaning energy. Long-term averages may be relevant but agglomeration of the effluent due to cleaning can yield misleading information.

Pulse jet collection efficiency for any given unit is a function of inlet concentration, dust characteristics, filter velocities, and the intensity duration and form of the pulse. Tests on fly ash and talc show that outlet concentration is proportional to inlet concentration to the nth power where $n \cong 0.5$ to 1. These tests were run with particles of roughly the same size, size distribution, and shape, using dacron and wool felt fabrics and outlet concentrations of 0.002–0.02 g/m³ (0.01–0.001 gr/ft³). Efficiency is in-

TABLE 5.3. Filtration Efficiency Factors for Needled Felt Fabric.

Decrease	Collection Efficiency Change Due to				
	Gravity	Sieving	Impaction	Interception	Diffusion
Filter velocity	I	NC	D	NC	I
Particle diameter	D	D	D	D	I
Particle density	D	NC	D	NC	I
Filter fiber diameter	NC	I	I	I	I

Note: D = decrease, I = increase, NC = no change.

versely related to filter velocity and in some tests gave up to 5 times *lower* outlet concentrations when velocities were reduced by 25% from high values of about 4.3 cm/sec. Efficiency decreases as more of the cake is removed by cleaning. For example, some tests showed a fivefold outlet concentration increase when cleaning pulse air intensity was increased from about 3 to 7 atmospheres.

Filter efficiencies are best obtained by actual tests using specific dusts, temperatures, fabrics, cleaning type, and cleaning sequence. Mechanical filters, shaker, and reverse air cleaning (except the pulse jet) are similar in collection efficiency. Often, several cleaning methods are combined. Table 5.4 shows measured particle collection efficiencies of specific particle sizes for several mechanical filters. Note that filters typically show, as these data do, a decreasing efficiency with decreasing size. Below a critical size, efficiency increases with decreasing size. This is due to the dominant removal of larger particles by impaction, interception and sedimentation and the dominant removal of the very small particles by diffusion.

Seepage also exists in filters. This is the movement of cake particles through the cake resulting ultimately in their escape. Properties of the dust such as size, shape, and cohesiveness as well as the operating conditions of filter velocity and type of fabric all influence amount of seepage. As seepage increases, efficiency decreases.

5.3.4 DESIGN AND OPERATING PARAMETERS

The filter must be operated dry and at a temperature sufficiently low to protect the filter fabric. Effect of temperature is the most important parameter in selection of a fabric. In addition, the fabric must also provide acceptable physical resistance to the mechanical action of the cleaning method (i.e., it must resist flex breakage) and to the abrasiveness of the particulate gas stream.

Other factors to consider in choosing a fabric include dust release and static-generation tendency. Coated and conductive fabrics have been developed for these purposes. Static generation can cause detonations in explosive combinations of dust and air as noted in Chapter 1.

Bag size varies with type of system. Reverse air collectors are mostly 11 1/2″ in diameter (29 cm) and are 30 to 34 feet long (9–10 m). One vendor uses bags 8″ in diameter (16 cm) by 22 ft (6.7 m) long. Filtration is on the inside of the bags and they are cleaned by collapsing them. To prevent complete closure of these bags, anti-collapse support rings are sewn into the bag at specific intervals. Starting at the bottom of the bag, the rings are installed at: 2 diameters up; 4 1/2 diameters up; 7 1/2 diameters up; and 11 1/2 diameters up. The typical U.S. bag length-to-diameter ratio as noted

TABLE 5.4. Mechanical Filter Efficiency [17].

Baler	Filtration Vel. (cm/sec)	Filter Collection Efficiency, Mass % of Particles						Filter Material	Cleaning Method
		Overall	<0.3 μm	0.5 μm	1 μm	2 μm	4 μm		
Pulverized	1	99.8	94.5	99.8	99.7	99.7		Glass/Teflon® [a]	Reverse air
Coker (high dust concentration)	1.5	99.8	99.9	99.2	99.4	99.6	99.6	Nomex	Reverse air
Coker (low dust concentration)	2.1	99.9	98*	98.9	98.3	99.3	99.3	Glass/Graphite	Reverse air plus shake

[a]Registered trademark of E. I du Pont de Nemours and Company, Inc., Wilmington, DE.

here is 30 to 33. European designs use a shorter ratio of about 16. This results in shorter and fatter bags with lower entrance gas velocities and possibly reduced fabric abrasion at the bag entrance. European reverse air filter velocities are about 2–3 cm/sec (4–6 ft/min) compared to U.S. reverse air and shaker velocities of about 1.25–1.5 cm/sec (2.5–3 ft/min).

Cage-type collector systems, such as the pulse jet, use bags from 4 1/2 to 6″ diameter (9–12 cm) by 8 to 19 ft long (2.4–5.8 m). Most coal-fired filter bags average 14 to 16 ft long.

Filter bags should be kept under proper tension, relative to bag diameter. This tension is estimated as 1.8 lb/in of circumference. Therefore, 8″ diameter bags would have a 45 lb tension and 11 1/2″ bags would have 65 lb tension. Adequate tension keeps the bags from rubbing on each other and other surfaces, from bending too much, and from closing and preventing proper cleaning. Too much tension would result in premature bag failure due to pulling apart.

Typical air-to-cloth ratios (filtration velocities) vary with both fabric construction and type of fabric. In general, woven fabrics operate at lower velocities (<3 cm/sec; <6 ft/min) to prevent driving the dust through the light material. Felts operate at higher velocities because the material is thicker and the dust does not penetrate as easily. Felts provide a greater filtration depth. Recommended air-to-cloth ratios (filtration velocities) may be

glass-woven	1:1 cm³/sec per cm²	(1 cm/sec; 2 ft/min)
Nomex-woven	1.5:1 cm³/sec per cm²	(1.5 cm/sec; 3 ft/min)
Nomex-felt	4:1 cm³/sec per cm²	(4 cm/sec; 7.9 ft/min)

Table 5.5 is presented to show typical filter units and fabrics used in the U.S. Other types of filter and fabric are used on the processes noted and the ranges of emission concentrations and operating conditions extend beyond the values noted in the table. Fabric filters should not be operated too hot, yet in some of these processes (such as municipal incinerators and boilers) it is important to keep the filter hot enough to keep it dry. The acid dew point of undiluted municipal effluent gas can be as high as 120–140°C.

A partial listing of some filter materials and their limitations is given in Table 5.6. New fabrics and fabric combinations are being developed for specific purposes. Some of the available and proven ones are listed in Table 5.7. Fiberglass fibers are unique in that all fibers are cylindrical and relatively straight and uniform in size. They can be made in extremely fine diameters. The combination of glass and Nomex, called Glamex, combines the uniform, small diameter glass fibers with Nomex to produce a felt fabric which is a more efficient filter than the plain Nomex felt. The advantage of smaller felt fibers can be seen from Table 5.3.

Table 5.5. Typical Filtration Units for Several Industries [14,16,18].

Industry	Filter Type	Filtration Velocity (cm/sec)	Filter Cloth(s)	Length-to Diam. Ratio	Filter ΔP (cm H$_2$O)	Approximate Temperature (°C)	Dust Concentration (g/m³)	
							Inlet	Emission Limit, Large New Source
Asphalt	Pulse jet	2.8–3.1	Felt-Nomex, Glass/Nomex		13–15	110–130		0.23
Cement	Reverse air	1.5	Felt-Nomex		10–15	130–190	2–20	0.23
Carbon black	Reverse air	0.5–1	Silicone, Teflon, glass	30	15–25	150–190	0.1–0.3	0.05
Cupola (foundry)	Reverse air	1	Silicone/glass, Nomex	23.5	8–15	290	2–4.6	0.115
Municipal incinerator	Reverse air	1	Felt-Teflon	30.6	2–3	230	<1	0.115
Glass furnace	Reverse air	1	Glass		13	200–290		0.46
Utility boiler	Reverse air-shake	1.5	Nomex, Teflon coated fiberglass, graphite, silicon	30	7.5–13	150–180	9	0.115

TABLE 5.6. Comparative Table of Filter Fabric Properties [19].

Material	Temp. Limit for Continuous Dry Heat (°F)	(°C)	Strength	Resistance to Acid	Alkali	Organic Solvents	Flex and Abrasion
Carbon (graphitized)	500	260	P	F	G	G	P
Cotton	176	80	G	P	F	G	G
Olefin-treated cotton	248	120	E	G	G	G	G
Dacron polyester	240	132	E	G	G	E	G
Dynel	212	100	G	E	E	F	G
Glamex	428	220	G	G	F	E	G
Glass (spun-yarn type)	752	400	F	E	P	E	P
Glass (continuous filament silicone treated)	500	260	F	G	F	E	F
Nylon ("Nomex")	400	204	E	F[a]	G	E	E
Nylon (6,6)	301	94	E	P	G	E	E
Orlon acrylic	270	135	G	G	F	E	G
Paper	176	80	P	P	F	G	P
Polyethylene	248	120	G	G	G	E	G
Polypropylene	221	105	E	E	E	G	E
Polyvinyl acetate	248	120	G	G	G	P	G
Stainless steel	842	450	G	E	E	E	G
Steel	806	430	G	F	F	F	G
Teflon	500	260	G	E	E	E	P
Wool	212	100	G	G	F	G	G

TABLE 5.6. (continued).

Material	Dust Release	Flame Resistance	Electrical Specific Resistance (ohms/cm)	Relative Static Genera- tion [11]	Cost Ratio to Cottonᵇ
Carbon (graphitized)	G	P		—	30.0
Cotton	F	P		+6	1.0
Olefin-treated cotton	G	G		—	2.7
Dacron polyester	E	E		0	2.7
Dynel	G	E		−4	3.2
Glamex	G	E		—	—
Glass (spun- yarn type)	F	E		—	6.0
Glass (continuous filament silicone treated)	E	E		+15	2.2
Nylon ("Nomex")	E	E	10^{16}	+10	8.5
Nylon (6,6)	E	E	4×10^{14}	+10	2.5
Orlon acrylic	G	P	5×10^8	+4	2.7
Paper	F	P		—	0.5
Polyethylene	G	P		−20	—
Polypropylene	G	P	10^{18}	−13	1.5
Polyvinyl acetate	E	P		—	—
Stainless steel	G	E	∼0	—	—
Steel	G	E		−10	—
Teflon	G	E	10^{18}	—	25.0
Wool	F	P		+11 to +20	3.7

ᵃIf temperature remains above acid dew point.
ᵇVaries with type of weave and weight of fiber. E = excellent, G = good, F = fair, P = poor.

TABLE 5.7. Recently Developed Filter Fabrics and
Relative Costs.

Product	Cost, Relative to Cotton
Woven glass	6
Woven homopolymer acrylic Dralon T	7
Permaguard	20
Ryton polyphenylene sulfide	24
Woven Teflon	36
Needled felt glass	36
Gore-Tex	50
Polybenzimidazole	80
Needled Teflon	85
Woven Nextel 312	150
Stainless steel needled felt	400

Most filtration devices are limited to temperatures below 300°C because of the fabric. Development of ceramic, metal, mineral, and other specialty filter material can extend filtration temperatures to 1,200°C. These filters include fibers of zirconia-silica, alumina-boria-silica, asbestos, boron nitride, and sintered chrome-nickel mats. Alumina, magnesia, and other ceramic honeycomb filters can also be used at very high temperatures. These materials are not fibers but are porous structures. The pores are one to several micrometers in diameter.

5.4 SCRUBBERS

Scrubbers are more universal control devices in that they can control either or both particulate and gaseous pollutants. There are numerous types of devices called scrubbers. These include wet scrubbers, wet-dry scrubbers, and dry-dry scrubbers. Scrubbers are a generic group of devices that use chemicals to accomplish the pollutant removal and the pollutants initially may be in either or both of gaseous or particulate form. In scrubbers, the gaseous pollutants are either absorbed or converted to particles. Then the absorbing liquid and all of the particulates must be separated and removed from the carrier gases. Wet scrubbers use liquids (usually water) for particle removal, and dissolved or suspended alkali material to react with absorbed acid gases. Oxidizing chemicals can be used for controlling organic pollutants. Dry scrubbers use alkali to combine with acid gases in gas-liquid or gas-solid phase reaction to form particles. The following sections discuss scrubbers (wet, wet-dry, and dry-dry) and some of the chemicals used. Although gas absorption in wet scrubbers occurs with particle removal, it will be covered alone in Section 5.7.

5.5 WET SCRUBBERS

Wet scrubbers cannot only control both particle and gaseous pollutants but they have the advantages of being able to handle hot gases, emissive particles, sticky particles, and liquids. Disadvantages are that they require more operating energy, discharge wet gases, and produce a wet product. A general method of categorizing is to call particulate scrubbers high-energy devices and gas absorbers low-energy devices, although gases and particulates are removed in both types. Wet scrubbers serve to quench or condition emission gas streams, and it is not uncommon to use prescrubbers for this purpose before passing the gas stream on to other scrubbers and collection devices. This is covered in Section 5.10.

Wet scrubbers are unique in that they can help to control emissions of toxic metals and metal compounds from incineration systems. Many toxic metals and metal compounds have vapor pressure above one atmosphere at incineration temperatures. Examples of materials completely volatilized at 1850°F (1010°C) are mercury, arsenic, and arsenic chloride. Hot spots that reach 2550°F (1400°C) would volatilize PbO during the combustion process. Wet scrubbers cause these materials to condense so they can be removed with the other particulates.

Wet scrubbers remove particulate matter directly by one or more of the particle collection mechanisms. Some examples are: inertial impaction—venturi; scavenging—spray tower; and centrifugal force—catenary grid. High-energy particulate scrubbers are so named because energy in the gas stream is used to accelerate the gas and the particles to velocities sufficiently high enough to achieve effective inertial impaction. An exception is the steam ejector type which uses steam ejection as the driving force. Good inertial impaction in wet scrubbers can be achieved with as little as 4″ (10 cm) water gauge pressure drop from some of the newer designs.

Inertial impaction is a primary collection mechanism but it is effective only on particles larger than about 0.5 μm in size. Remember that this is actually a centrifugal deposition. Particle collection is enhanced by agglomeration or condensation of liquid on a particle. (Gas conditioning is covered in Section 5.10.) Interception aids in the removal of particles by inertial forces. Gravitational settling is significant only for very large particles. Centrifugal deposition in a scrubber would be effective for >5 μm particles only if the device is constructed to take advantage of centrifugal force. Brownian or thermal diffusion of <0.3 μm particles is desirable, but many scrubbers have short holdup times and are not effective diffusional collectors. Use of phoretic forces, especially diffusiophoresis including Stephan flow, is extremely important in aiding in the removal of the <0.3 μm particles. Conditioning the inlet gases before the scrubber to achieve positive phoretic forces or at least to eliminate negative phoretic forces is very important, as noted in Section 5.10. Electrification of either or both the particles and the scrubbant liquid can be beneficial and is included with the hybrid devices.

5.5.1 ATOMIZATION

Many wet scrubbers utilize atomized liquid droplets as the collection targets for particle removal. A few scrubbers operate by passing the particles into a body of liquid or foam or onto a wetted collection surface. Atomized liquid droplets provide targets for particle collection and large surface areas for absorption. Liquids can be atomized by at least five different procedures: (1) spray nozzles, (2) pneumatic two-fluid atomization, (3) spinning disks or cups, (4) impingement, and (5) energy coupling. The names of the various atomization procedures simply indicate the physical mechanism by which the liquid is broken up.

Spray nozzles are designed to mechanically fracture the liquid by forcing it through a small nozzle under pressure. Some commercial nozzles are constructed to provide a centrifugal or swirl motion to the liquid which, when added to the pressure and flow forces, results in a greater total fracturing force. The liquid leaving the nozzle flies apart as it expands into the low-pressure region. Droplet direction depends on the configuration of the nozzle. Figure 5.13(a) shows a full cone spray nozzle containing internal offset, removable vanes for centrifugal swirl. This type of unit is recommended for gas scrubbing devices. The nozzles in Figures 5.13(b) and 5.13(c) produce flat spray patterns with the first spray being projected straight forward (in line) and the second spray deflected at angles from 17–84°. The (b) and (c) sprays are useful for washing down and keeping the surface of collectors free from dust buildup. Type (a) is good for providing large quantities of scrubbing spray droplets and will pass high concentrations of solids in solution.

Fogging nozzles are useful for conditioning gas streams by adding very small liquid droplets that evaporate to cool and partially saturate the gas. Nozzle (d) in Figure 5.13 produces droplets that are over 90% $<1{,}000$ μm at pressures of one-third atmosphere and greater. Nozzles such as the one shown as (d) are abrasion-resistant to slurry passage and can be obtained

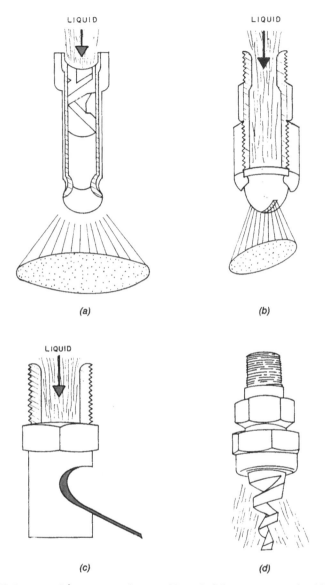

Figure 5.13 Some atomizing spray nozzles: (a) wide angle full cone spray nozzle with internal vanes; (b) in-line flat spray nozzle; (c) deflected flat spray nozzle; (d) fogging nozzle.

as an integral unit or as several pieces so that only a worn section need be replaced.

Pneumatic two-fluid atomization is accomplished by contacting one fluid stream (usually the liquid) with a high-speed jet of a second fluid (usually a gas). The kinetic energy of the high-speed stream is sufficient to overcome the surface force (surface tension) of the slow-moving liquid causing the liquid to fracture. The result is that finely divided droplets of liquid can be obtained in the gas stream by this mechanism. This type of atomization differs from the other types in that the material to be removed by the atomized droplets is contained in the atomizing fluid. The initial relative velocity difference between the particles, which are assumed to move at the velocity of the gas, and the atomized droplets is large. This velocity difference quickly decreases to zero as the targets are accelerated to the gas velocity. This acceleration is complete within just a few centimeters distance.

Several variations of throat spray atomization as commonly used in venturi scrubbers are presented in Figure 5.14. The wet approach shown in (a) provides a continuous flush on the inlet converger and helps keep it clean and prevent abrasion. This is a common old-style absorber scrubber. Atomization at the throat provides the greatest velocity difference between the droplets and the particles. Stationary or adjustable disks may be used at the throats of these high-velocity venturi scrubbers as shown in (b). The scrubbing liquid is atomized when it reaches the edge of the disk. Sometimes these units have deflecting cones over the disk to prevent breakup of liquid on top of the disk. Injection of spray liquid at the throat as shown in (c) minimizes the liquid forward velocity prior to atomization for maximum impaction. These systems have large open lines which reduce pluggage by solids in the liquid slurry.

Spinning disks, cups, arms, or other related devices are used to physically fracture liquid by discharging it at high velocities from the periphery of a rapidly rotating device. The disks are usually flat or curved plates and either solid or perforated. Rotating arms are not truly spinning disks, but are a combination of nozzles (drilled holes) in pipes that are mechanically rotated to help break up and distribute the spray. The centrifugal force developed by a spinning disk is sufficient to overcome the surface tension of the liquid flowing across the disk, causing it to break up into small droplets which are flung from the edges of the disk in a 360° arc.

Impingement is sometimes used to produce atomization by impinging two liquid jets upon each other or by forcing a single jet to impinge on a solid surface. Atomization by energy coupling is possible but the mechanism is not as well defined as it should be. The most common types of en-

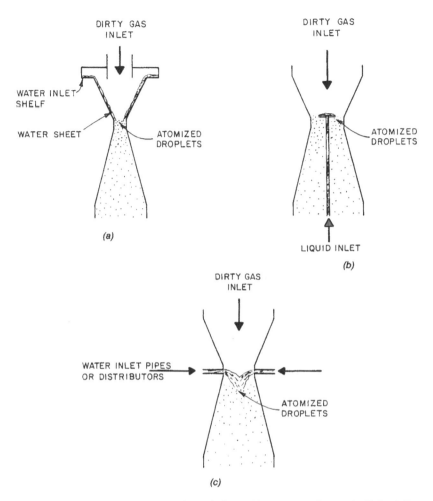

Figure 5.14 Throat pneumatic atomization techniques: (a) wet approach venturi; (b) flood disc venturi; (c) throat inlet venturi.

ergy used for this are sonic (supersonic and subsonic) vibrations and high-voltage electrical energy.

5.5.1.1 Atomized Droplet Sizes

It is possible to predict the size of the droplets created by atomization. Empirical data available in manufacturer's literature is useful for predicting the size of droplets produced from specific types of spray nozzles and rotating devices. These data are easily available in tabular form showing spray patterns, spray coverage, quantity of liquid consumed, and droplet size for specific nozzles, liquid pressures, and gas flow and pressures when used.

At least two different types of atomization can be made to occur by the pneumatic two-fluid atomization procedure. The first, called drop-type atomization, results when the liquid is introduced into the high-velocity gas stream from small diameter (< 1 mm ID) inlet nozzles, or then the liquid is introduced into the gas stream in drop form (for example, as from spray nozzles). Cloud-type atomization results in the formation of much smaller droplets, all of which appear to be less than 30 microns in diameter. Cloud-type atomization is reported by Hesketh [20] and verified by Behie [21].

Drop-type atomization has been studied extensively for many years. The most popular and thoroughly researched equation for predicting average droplet size produced by drop-type atomization is the Nukiyama-Tanasawa equation [22] which is

$$\overline{D}_x = \frac{585}{v_g' - v_l'} \left(\frac{\Upsilon}{\varrho_l}\right)^{0.50} + \left(\frac{\mu_l}{\Upsilon \varrho_l}\right)^{0.45} \left(\frac{1,000 Q_l}{Q_g}\right)^{1.5} \quad (5.19)$$

where

$\overline{D}_x$ = Sauter mean drop diameter (volume-to-surface ratio), μm
v_g' = velocity of gas, m/sec
v_l' = velocity of liquid, m/sec
ϱ_l = density of liquid, g/cm³
μ_l = viscosity of liquid, g/(cm sec)
Q_l = liquid volumetric flowrate, m³/sec
Q_g = gas volumetric flowrate, m³/sec
Υ = surface tension, erg/cm²

This equation simplified for air–water systems near normal temperatures and pressures and when liquid velocities are low is

$$\overline{D}_x = \frac{5 \times 10^5}{v_a} + 29.6(L_G)^{1.5} \quad (5.20)$$

where

v_a = velocity of air, cm/sec
L_G = ratio of liquid to gas flowrate, l/m³

As an example, the Sauter mean drop diameter of drop-type atomized water predicted by this equation is 290 μm when the air velocity is 4,575 cm/sec (150 ft/sec) and the liquid-to-gas ratio is 3.35 l/m³ (25 gal/1,000 ft³). It can be seen from Equation (5.20) that *both* increased air velocity and decreased quantity of liquid produce smaller atomized droplets by drop-type atomization.

Under the same conditions, if cloud-type atomization had been used (larger inlet nozzles), the mean diameter is estimated to be about 10 microns. The very small droplets formed by cloud-type atomization join together by hydrostatic force without coalescing to form clouds that move as a single system. These cloud systems then act like droplets which have much larger effective diameters. Using atomized droplet, size, velocity, and acceleration observations, Hesketh [23] shows that the $< 10\text{-}\mu$m droplets form clouds with effective diameters from 170 μm when the atomizing air velocity is 4,575 cm/sec (150 ft/sec) to 500 μm when the air velocity is 8,845 cm/sec (290 ft/sec).

5.5.1.2 Gas Velocity and Liquid Nozzle ID

Critical minimum velocities of the gas stream are required for satisfactory pneumatic atomization. Hesketh [23] presents the following equation for predicting critical velocities for cloud-type atomization

$$v_{g(\text{crit})} = 1.7 \left[\left(\frac{8 \times 10^6}{d_n} \right)^{1/2} + 460 \right] \qquad (5.21)$$

where

$v_{g(\text{crit})}$ = critical minimum velocity of gas stream, cm/sec
d_n = liquid inlet nozzle ID, mm

This equation is good for nozzles with an ID of approximately 1 mm or larger because below 1 mm ID drop-type atomization occurs. This equation cannot be extrapolated for the drop-type atomization critical velocities. Velocities above about 4,575 cm/sec (150 ft/sec) are the critical minimum velocity needed to obtain full atomization for both types of atomization at all nozzle diameters. The last term in the empirical Equation (5.21) is obtained from the fact that liquid streams are moving 460 cm/sec (15.3 ft/sec) at the time when most of the atomization occurs.

Pneumatic atomization cannot be effective if the liquid introduced is not projected into the gas stream, and then it will be atomized only when the gas velocities are above critical. If the liquid nozzles are too large, the liquid is not injected far enough. When this happens, the gas escapes through the open area and pushes most of the liquid streams against the walls where they are lost.

5.5.1.3 Atomization Efficiency

Very little of the available kinetic energy in high-velocity gas streams is actually utilized for atomization. Hesketh calculates that the atomization efficiency based on work to form new surface area is approximately 6.1% for cloud-type atomization of water by air. Marshall [24] shows that typical drop-type or pressure-nozzle atomization efficiency is 0.53% for water and approximately 0.17% for organic liquids with low surface tension (about one-third of water) regardless of the liquid viscosity. These efficiencies are calculated directly from the diameters of the droplets produced and indicate only the intensiveness of the atomization. The extent of atomization is unknown for both types, so it is generally assumed that all the liquid is atomized (which is not true).

Pneumatic atomization may consume several times the energy required for spray nozzle or spinning disk atomization depending on amount and pressure of the gas used. However, pneumatic atomization is the only method that can handle large quantities of liquids and liquid slurries for many scrubbing devices.

Some spinning disk atomizers utilize small amounts of air in addition to the mechanical force. Green [25] shows that small droplets are produced using 125 liters per minute of air at 1 cm³/min liquid rates in a spinning disk. Droplet size increased to 100 μm when air rate was reduced to 11 liters per minute.

5.5.1.4 Atomization Pressure Drop

Pressure drop or energy to produce atomization by pneumatic atomization in a wet scrubber is not significant. Most of the energy is expended in accelerating the gas and particles so that good impaction is achieved. This energy requirement is covered in Section 5.5.4. This pressure drop energy is considered the gas-phase energy requirement.

Atomization energy is required of the liquid phase to produce mechanically atomized droplets. For pressure spray nozzles, the nozzle pressure drop is added to the other liquid phase pressure drops resulting from liquid head, total friction losses, and flow. Total liquid pumping power input requirements assuming 85% efficiency for the motor-pump combination, can be shown by a mechanical energy balance for noncompressible fluids to be in metric units and in English units

$$kW_p = 2 \times 10^{-3}(\Delta p)(q) \qquad (hp)_p = 6.86 \times 10^{-4}(\Delta p)(q)$$

$$(5.22a)$$

where where

kW_p = input power to pump, kW $(hp)_p$ = input power to pump, hp

Δp = pressure drop across pump, atm Δp = pressure drop across pump, psig

q = liquid flowrate, l/min q = liquid flowrate, gpm

Using the 0.53% efficiency for pressure nozzle atomization given by Marshall [24], the theoretical power *actually* delivered to the scrubber for particle collection would be

$$kW_N = 8.9 \times 10^{-6}(\Delta p_N)(q) \qquad (5.22b)$$

where

kW_N = *actual* power delivered to scrubber for particle collection, kW

Δp_N = pressure drop across nozzle, atm

In terms of contacting power, the *liquid-phase contacting power* is obtained using Equation (5.22a) and the spray nozzle pressure drop.

Relative to pumping scrubbing-liquid slurries, velocities should be kept low to prevent eroding the pump and the slurry lines. For "thin" liquid slurries (up to about 25% solids) this would be

pump inlet = 1–3 ft/sec (0.3–1 m/sec)

pump discharge and process lines = 4–8 ft/sec (1.2–2.5 m/sec)

With these values in mind, one can calculate economic pipe diameter, D'_e, in English units by

$$D'_e = \frac{0.098\, W_s^{0.45}}{\varrho_s 0.31} \qquad (5.23)$$

where

D'_e = actual inside diameter, inches

W_s = slurry mass flow rate, lb/hr
ϱ_s = slurry density, lb/ft³

Example Problem 5.3

Determine the percentage of station power required to pump the scrubbing liquid slurry and the contacting power contribution of the liquid phase for particle collection if all the liquid passes through pressure spray nozzles. Pressure drop across the pumps is 2 atm (~30 psig) and across the nozzles is 1 atm (~15 psig). The scrubber uses 3.35 l of liquid per m³ of gas (~25 gal/1,000 ft³) and the liquid density is 1.2 g/cm³.

Solution

Typical utility emissions would amount to about 491,000 Am³/hr at 150°C per 100 MW of station capacity. Slurry flowrate per 100 MW is

$$q = \frac{(3.35 \ l/m^3)(491,000 \ m^3/hr)}{60 \ min/hr} = 27,400 \ l/min$$

Total pumping power is found using Equation (5.22a)

$$kW_p = 2 \times 10^{-3}(2)(27,400) = 110 \ kW$$

This is about 0.11% of the station power.

Liquid-phase contacting power to the scrubber would be half the total power for the pressure drops given or 55 kW or 0.06% of station power. Note that actual power delivered to the scrubber by the liquid for particle scrubbing as determined by Equation (5.22b) would be only

$$kW_N = (8.9 \times 10^{-6})(1)(27,400) = 0.24 \ kW$$

which is about 0.22% of the power delivered to the pump.

5.5.2 GAS ATOMIZED SPRAY—VENTURI

Gas atomized spray or pneumatic atomization wet scrubbers are particulate scrubbers with gas absorption capacity. These scrubbers are used as the primary example for wet scrubbers. The venturi scrubbers are the most common of this type and are used as the discussion example, although orifice scrubbers and other designs can be obtained. The accelerated gas stream provides most of the atomization energy. Spray nozzles are included to clean critical areas. These devices operate concurrently with respect to gas and scrubbing droplet motion. Pressure drops across these scrubbers range from about 5 to 350 cm H_2O. Gas absorption occurs at the lower contact velocities and more particle collection occurs at the higher. A good middle range that is effective for both collection mechanisms is 15 to 50 cm H_2O pressure drop.

Venturi scrubbers can be quite varied in configuration and as such have no true specifications as do venturi flow meters (e.g., standard flow meters have a 4:1 ratio of inlet-to-throat cross-section area, a 12.5° converging angle, and a 3.5° diverging angle). The gentle slopes in flow meters provide low permanent pressure loss. Venturi scrubbers may have converging and diverging sections, but usually the angles are greater to keep down the length. The throat-area reduction can also be achieved by placing obstacles, such as bundles of rods, in the throat of the scrubber. These venturi have straight sides as shown in Figure 5.15 and operate in a reverse manner in that the dirty gas enters at the bottom. Scrubbing liquid slurry is introduced through open pipes at the top. Some rod venturi have an increasing number concentration of rods toward the throat, then a decreasing number concentration beyond the throat to simulate converging/diverging sections.

A high throat gas velocity is desired yet many processes operate with varying inlet velocities. It is possible to use several scrubbers in parallel and to shut off some scrubbers to maintain desired flow in the others. Another procedure consists of using a variable throat opening. Scrubbers constructed in a conventional venturi converging/diverging style do this with movable plates, plumb bobs, and annular disks as shown in Figure 5.16. These venturi scrubbers show the more conventional shape. Gas and liquid enter at the top. The liquid enters as shown by Figure 5.14 or by any combination of these methods plus any additional spray nozzles. Many variations of venturi scrubbers exist. Some units have only one adjusting throat baffle and cylindrical systems may use an expandable rubber doughnut. Plumb bob and orifice plates may be adjusted from the top or bottom, but preference is from the bottom. The top of the plate is flushed with liquid. In the annular orifice scrubber shown as Figure 5.16(c), most of the scrubbing liquid is injected at the orifice.

Scrubbers saturate the inlet gases and cool them to the adiabatic saturation temperature. This occurs within a few centimeters from the point of scrubbing liquid injection for low-temperature gases, but takes a considerably longer time (distance) for higher-temperature dry gases. The saturation water leaves with the scrubbed gases, but the bulk of the scrubbing liquid is entrained as liquid and must be separated from the gas. This mist elimination stage is considered part of the scrubbing unit.

Good atomization is mandatory for successful scrubber operation. Ve-

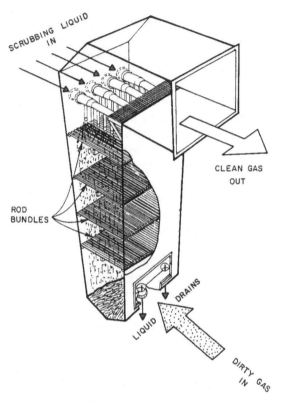

Figure 5.15 Venturi-type scrubber with throat-area reduction achieved using bundles of freely rotating rods.

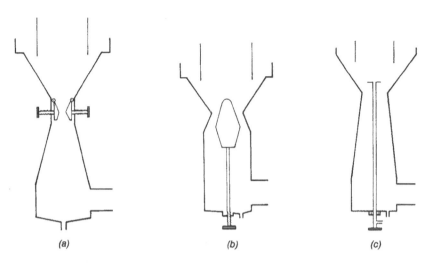

Figure 5.16 Throat-area adjustment mechanisms for conventional venturi scrubbers: (a) baffles; (b) plumb bob; (c) annular orifice.

locities of >4,575 cm/sec (150 ft/sec) have been noted as necessary for complete atomization, yet velocities of half this give effective particle removal and high gas absorption. It is assumed that the scrubbing liquid is injected in such a manner and quantity so as to assure that the throat is continuously permeated with atomized droplets. The maximum size throat inlet nozzles would be about one-ninth the scrubber throat diameter (or width). Throat nozzles should be evenly spaced around the throat and directed toward the center for inward-directed liquid feed. Conventional venturi scrubbing units require a *minimum* of about 0.7 l/m³ (5 gal/1,000 ft³) of liquid to provide for losses due to ineffective atomization and to cover the throat with a minimum of scrubbing droplets.

5.5.3 CUT DIAMETER

Well-designed, operated, and maintained venturi scrubbers are capable of high particle-removal efficiencies. Aerodynamic cut diameters approaching 0.2 μm can be obtained. This means that by mass, half of the 0.2-μm particles are captured and half pass through. Less than half of the smaller particles are captured and more than half of the larger particles are captured. Penetration of particles through the scrubber is the opposite of collection efficiency. Overall fractional penetration, Pt_o, equals one minus overall fractional efficiency, ϵ_o. Aerodynamic cut diameter, d_{50}, is related to penetration but depends on the size distribution of the particulate matter and the collecting device. Calvert et al. [26] have developed a plot of penetration and the ratio cut diameter divided by aerodynamic mass mean diameter, $\bar{d}_a$, for venturi scrubbers. This is given as Figure 5.17.

The lower the aerodynamic cut diameter, the more effective particle removal is for a particular system. However, penetration, cut diameter, mean diameter, and size distribution are all related to the overall operating effectiveness. This is useful for comparing various devices and is shown by the following example.

Example Problem 5.4

Compare the following data for three venturi scrubbers and rank each device in order of particle collection effectiveness.

| | Dust | | Overall Fractional |
Scrubber	$\bar{d}_a$, μm	σ_g	Efficiency, ϵ_o
1	1.0	2.0	0.92
2	1.2	1.5	0.99
3	0.9	2.5	0.94

Solution

Assuming all data are valid and using Figure 5.17, which applies only to venturi scrubbers, the ratios of $d_{50}/\bar{d}_a$ for scrubbers 1, 2, and 3, respectively, are approximately 0.30, 0.24, and 0.20. The aerodynamic cut diameters in respective order are then 0.30, 0.29, and 0.18. It is concluded that scrubber 3 is best and scrubbers 1 and 2 are very similar in effectiveness.

5.5.4 CONTACTING POWER

A common expression is "You get what you pay for." To some extent, this is true of wet scrubbers in that the scrubber particle collection efficiency for a specific particulate matter is related to the energy expended in the gas-liquid contacting process. That is, collection efficiency is a function of total power dissipated. However, scrubber collection is also dependent on scrubber geometry, scrubber size, and the manner in which contacting occurs.

Gas-liquid contacting energy is supplied by gas stream energy, liquid stream energy, mechanical energy, and combinations of these. In addition, there is the transfer of thermal energy and mass as related to application of phoretic forces. Most scrubbers use all of these mechanisms for con-

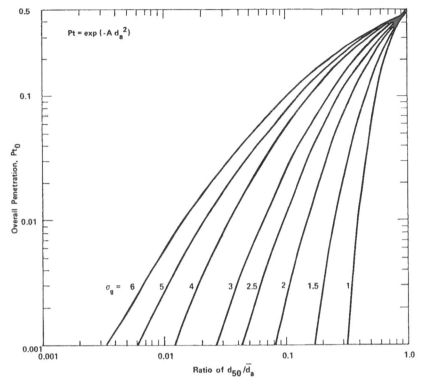

Figure 5.17 Venturi-type scrubber overall penetration versus the ratio of aerodynamic cut diameter to aerodynamic mass mean diameter as a function of σ_g for log-normally distributed particles [26].

tacting but have one dominant source of energy. For example, most venturis use predominantly gas stream energy, but also use liquid stream energy from spray nozzles. High-velocity preformed spray and ejector jet scrubbers utilize significantly more liquid stream energy. Centrifugal scrubbers use higher amounts of mechanical energy. Gas phase contacting power appears to be the most effective for particle removal.

Contacting power as used here does not represent the true energy supplied to the scrubber but includes energy supplied to overcome inefficiencies in pumps, blowers, atomization devices, and other mechanical losses. Liquid-phase contacting power is shown by Example Problem 5.3. Gas-phase contacting power is considered to be the friction loss across the wetted scrubber when energy is from the gas stream. This would equal the venturi inlet-outlet pressure drop *when* inlet and outlet measurements are made across equal size ducts. In an ejector device, contacting power is the kinetic energy rate of the liquid at the nozzle which equals volumetric flowrate times upstream pressure. Mechanical contacting power would be motor input power. The energy consumed by the scrubbing device is dissipated in fluid friction and turbulence and is the same whether particles are present in the gas stream or not.

Semrau [27] expresses fractional overall collection efficiency, ϵ_o, for a scrubber as

$$\epsilon_o = 1 - \exp - [f(P_T)] \qquad (5.24a)$$

where $f(P_T)$ means a function of total contacting power, P_T. Rearranging this equation gives

$$f(P_T) = - \ln (1 - \epsilon_o) = \ln \left(\frac{1}{1 - \epsilon_o} \right) \qquad (5.24b)$$

Semrau plots $\ln [1/(1 - \epsilon_o)]$ versus gas stream energy, ΔP, for a gas atomizing spray scrubber. These data show that liquid-to-gas ratio alone has no significant effect on collector effectiveness. The relationship between $f(P_T)$ and $\ln [1/(1 - \epsilon_o)]$ is characteristic of the aerosol only. For example, it is shown by Semrau that

$$f(P_T) = AP_T^B$$

where A and B are empirical constants for the given particulate matter. In a series collection device, the numerical values of A and B change as particles are removed at each stage.

Calvert [28] extends the contacting-power theory by using the cut-diameter concept to show that there are differences in energy consumption and scrubbing effectiveness for different types of wet scrubbers. Figure 5.18 shows the Calvert data supplemented with data for the Ceilcote ionizing wet scrubber (obtained from the October 30, 1980 McIlvaine Wet Scrubber Newsletter) and the Calvert collision scrubber (obtained from Calvert Corp. company literature). Without the aid of electrification, only the gas-atomizing type of scrubbers are capable of obtaining aerodynamic cut diameters <0.6 μm at reasonable expenditures of energy. The venturi is a gas-atomizing scrubber. The collision scrubber achieves a d_{50} of 0.2 μm at a gas phase pressure drop of about 140 cm (55 inch) water gauge and the venturi does this at about 300 cm (120 inch) water.

Total power input to the gas phase to overcome all pressure losses, both before and after the scrubber and including the scrubber, can be obtained from the pressure drop and volumetric flowrate, assuming 60% motor-

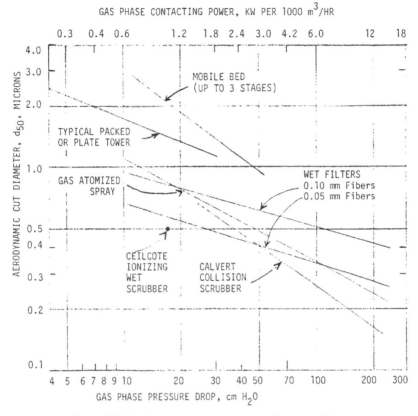

Figure 5.18 Cut diameter versus contact power for wet scrubbers.

blower efficiency, by the following equation in metric and English units

$$kW_B = 6 \times 10^{-5}(\Delta P)(Q) \qquad HP_B = 3 \times 10^{-4}(\Delta P)(Q)$$

$$(5.25)$$

where   where

kW_B = input power to blower, kW

ΔP = blower pressure drop, cm H_2O

Q = gas volumetric flowrate, Am³/hr

HP_B = input power to blower, hp

ΔP = blower pressure drop, inches H_2O

Q = gas volumetric flowrate, acfm

Note that radial tip fans can be up to 75% efficient.

Gas phase contacting power input to the scrubber would be considered as Equation (5.25) using pressure drop across the wetted scrubber when pressure drop is obtained with inlet and outlet ducts the same size. These values are shown at the top of Figure 5.18.

Example Problem 5.5

Determine the gas phase contacting-power input for the data in Example Problem 5.3 for a scrubber operating at a pressure drop of 25 cm H_2O.

Solution

Adapting Equation (5.25) for the scrubber only, this would be per 100 MW of capacity.

$$kW_s = (6 \times 10^{-5})(25)(491,000) = 737 \text{ kW}$$

That is, the scrubber alone would require 0.74% of the station power. Associated ductwork, etc. would add to this. This shows that contacting power in gas-atomized spray scrubbers comes predominantly from the gas stream.

5.5.5 PRESSURE DROP

Pressure drop is one of the most important parameters. As can be seen from Figure 5.18, gas phase pressure drop across a scrubber is related to aerodynamic cut diameter. Pressure drop can be easily and accurately measured on operating units if reasonable care is used. The need exists, however, for estimating pressure drop in the designing of units and for predicting pressure drops at various operating conditions. Numerous studies have been made to provide data and procedures for predicting scrubber pressure drops [29–34].

It has been recognized, since the studies on venturi scrubbers in the 1930s by Johnstone, that turbulence is important in pressure drop and particle-collection efficiency. As a result of the follow-up work by Eckman and Johnstone [29], scrubber designs were modified to incorporate more turbulence. Flow turbulence is a function of Reynolds flow number and this number is a direct function of venturi throat size (diameter), all other factors constant. One of the simplest equations for predicting venturi scrubber pressure drops, and one of the few to incorporate throat size, is given by Hesketh [30] in metric and English units

$$\Delta P \cong \frac{v_t^2 \varrho_g A_t^{0.133} L_G^{0.78}}{3870} \qquad\qquad \Delta P \cong \frac{v_t^2 \varrho_g A_t^{0.133} L_G^{0.78}}{1270}$$

$$(5.26)$$

where

ΔP = venturi scrubber pressure drop, cm H_2O

v_t = throat velocity of gas and particles, cm/sec

A_t = throat cross-section area, cm^2

L_G = liquid-to-gas ratio, l/m^3

ϱ_g = throat gas density, g/cm^3

where

ΔP = venturi scrubber pressure drop, inches H_2O

v_t = throat velocity of gases and particles, actual ft/sec

A_t = throat cross-section area, ft^2

L_G = liquid-to-gas ratio, gal/1000 acf

ϱ_g = throat gas density, lb/ft^3

The effect of venturi throat area can be seen by plotting the dimensionless pressure ratio, $\Delta P/\Delta H$, against the liquid-to-gas ratio. The gas velocity head, ΔH, is

$$\Delta H = \frac{v_t^2 \varrho_g}{2} \qquad\qquad (5.27)$$

Figure 5.19 shows this for pressure drops calculated by Equation (5.26) at various venturi throat areas. Included for comparison are pressure drop data as predicted by Yung et al. [31] and Boll [32]. These equations have no factor to account for throat area. Boll's test unit had a 1.1×10^3 cm^2 throat area. A venturi with a 23×10^3 cm^2 throat is equivalent to typical industrial size and a unit with 1.1×10^3 cm^2 throat is large pilot size.

5.5.6 COLLECTION EFFICIENCY

Particle collection efficiency in wet scrubbers has been shown to vary directly with contact power, and this is related to gas phase pressure drop across the wetted scrubber if ΔP is measured using the same size ducts. For a venturi scrubber, overall collection efficiency could be estimated using the measured pressure drop, Figures 5.17 and 5.18 and the size distribution properties of the particulate matter as demonstrated by Example Problem 5.6 which follows. Conversely, knowing a required overall collection

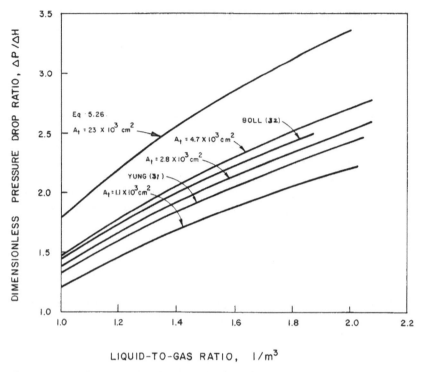

Figure 5.19 Scrubber throat head loss as a function of liquid-to-gas ratio for various venturi scrubber throat areas.

efficiency the required pressure drop can be established, and by use of Equation (5.26) operating parameters can be established.

Example Problem 5.6

Determine the collection efficiency for a venturi scrubber operating at a 20-cm H_2O pressure drop. The dust has a mass mean aerodynamic diameter, $\bar{d}_a$, of 10 μm and a standard geometric deviation, σ_g, of 3.

Solution

The curve for a venturi (gas-atomized spray) on Figure 5.18 shows that an aerodynamic cut diameter, d_{50}, of about 0.62 should be expected. Using the ratio $d_{50}/\bar{d}_a$ of 0.062 and Figure 5.17, an overall penetration of about 0.01 would be obtained at a σ_g of 3. Therefore, overall efficiency is about 99%.

5.5.7 VENTURI ABSORPTION

Although venturi scrubbers are used predominantly for control of particulate air pollutants, these devices can simultaneously function as absorbers. At low velocities there is more holdup time which permits more gas molecules to pass through the gas and liquid films and be removed with the liquid. Using the Chilton-Colburn concept of transfer units [35], the fractional absorption efficiency, ϵ_o, in a venturi scrubber is

$$\epsilon_o = 1 - \exp(-N_G) \qquad (5.28)$$

where N_G is number of overall gas transfer units assuming an irreversible chemical reaction occurs between the absorbed gas molecules and the chemical scrubbant. This is normal for scrubbing systems, and depending upon the specific chemicals absorbed and the absorbents, the value of N_G for venturi scrubbers can be expressed in a variety of forms.

Most empirical expressions show N_G to be a function of scrubber pressure drop. As pressure drop increases at low gas velocities, absorption efficiency increases. Note from Equation (5.26) that pressure drop for a particular device and fluids is dependent on throat velocity and liquid-to-gas ratio. This suggests that to raise pressure drop while operating at low velocities the liquid-to-gas ratio must be increased. Other types of scrubbers behave similarly in that increasing liquid quantity can improve absorption.

Venturi scrubbers are cocurrent absorbers. As gas molecules become absorbed, equilibrium concentrations are approached until no more absorption occurs. Galeano [36] suggests that the venturi length actually involved in absorption is about 6 cm for liquid film-controlling absorptions and about 30 cm for gas film controlling. As a general rule, a venturi provides absorption capacity equivalent to one theoretical stage of separation on a vapor-liquid equilibrium diagram (these diagrams are discussed under absorbers).

Venturi scrubbers are good fly ash collectors and at the same time remove substantial amounts of sulfur dioxide (SO_2). Chemical scrubbants used in venturi and other types of scrubbers during flue gas desulfurization (FGD) contain alkaline cations such as sodium, magnesium, and calcium. These are listed in order of decreasing reactivity. The calcium compounds such as lime (CaO) and limestone ($CaCO_3$) are less expensive so products produced using these materials are usually discarded. The more reactive sodium and magnesium chemicals are usually regenerated either within the scrubbing system or externally.

Specific relationships for the absorption of SO_2 from flue gas in venturi scrubbers using calcium and magnesium are presented here to show some of the significant relationships. These equations are specific to certain operating conditions and systems. For MgO scrubbing [37] at a liquid-to-gas ratio, L_G, equal to 5.4 l/m³ (40 gal/1,000 ft³) the SO_2 removal fractional efficiency is

$$\epsilon_o = 1 - \exp[(6.86\Delta P^{-1.014}y^{-3.75+0.27 \ln y}10^{6-0.031\text{pH}}) - 3] \qquad (5.29)$$

where ΔP is in cm H_2O, y is inlet SO_2 concentration in ppm and pH is actual chemical pH of recycle loop scrubbing slurry which is about 7.0. Efficiency drops off with inlet SO_2 concentration at low pHs and at low SO_2 concentrations (<500 ppm) more than is predicted by this equation.

The following expression has been given [38] for CaO scrubbing of SO_2 in a venturi scrubber

$$\epsilon_o = 0.307 + 0.0457(S) + 3.75 \times 10^{-3}(\Delta P) + 0.048(L_G)$$

$$+ 0.1516(I) - 0.0258(I^2) - 5.98 \times 10^{-3}(C_s) \qquad (5.30)$$

where

S = stoichiometry, moles CaO/mole SO_2 input
ΔP = venturi scrubber pressure drop, cm H_2O
L_G = liquid-to-gas ratio, l/m³ (0.134 l/m³ = 1 gal/1,000 ft³)
I = ionic strength of Na^+ present, molarity
C_s = slurry solids concentration, percent by weight

Example Problem 5.7

Predict SO_2 removal efficiency by FGD in a single-stage MgO venturi scrubber and a two-stage CaO scrubber. Pressure drop for each stage is 17 cm H_2O. Inlet SO_2 concentration is 1,600 ppm, scrubber slurries have a pH of 7, contain 5% by weight solids and a ratio of 5.4 l liquid per m³ gas. Consider that adequate chemical is available for absorption and that the lime system stoichiometry is 1 mole CaO/mole SO_2 input and no sodium ions are present. (Note that normally stoichiometry is expressed as mole CaO/mole SO_2 absorbed.)

Solution

Absorption efficiency of SO_2 in the single-stage MgO system using

Equation (5.29) is

$$\epsilon_o = 1 - \exp[(6.86)(17)^{-1.014}(1{,}600^{-3.75+0.271 \ln 1{,}600})(10^{6-0.031(7)}) - 3]$$

$$= 0.911$$

Absorption for the first stage of the CaO system using Equation (5.30) is

$$\epsilon_1 = 0.307 + 0.0457(1) + 3.75 \times 10^{-3}(17)$$

$$+ (0.048)(5.4) - 5.98 \times 10^{-3}(5)$$

$$= 0.646$$

Efficiency for two stages in series under these conditions would be

$$\epsilon_o = 0.646 + (0.646)(1 - 0.646) = 0.875$$

These data suggest that with the less-reactive calcium solution a single theoretical stage of separation as achieved in a venturi scrubber is not very effective. Coupled with another type of scrubber as the second stage to provide additional theoretical absorption stages (e.g., 4 or 5 total), overall efficiency of SO_2 absorption should be increased to 98.5–99.5%, respectively, under these conditions. Note that the actual stoichiometry of lime to SO_2 *absorbed* in this system is high. In the first stage it is [1/(1)(0.646)] or 1.5. Use of more-reactive lime (i.e., thiosorbic lime) also improves absorption.

5.5.8 VENTURI DESIGN PARAMETERS

Some of the venturi scrubber parameters listed in the sections on specifications, cut diameter, contacting power, pressure drop, collection efficiency, and absorption are summarized in this section. For example, if a high-efficiency particulate scrubber is required, then venturi throat velocities of 4,575 cm/sec (150 ft/sec) and higher would be valuable. Venturi throat velocities as high as 20,000 cm/sec (656 ft/sec) are used [39]. If absorption is desired with some particulate removal, then lower velocities such as 2,000–2,300 cm/sec (65–75 ft/sec) should be considered. Velocities coupled with liquid-to-gas ratio and the other factors as given by Equation (5.26) would be useful in determining the expected scrubber pressure drop.

Scrubber pressure drops are needed to estimate particle collection efficiency for a given dust size and size distribution. Pressure drop readings for all portions of the system are needed with volumetric gas flowrates for determining blower requirements. Pressure drops as high as 350 cm (140 in.) H_2O are used in some high-energy venturi scrubbers, although about 17 cm (6.5 in.) H_2O can give good absorption and scrubbing.

Liquid-to-gas ratio significantly influences the scrubber pressure drop. Minimum values of about 0.7 l/m³ (5 gal/1,000 ft³) should be used for particulate scrubbing, but values as low as 0.3 l/m³ (2 gal/1,000 ft³) are reported [40] for some new highly effective devices. Absorption requires higher rates of liquid such as 5.5 to 11 l/m³ (40–80 gal/1,000 ft³). This assumes that liquid is introduced evenly around the throat and the atomized liquid thoroughly covers the throat cross section.

Venturi scrubber efficiency is dependent on the throat length. If the throat is too small, efficiency is less than predicted. If the throat is longer than necessary, the pressure drop will be higher than predicted with no significant increase in efficiency. Pressure drops higher than necessary waste energy and money.

Hesketh and Mohan [41] have developed a procedure for specifying optimum venturi throat length based on a dimensionless velocity ratio, v_r

$$v_r = \frac{v_{de}}{v_t} \tag{5.31}$$

where

v_{de} = droplet velocity at throat exit, ft/sec
v_t = actual gas velocity in throat, ft/sec

Ratios <0.8 result in excessive throat length and high pressure drops. Ratios <0.3 give short throats with low particle collection efficiency. Using the optimum ratio of 0.5, throat length can be calculated in English units by

$$X_t = 328.582v_t^{(0.02343L_G - 0.8657)} \exp(-0.063L_G) \tag{5.32}$$

where

X_t = throat length, inches
L_G = liquid-to-gas ratio, gal/1,000 acf

5.5.9 MIST ELIMINATION

No matter how effectively the scrubber operates, it is worthless if the liquid containing the collected pollutants cannot be removed from the exiting gas stream before it enters the atmosphere. Many of the larger droplets fall out by gravity, centrifugal force, and/or coagulation. The remaining entrained drops must be removed. For very high removal efficiencies, single-stage wet ESPs can be used. However, the more common technique is to use mist eliminators. These devices are normally included as part of wet scrubbing devices. Some mist eliminators are such an integral part of the scrubber that pressure-drop measurements include the eliminators. The eliminators are usually flushed with water or fresh liquid to keep them clean. The washing is done from the dirty side. This means they can be effective gas absorbers also. Mist eliminators must remove essentially 100% of the entrained liquid and must keep reentrainment from the eliminator to a minimum. If anything escapes, it should be water with essentially no dissolved or suspended solids.

The initial entrainment from scrubbers entering mist eliminators amounts to about 1% of the scrubber recycle liquid. If this liquid, containing scrubbing-system solids, were not controlled, particulate emissions would be enormous. For example, a 500 MW scrubbing system with 5% solids could release about 1,250 g/sec of solids.

Although the initial entrainment from scrubbers consists mostly of 50- to 500-μm droplets, this varies as shown in Figure 5.20 for water mists. Data for this figure were obtained from McNulty and Monat [42]. This diagram shows that mist droplet size can vary considerably depending on the type of scrubber and the placement and type of nozzles. Note that ammonia and acid mists result in smaller droplets than water mists.

Reentrainment is the stripping-off of liquid in the mist eliminator by the gas stream. Reentrainment depends on separator configuration and position, gas velocity, entrained wash liquid, and liquid drainage rates. There is confusion as to whether reentrained droplets are large (80–750 μm) or small (< 10 μm). Calvert [43] reports that the average reentrained droplet is about 250 μm in size.

Reentrainment mechanisms consist of transition of liquid from liquid phase to entrained flow, rupture of bubbles, creeping of liquid on separator surface, and shattering of drops. Pneumatic two-fluid atomization causes drops or sheets of liquid to atomize and occurs at lower gas velocities than in straight tubes because gas jets are developed by the separator configuration. Reduced sharp angles should reduce reentrainmnent. For example, zigzag baffles as shown in Figure 5.21, inclined at an angle of 30° from the

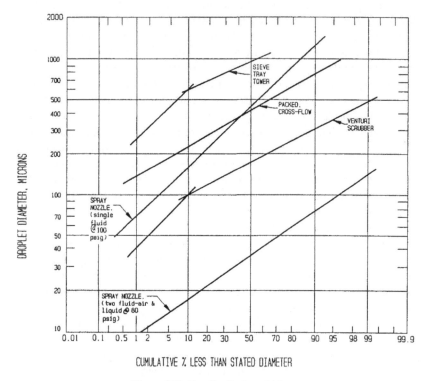

CUMULATIVE % LESS THAN STATED DIAMETER

Figure 5.20 Size distribution of mists.

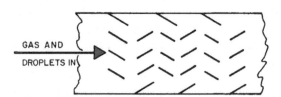

Figure 5.21 Top view of zigzag baffle mist eliminator.

horizontal gas flow direction should give less entrainment than baffles inclined 45°.

5.5.9.1 Entrainment Separators

The droplets entering the mist eliminators (entrainment separators) are large (50–500 μm), so diffusional collection is not significant. The mechanisms for removal of droplets are then impaction, interception, sedimentation, centrifugation, and electrostatic precipitation. All but impaction and interception can be treated just like the respective particulate control device previously discussed. Impaction and interception-plus-sedimentation are common mist elimination techniques and are presented here.

Effective droplet removal in separators is obtained at velocities of 1–5 m/sec (3–16 ft/sec) depending on design, although a few operate at velocities as high as 12 m/sec (39 ft/sec). The danger at higher velocities is that although entrained-droplet removal by impaction increases, reentrainment increases and eventually dominates. Reentrainment starts at about 18 m/sec in straight ducts. Bell [44] notes that reentrainment in vertical louvers starts at about 5 m/sec. Studies with "W" and "Z" shaped zigzag baffles [45] show that impaction drops off significantly below 25 m/sec and reentrainment increases above 3 m/sec. York [46] suggests a form of the Souders-Brown equation for predicting maximum allowable gas velocities in wire mesh separators to prevent reentrainment. This is

$$v = K_m \left(\frac{\varrho_l - \varrho_g}{\varrho_g} \right)^{1/2} \qquad (5.33a)$$

where K_m is an empirical constant $= 35$ with r in lb/ft^3 and v in ft/sec. Knitted mesh separators with horizontal flow operate at about 4.5 to 7 m/sec gas velocities.

Maximum allowable gas velocities in chevron mist eliminators can be determined by:

$$v = \frac{K_c}{\sqrt{\varrho_g}} \qquad (5.33b)$$

where

K_c = critical flow factor constant from 2.3 to 4.4 (see Table 5.8)
v = maximum gas velocity approaching chevrons, ft/sec
ϱ_g = gas density, lb/ft^3

TABLE 5.8. Commercial Chevron Mist Eliminators Flow Factor (Data Source Reference [42]).

Number	Chevron Description	Flow Factor, K_c	Critical Pressure Drop, inches H_2O
1.	4-pass sinusoid with hooks, 1 1/4 in. spacing, plastic	3.93	1.10
2.	3-pass modified zigzag, 1 in. spacing, 45° angle to flow, plastic	3.29	0.43
3.	4-pass sinusoid with hooks, 1 3/8 in. spacing, plastic	2.97	0.40
4.	4-pass zigzag with hooks, 2 in. spacing, 45° angle to flow, stainless steel (SS)	3.10	1.10
5.	4-pass zigzag, no hooks, 2 in. spacing, 45° angle to flow, SS	3.00	0.57
6.	Corrugated sheet metal packing, 1/2 in. corrugation height, 12 in. thick, 45° angle to flow	4.10	1.40
7.	Corrugated sheet metal packing, 1 in. corrugation ht., 12 in. thick, 30° angle to flow	4.30	1.50
8.	Corrugated sheet metal packing, 1 in. corrugation ht., 12 in. thick, 45° angle to flow	3.70	0.53
9.	Corrugated sheet metal packing, 1/2 in. corrugation ht., 6 in. thick, 45° angle to flow	3.20	0.70

TABLE 5.8. (continued).

Number	Chevron Description	Flow Factor, K_c	Critical Pressure Drop, inches H_2O
10.	3-pass modified zigzag, 1.5 in. spacing, 45° angle to flow, plastic	3.39	0.27
11.	3-pass modified zigzag, 1 in. spacing, 45° angle to flow, SS	4.00	0.20
12.	2-pass modified zigzag, 0.75 in. spacing, SS	4.40	0.51
13.	3-pass zigzag with hooks, 2 in. spacing, 45° angle to flow, SS	3.20	0.94
14.	3-pass zigzag, no hooks, 2 in. spacing, 45° angle to flow, plastic	3.00	0.39
15.	2-pass separated zigzag, 1 3/8 in. spacing, 30° angle to flow, SS	3.60	0.23
16.	2-pass modified zigzag, 0.75 in., spacing, plastic	3.80	1.00
17.	2-pass wing-shaped blade, 3 3/4 in. spacing, plastic	2.30	0.30
18.	3-pass zigzag, no hooks, 2 in. spacing, 45° angle to flow, SS	3.20	0.51

Values of K_c were determined by McNulty et al. [42] for 18 commercially available chevrons as listed in Table 5.8. In this table, "angle to flow" is the angle between main gas stream direction and the chevron blade and "SS" means stainless steel construction.

Tube banks operate at about 5.5–7 m/sec velocities with horizontal flow. Sedimentation rate for droplets of the size entering separators range from 10 to 200 cm/sec so sedimentation can be significant at these gas rates.

Reentrainment is reduced as liquid drainage rate increases, so orientation of the separator is important. Best drainage occurs when baffles are vertical and gas flow is horizontal as in Figure 5.21. In this arrangement with 30° baffles, reentrainment becomes heavy when liquid-to-gas ratio exceeds 0.5 1/m³ at gas velocities over 500 cm/sec [47]. The initiation of reentrainment occurs at similar velocities and inlet liquid-to-gas ratios for fibrous packing and tube bank separators operating with horizontal gas flow and vertical elements. These and several other impaction-type entrainment separators are shown in Figure 5.22. Many variations of designs exist. Already noted are the "W" and "Z" zigzag baffles. In addition, some zigzag units have small hooks at the end of each baffle to reduce reentrainment due to liquid creeping. Although eight gas turns are shown in Figure 5.22(d), three are usually adequate. Cyclonic mist eliminators show no reentrainment at inlet velocities of 40 m/sec and about 0.5% reentrainment at 60 m/sec.

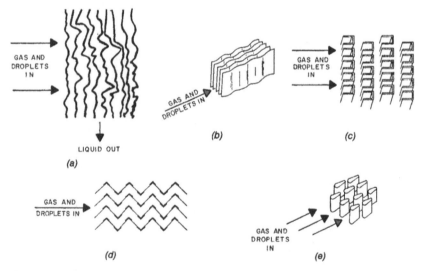

Figure 5.22 Some impaction-type mist eliminators: (a) knitted wire mesh; (b) wave plates; (c) staggered channels; (d) line separator; (e) streamline tube bank.

Gas velocity ranges for well-designed entrainment separators (i.e., mist eliminators) are typically:

Zigzag with up flow gases and horizontal baffles	12–15 ft/sec	365–455 cm/sec
Zigzag with horizontal flow gases and vertical baffles	15–20 ft/sec	455–610 cm/sec
Cyclone inlet velocity	100–130 ft/sec	3,050–4,000 cm/sec
Knitted mesh with vertical gas flow	10–15 ft/sec	305–455 cm/sec
Knitted mesh with horizontal gas flow	15–23 ft/sec	455–700 cm/sec
Tube bank with vertical gas flow	12–16 ft/sec	365–490 cm/sec
Tube bank with horizontal gas flow	18–23 ft/sec	550–700 cm/sec

Mist eliminators become plugged as a result of solids buildup. This can occur by impaction, settling, and/or evaporation of liquid with subsequent deposition of all solid matter. This is minimized by using vertical separator elements, maintaining substantial film thicknesses on all wetted surfaces, and washing periodically.

The release of droplets from mist eliminators has three significant effects: (1) moisture with dissolved and suspended solids consists of particulate matter emissions; (2) moisture with its chemical components increases corrosion; and (3) the droplets significantly increase reheat requirements to overcome the latent heat of evaporation if reheat were used. A method of reducing the first two of these problems is to use staged mist eliminators with the unit closest to the exit washed with clean liquid so if droplets are reentrained they transport a minimum of solid material.

5.5.9.2 Pressure Drop

Mist eliminators usually operate at low-pressure drops of about 1 cm H_2O or more. Pressure drops increase rapidly if the unit becomes plugged. Pluggage affects operation in that higher gas velocities result in the open portion, causing more reentrainment. Pressure drops do not seem to be influenced by orientation and amount of liquid entrainment, as related to scrubber emissions, except for the knitted mesh separators. In mesh separators, pressure drop is

$$\Delta P = 1.7^{\sqrt{L/A}} \Delta P_{dry} \tag{5.34}$$

where

L = liquid rate, m³/min

A = mesh cross-section area in the liquid flow direction, m²

ΔP_{dry} = pressure drop of dry mesh at that gas velocity

In mesh separators, pressure drop varies directly with gas velocity to the 1.65 power.

Pressure drops at maximum gas flow rates before entrainment occurs are listed in Table 5.8 as the critical pressure drops. The pressure drop for most chevrons is relatively independent of liquid loading, at least for clean conditions. Therefore, this critical pressure drop can be used to determine whether the chevron is operated at gas velocities above the entrainment point.

5.5.9.3 Overall Efficiency

Wet scrubber efficiency is dependent on how well the mist separator works. Properly designed and operated separators can have 99.5% or higher efficiency. This includes both penetrated entrainment plus reentrainment. Series of separators can promote droplet coalescing and removal plus reducing amounts of solids as previously noted. For comparison, the 500-MW example noted as losing 1,250 g/sec of solids due to scrubber entrainment of a 5% solids liquid should not lose more than about 6 g/sec of solids due to mist elimination liquid entrainment penetration plus reentrainment. This, plus the uncollected particulates in the gas stream would be the combined scrubber particulate emissions. The following example shows how reheat is affected by the presence of liquid droplets.

Example Problem 5.8

Emissions from a scrubber are to be reheated to eliminate the visible moisture plume. Determine what effect bypassed gases would have in providing reheat if a high-energy scrubber is used. Combustion flue gases at 140°C containing 6% water vapor are to be scrubbed at a high liquid-to-gas ratio of 11.5 l/m³.

Solution

Assume adiabatic saturation produces scrubbed gases at 50°C containing 12% water vapor. Scrubber entrainment equals 1% of the recycle liquid or 0.115 l/m³. The separator reduces this by 99.5% to 5.75 × 10⁻⁴ l/m³. Assume no heat losses and include sensible heat for the gases (air) and latent plus sensible heat for the water. A heat-balance on the system shows the bypass-scrubbed gas mixture reheat would be

Bypass	Scrub	Resultant Scrubbed Gas ΔT
10%	90%	8°C
15%	85%	12°C
20%	80%	15°C

5.5.10 CHEMICAL ADDITIVES

Numerous beneficial effects can be obtained by adding chemicals to the scrubbing liquid. Use of alkali cations (sodium, magnesium, and calcium) to tie up absorbed SO_2 as sulfites and sulfates has been discussed. This increases SO_2 removal *and* prevents the resulting liquid from becoming a strongly corrosive acid. Other materials serve as oxidants, inhibitors, and catalysts to improve the scrubbing. Magnesium alkali has the unique effect of forming flat crystal precipitates which bind strongly with heavy metals present in the slurry.

Particle removal may be improved by use of surfactants (wetting agents). Very small amounts of these materials reduce surface tension of the liquid phase. This can make atomization easier, enhance particle wettability, and reduce deposition of solids. Typically, nonionic, low-foaming surfactants are best for the high-energy scrubbing systems. Care must be exercised to avoid excessive buildup of surfactants in recycle loops as this could result in foaming within the scrubber. Surfactants may reduce absorption while improving particle collection.

5.5.11 RECENT INNOVATIONS IN WET SCRUBBERS

At least four new and unique wet scrubbers have been developed recently. These include the Calvert Collision Scrubber, the Ceilcote Ionizing Wet Scrubber (IWS), the Otto York Catenary Grid Scrubber, and the Turbotak Waterloo Scrubber. Each of these is patented and commercially available, and the names are registered trademarks. The collision scrubber and the ionizing wet scrubber (IWS) have been noted in previous sections and are included in Figure 5.18 cut power relationships. The IWS could also be considered as a hybrid device.

5.5.11.1 Collision Scrubber

The Calvert collision scrubber operation is described in Section 3.7.3.4. Parts of the system are venturi scrubber sections with cocurrent gas-liquid flow, yet there is the unique collision chamber. The cut power data in Figure 5.18 suggests that, at overall pressure drops greater than about 18 cm (7 inch) water, the collision scrubber has a particle collection efficiency advantage over that of conventional venturi scrubbers. The total Calvert system includes gas conditioning to achieve "Flux Force" enhancement. This is discussed in Section 5.10.

5.5.11.2 Ionizing Wet Scrubber

The Ceilcote ISW is a cross-flow scrubber with Tellerette packing. The gases flow horizontally while the liquid falls vertically. It utilizes a high-voltage ionizer to charge inlet particles before they reach the packing. The charging section is flushed with liquid by sprays at the top and the packing is washed by liquid from spray headers. The unique type of packing and the good washing enable this packed scrubber to operate very successfully as a particle-control device.

The larger particles (>3 μm) are removed by inertial impaction and the smaller particles are attracted to the neutral packing and droplet surface by image force attraction. The washing liquids can also be given a positive or negative charge. The IWS has a low pressure drop and charging energy is only about 0.3 KVA per 1,000 acfm.

The system operates at essentially constant pressure drop and the fifty percent cut power efficiency is given in Figure 5.18. This system is staged by placing two or more in series to achieve higher overall efficiencies.

5.5.11.3 Catenary Grid Scrubber

The catenary grid scrubber is discussed in Section 3.7.2.1. This device operates like a spray tower in that the gases rise and the liquids fall in a countercurrent manner. The unique collection mechanism makes this a very efficient, low pressure drop device. The limited data on collection efficiency, if placed on Figure 5.18, would suggest cut power efficiencies greater than any device shown. However, these data are only preliminary and need to be verified.

The catenary grid is smaller in size than other countercurrent scrubbers because the gas velocity average is 18 ft/sec (5.5 m/sec), whereas spray tower velocity is about 12 ft/sec (3.7 m/sec). The catenary can operate at gas velocities in excess of 30 ft/sec (9.2 m/sec). The unit operates with liquid-to-gas ratios as low as 0.5 gal/1000 ft³ (67 1/1000 m³) which makes it attractive for once-through liquid operation.

5.5.11.4 Waterloo Scrubber

The Turbotak Waterloo scrubber is similar to a venturi in that it is a co-current flow device. However, there are no converging or diverging sections. The system utilizes highly efficient air-atomizing nozzles to condition the gas as it passes through an inlet duct toward a fan-turbulent mixing zone. This gas-conditioning nozzle uses 70 psig air. Before reaching the fan, patented low energy CALDYN nozzles are used to provide collecting droplets. Turbotak reports [48] that the collection droplet size for optimum inertial impaction should be 15–20 times the diameter of the particle to be collected. Larger collector droplets permit the particles to follow the gas streamline and miss the targets while smaller collectors move in concert with the particles. Using the Caldyn nozzles the Waterloo scrubber can be "tuned" to optimize collection for specific sizes of particles.

The fan produces turbulent mixing for the particle–droplet interactions to occur. The agglomerated liquid is removed from the gas in downstream entrainment separators. Liquid-to-gas ratios as low as 0.5 gal/1000 ft³ (67 1/1000 m³) can be used in this system also.

5.5.12 OTHER CONVENTIONAL WET SCRUBBERS

The atomized spray wet scrubbers have been discussed in detail, with the venturi scrubber as the primary example. Many of the concepts covered thus far apply to other categories of scrubbers. A few other scrubber types are noted in this section. Combinations of scrubbers and other devices are included in Section 5.9 as Hybrids.

5.5.12.1 Preformed Spray Scrubbers

These scrubbers consist of spray towers or spray chambers and ejectors which cover the gas-liquid flow modes of countercurrent, crosscurrent, and cocurrent, respectively, as shown in Figure 5.23. Spray towers and chambers can operate with either gravity or high-pressure spray systems. The gravity systems can obtain effective aerodynamic cut diameters, d_{50}, of about 2 μm while the pressure spray systems (up to 27 atmospheres) can achieve a d_{50} of about 0.7 μm. Large amounts of liquid are used in these devices. Some operate as high as 16 l/m³ (120 gal/1,000 ft³) or more on an overall basis. Spray scrubbers have an open unobstructed interior which enables them to operate at low-pressure drops.

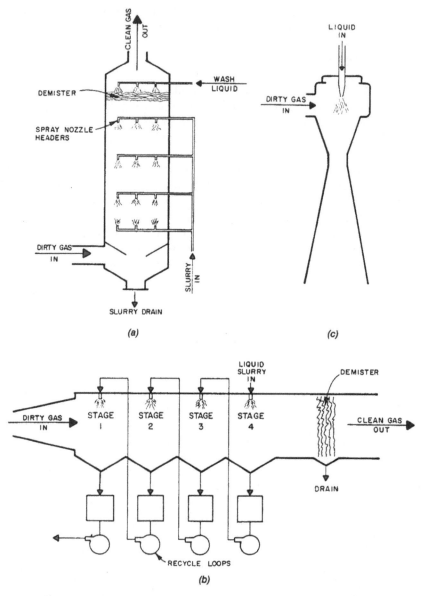

Figure 5.23 Preformed spray scrubbers: (a) spray tower; (b) horizontal spray chamber; (c) ejector scrubber.

344

The horizontal spray scrubber has been providing good particle and SO_2 removal in flue gas scrubbing applications. In order to meet particulate emissions, the scrubber is usually placed after a low-efficiency ESP. Particulate removal is in the mid-90% range for typical operating conditions. SO_2 removal of 99% on low-sulfur coal has been obtained in four stages by lime scrubbing at the high liquid-to-gas ratio of 6.7 l/m^3 (50 gal/1000 ft^3) per stage. Design operating conditions for the horizontal scrubber are given in Table 5.9. SO_2 fractional removal efficiency, ϵ_o, by sodium carbonate in the horizontal scrubber, can be predicted by [50]

$$\epsilon_o = 1 - \exp\phi \qquad (5.35)$$

For 5% Na_2CO_3 solutions with gas phase controlling absorption

$$\phi = 1.2 \times 10^{-2} \sqrt{\frac{D_{AB}u}{\pi D_c}} \left(\frac{lNL_G}{v_l D_c}\right) \qquad (5.36)$$

where

D_{AB} = SO_2 diffusivity in gas phase, cm^2/sec
u = relative droplet-gas velocity difference, cm/sec
l = mean distance traveled by droplets, cm
N = number of stages
v_l = liquid droplet velocity, cm/sec
L_G = liquid-to-gas ratio per stage, l/m^3

5.5.12.2 Moving Bed Scrubbers

Moving bed scrubbers are towers packed with mobile packing. They op-

TABLE 5.9. Operating Conditions at
Design Gas Flowrate for Some Flue Gas Scrubbers [49].

	Horizontal Spray	Vertical Spray TCA	Vertical PPA
Inlet gas flowrate, Nm^3/hr	765,000	765,000	765,000
Gas velocity, m/sec	6.7	3.7	3.7
Number of stages	4	4	3
Liquid-to-gas ratio per stage, ℓ/m^3	2.9	2.1	8.0
Pressure drop, cm H_2O	10	36	31
Nozzle pressure, atm	2.4	1	2
Power consumption, kW	2.6	3.4	3.9

erate in a countercurrent manner as shown in Figure 5.24. The Mobile
Ball Scrubber shown can have hollow balls of plastic, thermoplastic rubber, or other materials. They are also known by names such as the ping-
pong ball scrubber and UOP's Turbulent Contact Absorber (TCA). The
Marble Bed Scrubber contains several layers of glass marbles. Each scrubber can have various numbers of stages and slurry introduction arrangements. Table 5.9 shows design data for a TCA with four stages. Three-
stage TCAs can obtain $d_{50}s$ of about 1 μm. Good absorption can be attained
with these units.

5.5.12.3 Tower Scrubbers

Packed and plate tower scrubbers are used more as absorbers than particulate control devices as they are susceptible to plugging by solids partic-

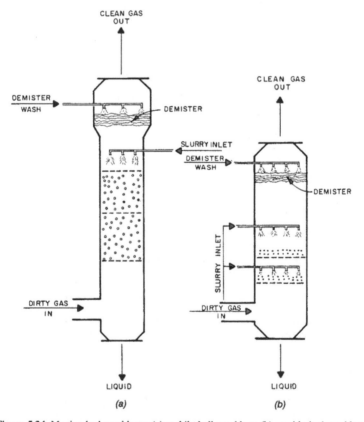

Figure 5.24 Moving bed scrubbers: (a) mobile ball scrubber; (b) marble bed scrubber.

ulates. These units are discussed in Section 5.7 under absorbers. The three-stage PPA scrubber noted in Table 5.9 is a Polygrid Packed Absorber which has fixed "egg crate" shaped packing 3 cm thick with 5 cm square openings.

5.5.12.4 Variations

Numerous other scrubbers exist, each with certain advantages for particular applications. Few are more effective than venturi scrubbers for particulate removal but some may have advantages in other aspects such as low cost, low pressure drop, portability, and easy maintenance. These devices include centrifugal scrubbers, baffle scrubbers, disintegrators, and others. Usually the name of the device implies the manner of operation. Comparisons of these systems can usually be made using the procedures outlined in the previous portions of Section 5.5.

5.6 DRY SCRUBBERS

Dry scrubbers include both dry/dry scrubbers and wet/dry scrubbers. Neither of these actually uses wet scrubbing-type contacting. Both types consist of a chemical-injection system, a reaction zone, and a particle-removal device. The chemicals injected are usually active alkali for reacting with the acid pollutants. The dry/dry systems inject dry alkali which undergoes a gas-solid phase reaction. The wet/dry systems spray a concentrated slurry into a spray drier reaction chamber. The reaction occurs in both gas-liquid and gas-solid phase interactions. In this system, all the liquid evaporates. The reaction products, as well as the unreacted alkali, are solid particles which are removed by ESPs or bag filters.

Dry scrubbers have been successfully used in flue gas desulfurization (FGD) for coal-fired combustion processes. The recent trend to increased waste incineration has provided a greatly increased market for dry scrubbers.

All dry scrubbers have the advantage of producing a dry product. The composition of the waste differs between systems, as some utilize the alkali more efficiently. Wet/dry scrubbing is a more efficient process for utilization of alkali. Based on wet scrubbing alkali utilization equal to 1, the dry scrubbing process alkali utilization in typical relative terms for hydrated lime is [51]:

$$
\begin{aligned}
\text{Wet scrubber} &= 1.0 \\
\text{Wet/dry spray dryer} &= 1.13 \\
\text{Dry/dry sorbent injection} &= 1.71 \\
\text{Dry/dry furnace injection} &= 3.37
\end{aligned}
$$

These data show the major disadvantage of this technique. The actual values will vary depending on whether system recycling is used and other operating parameters.

5.6.1 DRY/DRY SCRUBBERS

The two basic systems are the dry furnace injection and the dry sorbent injection (DSI).

5.6.1.1 Dry Furnace Injection

This system involves injecting dry sorbent directly into the furnace. It is the cheapest but least effective method, and has only been used for SO_2 control on coal combustion processes. Removal efficiencies of about 50% have been reported with hydrated lime. The reaction is both time and temperature sensitive. It appears that the temperature window for optimum SO_2 removal is from 1600 to 2300°F (870–1260°C). Below 1600°F the reaction rates are too low, and temperatures above 2300°F result in sintering of the lime. The LIMB (lime injection multiphase burner) systems usually operate at 2200–2300°F. In addition to the high alkali requirements and the low SO_2 removal, these systems can plug up and erode the boiler heat-transfer surfaces.

5.6.1.2 Dry Sorbent Injection

The DSI systems inject alkali into the flue gas stream by pneumatic injection. The alkali is directed downward into the rising gas at the bottom of a reaction vessel. The gases then continue to flow upward carrying the alkali. Some commercially available systems condition the gases first by spraying water to increase the gas humidity. This also cools the gas. Acid gas removal and alkali utilization are affected by location of alkali injection, acid gas concentration, moisture content, alkali concentration, alkali type, and whether the alkali is recycled. Normally, removal efficiencies increase directly with more chemical, greater contact time, higher inlet gas concentration, and moisture to within about 20–30°F of the saturation temperature. Acid gas removal is not as good as in wet scrubbers.

5.6.2 WET/DRY SCRUBBER

These are conventional spray dry systems where the alkali is introduced into the top of a reaction vessel as a concentrated wet slurry by either two-fluid atomizing nozzle or rotating disks. The slurry droplets are 50–90 μm in diameter. The spray dryer is a large reaction vessel with the gases entering at the top and flowing cocurrent with the alkali droplets. Acid gas removal in the high 80 and low 90 percent have been achieved depending on the type of particle removal system used. A problem with the spray dry absorbers (SDA) is that the spray nozzles can erode rapidly. This results in large droplets being formed which reduces gas removal efficiency and causes unevaporated liquid to accumulate in the bottom of the reactor.

If the gases entering the reactor are from a boiler they may have temperatures up to 220°C (430°F). The evaporating liquid cools the gas to about 140°C (285°F). Gas residence time in the reactor is 10–12 seconds. With this residence time and with proper reactor geometry, evaporation is complete in the gas phase and no liquid droplets reach the reactor walls. Also, about 25% of the total dry reaction products and ash settle out of the gas while in the reactor and are discharged through the bottom. Typically, about 40 cm³ of liquid slurry spray is required per m³ of gas.

5.6.3 PARTICLE REMOVAL

Dry scrubbers are designed to cause gaseous pollutants to react and produce solid particulate matter. In addition, there are particles that enter with the flue gas plus particles resulting from the excess alkali used to drive the reaction. All of this material results in a considerable particulate removal requirement. For coal combustion operations the total dry scrubber particulates are easily 3 times the original flue gas dust loading. As noted in the previous section, up to 25% of this may fall out in the reaction vessel but high-efficiency dust collection devices are also required.

Normally either fabric filters or electrostatic precipitators are used for particulate control following dry scrubbers. Most systems use fabric filters because of the greater removal efficiency *and* because the gaseous pollutants continue to react in the dust cake resulting in higher overall acid gas removal efficiencies. For example, EPRI pilot plant data show [52] that with high-sulfur coal combustion, a spray dryer-baghouse combination is capable of achieving over 90% SO_2 removal and meeting the 0.03 lb/10⁶ Btu federal particulate standard for new sources. The spray dryer-electrostatic precipitator arrangement removed about 80% of the SO_2 and emitted about 0.11 lb of particulates per 10⁶ Btu. The tests were conducted with inlet flue gases at 300°F and containing 2500 ppm SO_2. Lime at a stoichiometric ratio ≥ 1.45 was used and the gases were cooled to within 20°F of saturation.

In municipal solid waste (MSW) mass burn, the acid gas inlet concentration is much lower and the effects of pollutant removal differ. It is reported [53] that for full-scale lime spray dryer systems, there are various advantages for each type of particle collector. The spray dryer-fabric filter more effectively utilizes the reagent (therefore, higher acid gas removal efficiencies) and also has the ability to dampen any adverse effects of fluctuating inlet gas concentration and other gas condition changes. A spray dryer-ESP may be less expensive to install and may occupy less space. For both arrangements, HCl removal efficiencies can reach 99%. For SO_2, the ESP system achieves removals in the 70 percent range and the filter system gives removals in the 90s at a lime stoichiometry of 2.2 to 3.0. Inlet gases were 240–280°F with 3.2–21.5 ppm HCl and 3.2–57 ppm SO_2.

5.6.4 CHEMICALS

Most dry scrubbers use calcium compounds as the alkali for acid gas removal. Large systems can slake and use pebble lime because the operating ports are large enough to pass the coarse grit contained in this lime. Smaller systems often use hydrated lime. Operations with spray dry-baghouse arrangements provide extra time for the gas/solid phase reactions. In some systems, fly ash is being tested as a reagent with lime. Coal combustion fly ash contains much calcium which can be attractive for FGD. Certain fly ash, when the loading is increased up to 20 g fly ash/g $Ca(OH)_2$ increased the $Ca(OH)_2$ utilization from 17 to 78% [54]. Addition of trace quantities of NaOH to fly ash/$Ca(OH)_2$ reaction slurries increased the reactivity of both SO_2 and NO_x [55].

Very reactive chemicals can be used to precoat filters so that when acid gases pass through the precoat they are converted to particulate material and captured. Trona, nahcolite, and other natural sodium carbonate-bicarbonate ores as well as lime have been used as precoat material for SO_2 removal. This precoat or bag conditioning can perform several functions: it is effective in removing reactive gases; it can increase particulate size, which increases particulate collection efficiency; it can significantly reduce operating pressure drop; and it can prolong filter fabric life.

5.7 ABSORBERS

Absorbers, as related to air pollution control, are a special category of wet scrubbers. Absorption can occur in any wet scrubber and has been specifically mentioned in Section 5.5.7. Notes related to absorption, especially SO_2 absorption, are given in other parts of Section 5.5 where absorption efficiencies in the high 90% are obtained. Absorption towers are presented in this section, as they work better for gaseous absorption than for particulate collection.

The two common types of towers, plate and packed, make effective pollution control devices. They utilize simple mechanical methods of achieving good contacting between the gas and liquid phases to provide favorable overall mass transfer. Towers or columns have been used industrially for many years in chemical operations such as distillation, rectification, absorption, and extraction. As such they have an advantage of being familiar devices to many engineers. The three general types of towers are packed, bubble cap, and perforated-plate towers.

5.7.1 ABSORPTION TOWERS

5.7.1.1 Packed Towers

These are cylinders filled with a packing material and supported at the top and bottom to permit phase separation. Supports may also be in the midsections of the tower to relieve the load on the packing. The packing in these towers can be anything from broken bottles or pieces of coke to ingenuously prepared shapes such as fritted plates, interlocking saddles, spheres, and cylinders. Packings are made of carbon, ceramic, glass, plastic, Teflon, stainless steels, and other metals. Some of the commercially available forms of packing are the Berl saddles, Intalox saddles, Raschig rings, Pall rings, and Cannon packing. A few of these are shown in Figure 5.25. The Raschig rings and Intalox saddles shown are made of ceramic and the Pall rings are plastic and stainless steel. Note the variations in sizes. In general, packings < 2.5 cm (1 in.) are used in towers < 30 cm (1 ft) diameter; packings < 3.8 cm (1.5 in.) for towers < 91 cm (3 ft) diameter; and packings > 5 cm (2 in.) for towers > 61 cm (2 ft) diameter.

The requirements of packings are to (1) provide good surface contact area, (2) give low pressure drop to gas flow, (3) provide even distribution of both fluid-phases (i.e., minimize channeling), (4) be unreactive with either phase (unless catalytic influence or chemical reaction is desired), (5) be sturdy enough to support themselves in the column, (6) have abrasion resistance to prevent being worn away by attrition, and (7) be economical for the desired operation (low cost, available, and easily handled).

Figure 5.26 is a schematic of a packed tower operation with countercurrent flow. The liquid enters at the top of the tower and must be distributed in some manner such as by a spray system, overflow weirs, or a perforated distribution plate. The liquid should pass uniformly throughout the packing where it contacts the rising gas. Sometimes it is necessary to add intermediate packing support plates if the packing is not capable of supporting the entire column weight of liquid and packing. These supports can also be used as liquid redistributors to try and reduce liquid channeling. A

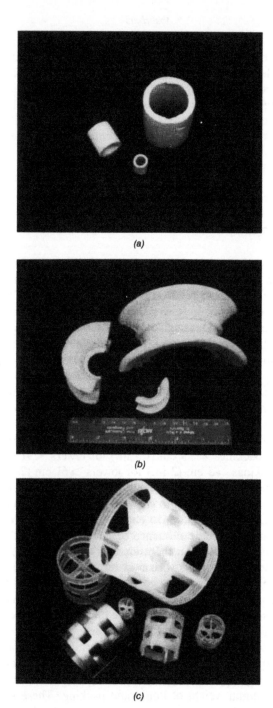

Figure 5.25 Examples of absorber tower packing: (a) Raschig rings; (b) Intalox saddles; (c) Pall rings.

352

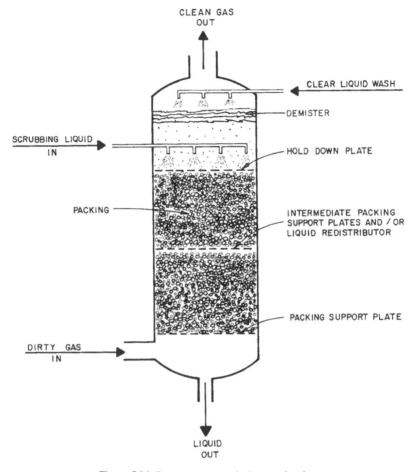

Figure 5.26 Countercurrent packed tower absorber.

packing support is required at the bottom of the tower. This screen or plate support grid provides a receptacle for draining the dirty liquid, and also distributes the entering gas before the gas starts upward through the packing. It is usually necessary to utilize a gas demister (mist eliminator) at the top of the tower to prevent entrainment of liquid droplets in the exit gas stream.

Tower packing must be installed with considerable care and not simply dumped into the shell. If it is not added in small quantities and spread evenly to assure uniform distribution, excessive channeling will occur during operation. Towers can be filled with water before adding the packing to prevent fracturing the packing as it is charged to the column and to aid in distributing the packing.

Use the following nine rules of thumb as guidelines for selecting packed towers:

(1) Structured and random packings are suited to packed towers under 3 ft dia. and when low ΔP is desirable. In such cases, volumetric efficiencies, given adequate liquid distribution throughout the column, can exceed those of trayed towers.

(2) Replacing trays with packing allows greater throughput or separation in existing tower shells.

(3) For gas rates of 500 ft³/min, use 1-in. packing; for gas rates of 200 ft³/min or more, use 2-in. packing.

(4) The ratio of the diameter of the tower to the packing should be at least 15:1.

(5) Because of deformability, plastic packing is limited to an unsupported depth of 10–15 ft; metal to 20–25 ft.

(6) Liquid redistributors are required every 5–10 tower diameters with rings, and at least every 20 ft for other types of dumped packings. The number of liquid streams should be 3–5/ft² in towers larger than 3 ft dia. Some experts advocate 9–12/ft².

(7) The Height Equivalent to a Theoretical Plate (HETP) for vapor-liquid contacting is 1.3–1.8 ft for 1-in. rings, 2.5–3.0 ft for 2-in. rings.

(8) Packed towers should operate near 70% of the flooding rate.

(9) Limit the tower height to about 175 ft maximum because of wind load and foundation considerations.

5.7.1.2 Bubble Cap Towers

Bubble cap towers have no support grids or packing, but instead have evenly spaced plates with specially designed risers and downcomers as shown in Figure 5.27. This figure is a sectional view showing two plates of a typical bubble cap tower. The gases enter each plate through the gas risers which are spaced according to a consistent geometrical arrangement. The gas is forced through the liquid and out slots in or underneath the bubble caps. This causes good turbulence and provides a large surface area for contact between the gas and liquid phases.

The liquid on each plate flows to the next lower plate by means of downcomers, which are pipes containing various levels of overflow weirs. The actual height of liquid on each plate depends on the height and design of

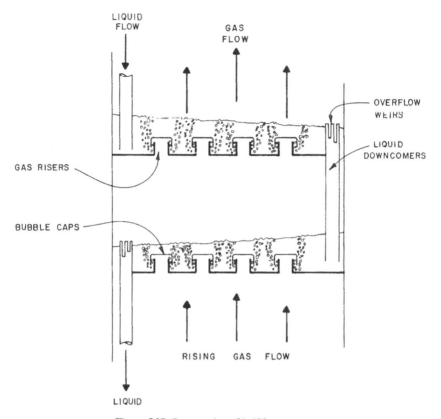

Figure 5.27 Cross section of bubble cap tower.

the liquid overflows as well as on the liquid and gas flowrates. The gas pressure drop across each plate increases with liquid height on the plate and gas flowrate. Increasing the number of downcomers per plate as column diameters increase reduces the tendency for an uneven liquid head on each plate as shown in Figure 5.27. Any liquid addition onto a plate, whether from a downcomer or an external feed source, must be initiated in a manner that provides least disturbance to the liquid on the plate.

5.7.1.3 Perforated (Sieve) Plate Towers

These towers are designed similar to bubble cap towers except that instead of risers or downcomers, the plates simply have holes in them. The number of holes, their shape, and their arrangement on the plates varies from column to column, but they are often 3-mm (1/8-in) holes on 1-cm

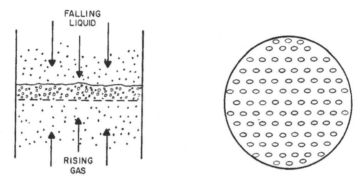

Figure 5.28 Typical sectional views of a sieve-plate tower.

(3/8-in) triangular center in 0.64-cm (1/4-in) thick plates. Figure 5.28 shows a section of a sieve plate tower and the top of a plate. The rising gas provides a resistance to liquid flow so that a liquid head can be maintained on each plate. The two phases contact each other as the gas bubbles up through the liquid as well as when the liquid falls through the gas. The liquid level on the plate is a function of both liquid and gas flowrates. It is necessary to have level plates with consistently sized and arranged holes in order to maintain an even liquid height on the plate. Uneven liquid height would result in channeling or bypassing of the gas because the gas would pass through the section having least liquid resistance to gas flow. Perforated plate towers with slotted plates similar to the plate shown in Figure 5.28 are called Turbogrid towers.

5.7.2 ABSORPTION TOWER CAPACITY

Absorber tower capacity is usually considered to be limited by (1) a maximum vapor rate above which liquid would be carried upward by entrainment in the rising vapor concluding effective countercurrent separation and (2) a maximum liquid rate above which the column would flood and dump, terminating the operation. Using these rate-limiting restrictions, capacities for the various type towers are given in the following subsections. However, it should be noted that new operational methods discussed in Section 5.7.3 may make it desirable to alter these conditions for countercurrent tower operations.

The variables affecting absorption tower performance can be considered as consisting of three groups: those affecting capacity, those affecting efficiency and those affecting both. Basically capacity is set by tower diameter and efficiency is set by height. It is best to determine tower diameter and hence capacity first.

5.7.2.1 Packed Towers

Packed tower capacities can be found by fixing the desired pressure drop, using the physical properties and flowrates of the liquid and gas streams involved and the generalized pressure drop correlation given in Figure 5.29 for countercurrent packed towers with gas-liquid phases. The value of X for Figure 5.29 is determined by

$$X = \frac{LM_l}{GM_g}\left(\frac{\varrho_g}{\varrho_l}\right)^{0.5} \tag{5.37}$$

where

L = liquid flowrate, g mole/(cm^2 hr)
G = gas flowrate, g mole/(cm^2 hr)
M_l = molecular weight of liquid
M_g = molecular weight of gas

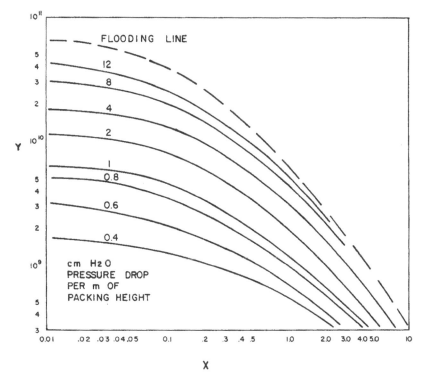

Figure 5.29 Generalized pressure drop correlation for countercurrent packed towers with gas-liquid absorption [55].

Note that this requires an assumed value for G and a trial and error solution. In this equation, ϱ_l is assumed essentially equal to $\varrho_l - \varrho_g$. Using this value of X and the desired pressure drop per height of packing, obtain a value for Y from Figure 5.29. Packed tower scrubbers normally operate in the midrange of pressure drops given, i.e., from about 1.5–3.3 cm H_2O/m packed height. (See also Section 5.7.4 for packing relative ΔP.) High-pressure absorbers operate above this and vacuum stills are below. Packed beds flood when the ΔP is much higher than about 10 cm H_2O per m of packing height. Gas flowrate is obtained by

$$G = \left[\frac{Y \varrho_g \varrho_l}{M_g^2 F \Psi (\mu_l)^a}\right]^{0.5} \qquad (5.38a)$$

where

Y = ordinate of Figure 5.29
F = packing factor from Table 5.10
Ψ = ϱ_{H_2O}/ϱ_l dimensionless
μ_l = viscosity of liquid, g/(cm sec)
a = 0.2 when F is ≥ 90 and 0.1 when F is < 90

When packing factor, F, is less than 90, use $(\mu_l)^{0.1}$ in Equation (5.38a). Correct the original value of G used in Equation (5.37) and repeat as necessary.

Tower diameter can be determined using the ideal gas relation

$$D_T = 1.25 \left[\frac{Q}{G} \frac{P_T}{273 + t}\right]^{0.5} \qquad (5.39)$$

where

D_T = tower diameter, m
Q = gas volumetric flowrate, Am^3/hr
P_T = tower pressure, atm
t = temperature, °C

Packed tower flooding velocities may be as high as 120 to 180 cm/sec at atmospheric pressure, although flooding can occur when the rising gas velocity is greater than 1.5 times the velocity which results in a pressure drop of 4 cm of water per meter of packing height.

A packed tower of a specific size operating with a given type of packing

TABLE 5.10. Packing Factors, *F*, for Packed
Tower Absorbers for Wet and Dump Packed [56].

Packing	Material	Nominal Packing Size (cm)							
		0.6	1.0	1.6	1.9	2.5	3.8	5.1	7.6
Intalox saddles	Ceramic	725	330		145	98	52	40	22
Intalox saddles	Plastic					33		21	16
Raschig rings	Ceramic	1,600	1,000	380	255	155	95	65	37
Raschig rings,									
0.08-cm wall	Metal	700	390	170	155	115			
Berl saddles	Ceramic	900			170	110	65	45	
Pall rings	Plastic			97		52	32	25	
Pall rings	Metal			70		48	28	20	
Tellerettes	Plastic							45	

can operate at a different gas mass velocity by changing the packing. This difference can be estimated using the packing factors obtained for the original and final conditions as denoted by 1 and 2

$$G_2 = G_1 \left(\frac{F_1}{F_2} \right)^{0.5} \tag{5.38b}$$

Example Problem 5.9

Estimate the size packed column that should be used to provide a pressure drop of 2.0 cm H_2O per m of packing height when the scrubbing liquid rate is 41.5 g mole/(cm^2 hr). The gas flowrate is 3,000 Am^3/hr. Assume the scrubbing liquid is water and the gas is air at standard conditions. Plastic Pall rings, 3.8 cm in size are suggested.

Solution

Assume a value of *G* equal to 26 g mole/(cm^2 hr). Then the value of *X* from Equation (5.37) would be

$$X = \left(\frac{41.5}{26} \right) \left(\frac{18}{29} \right) \left[\left(\frac{1.2 \times 10^{-3}}{1} \right) \right]^{0.5} = 0.034$$

Obtain from Figure 5.28 at $\Delta P = 2$ cm H_2O per meter packing height the value of $Y = 1.03 \times 10^{10}$.

Solving for G using Equation (5.38a) gives

$$G = \left[\frac{(1.02 \times 10^{10})(1.2 \times 10^{-3})(1)}{(29)^2(32)(1)(0.01)^{0.1}}\right]^{0.5} = 26.9 \ \frac{\text{g mole}}{\text{cm}^2 \ \text{hr}}$$

This is close enough to the original assumed value of G, so no iteration is made.

Tower diameter from Equation (5.39) is

$$D_T = (1.25)\left[\left(\frac{3,000}{26.9}\right)\left(\frac{1}{273 + 20}\right)\right]^{0.5} = 0.77 \ \text{m}$$

Note: A larger packing size may be desired.

5.7.2.2 Plate Towers

The maximum superficial vapor velocity before flooding in a bubble cap tower is generally about 90 cm/sec at 1 atmosphere pressure. A more exact value of bubble cap maximum vapor velocity, v_b, in cm/sec can be estimated using

$$v_b = K_v \left(\frac{\varrho_l - \varrho_g}{\varrho_g}\right)^{0.5} \tag{5.40}$$

where K_v is a constant dependent on plate spacing and liquid seal depth on the plate. Values for this are given in Table 5.11.

Tower diameter can be calculated from gas velocity using

$$D_T = \sqrt{\frac{Q}{9\pi v_b}} \tag{5.41}$$

where

D_T = tower diameter, m
Q = gas volumetric flowrate, Am³/hr
v_b = gas superficial velocity, cm/sec

5.7.3 ABSORPTION TOWER EFFICIENCY

There are two efficiencies that can be considered for absorption devices. The first is absorption efficiency as related to gas removal, and the second

TABLE 5.11. Values of K_v for Use
in Equation (5.40) [57].

Plate Spacing (cm)	Liquid Seal Height (cm)			
	1.3	2.5	5.1	7.6
15	0.9			
30.5	3.1	2.4	1.8	
61	5.5	5.2	4.9	4.6
91.5	6.1	5.8	5.8	5.5

is the actual absorber-stage efficiency of the unit compared to an ideal absorber. Gas absorption efficiency is presented first. Stage efficiency is discussed in Section 4.3.8 for absorption in general and is presented here as related to specific absorption devices.

Gas absorption efficiency is the number of moles of gaseous pollutant absorbed per mole of inlet gaseous pollutant. This is the same as a volumetric measurement of pollutant in and out, e.g., in ppm. Estimations of absorption efficiency can be made using Equation (4.18), but values of $K_G a$ are best obtained by empirical data. Equation (4.18) is rearranged and presented here for the mass transfer coefficient.

$$K_G a = \frac{\Delta N_A}{V(p_{AG} - p_{AL})_{ln}} \qquad (5.42)$$

Fortunately, once values of $K_G a$ have been obtained empirically or from manufacturers' literature for a given device and system, $K_G a$ values for different size and type of packing can be estimated. This is commonly done using tabulated data on CO_2 absorption in NaOH solutions. The procedure for this is

$$(K_G a)_2 = (K_G a)_1 \frac{(K_G a)_{T_2}}{(K_G a)_{T_1}} \qquad (5.43)$$

where

$(K_G a)_2$ = value to be determined at conditions 2
$(K_G a)_1$ = known value at conditions 1
$(K_G a)_{T_1}$ = value from Table 5.12 at conditions 1
$(K_G a)_{T_2}$ = value from Table 5.12 at conditions 2

TABLE 5.12. Tabulated $K_G a$ Values in 10^4 g mole/(hr m³ atm) for a Consistent System of CO_2 and 1% NaOH [58].

Packing	Material	Nominal Packing Size (cm)							
		2.5		3.8		5.1		7.6	
		a	b	a	b	a	b	a	b
Intalox saddles	Ceramic	4.2	5.1	3.7	4.3	2.9	3.4	2.2	2.6
Intalox saddles	Plastic	5.1	5.6			3.2	3.7	2.4	2.7
Raschig rings	Ceramic	3.7	4.8	3.0	3.4	2.7	3.2		
Raschig rings	Metal	3.8	4.5	3.2	3.5	2.4	2.9		
Berl saddles	Ceramic	4.0	4.5	2.9	3.5	2.4	2.9		
Pall rings	Plastic	3.8	4.5	3.5	4.2	3.0	3.6		
Pall rings	Metal	4.8	5.6	4.2	5.1	3.5	4.2	2.7	3.0
Tellerettes	Plastic					3.0	3.4		

Note: a = Liquid rate of 2,440 g/(cm² hr).
      b = Liquid rate of 4,880 g/(cm² hr).

Values of $K_G a$ can also be extrapolated to correct for changes in concentrations using the log mean pressure drop

$$(K_G a)_2 = (K_G a)_1 \left[\frac{(p_{AG} - p_{AL})_{ln2}}{(p_{AG} - p_{AL})_{ln1}} \right]^{0.32} \tag{5.44}$$

$K_G a$ increases with temperature and for CO_2—NaOH systems this is about 1% per °C. As equilibrium is approached values of $K_G a$ decrease.

Stage efficiencies for packed and plate towers can be obtained in a similar manner. This can be done by either a mathematical or a graphical procedure, although the graphical procedure is more easily understood and just as accurate. The McCabe-Thiele-type graphical procedure consists of stepping off the number of theoretical contact stages or number of theoretical plates (NTP) on a vapor-liquid equilibrium diagram as discussed in Section 4.3.8. The overall fractional absorption stage efficiency of a plate-type tower then becomes

$$\epsilon_{\sigma_{stage}} = \frac{NTP}{\text{number of actual tower plates}} \tag{5.45}$$

This is the same procedure used in the calculation of distillation efficiency, except in distillation the stillpot counts as one extra actual plate.

There are no plates in a packed tower, so it is necessary to estimate the number of plate equivalents which is the number of transfer units (NTU). This can be done for a packed tower using the Chilton-Coburn equation:

$$\text{NTU} = \int_{y_1}^{y_2} \frac{dy}{y^* - y} \qquad (5.46)$$

where

y_1 = vapor composition entering the bottom of the countercurrent column, mole fraction
y_2 = vapor composition leaving, mole fraction
y^* = equilibrium vapor composition at the point where y is determined, mole fraction

The overall efficiency of the packed tower can then be found by replacing the number of actual tower plates in Equation (5.45) with the NTU.

The height equivalent to a theoretical plate (HETP) or contact stage can be applied to both plate and packed towers. It is obtained by dividing actual tower height by the number of theoretical stages for a given type separation, tower design, and operation condition. If the HETP value is available from either actual measurements, the equipment, or packing manufacturer's literature or pilot-plant studies, it can be used to calculate the approximate height, Z, of a proposed absorber by

$$Z = (\text{NTP})(\text{HETP}) \qquad (5.47)$$

The value of HETP will vary depending on type of plate or packing, column diameter, height of tower (to a minor extent), mass flowrates and absorbents and absorbates used.

An alternate procedure for estimating the height of an absorbing tower is to use the equation

$$Z = N_G H_G \qquad (5.48)$$

where

N_G = number overall gas transfer units
H_G = height of a transfer unit

Values of N_G can be estimated using the procedure from Section 4.3.9 and the approximate value of H_G of 0.6 m can be used for some preliminary estimates (N_G can be calculated using AIChE design manual procedures).

Typical absorber-stage efficiencies and operating characteristics of common absorption towers are listed for comparison in Table 5.13. All but the cycled tower are operating below the standard maximum vapor velocity and maximum liquid throughput limitations.

Contact efficiencies for absorption and other mass transfer operations can be vastly improved by a practice called controlled cycling. Controlled cycling, which is not to be confused with pulsation, can be used on countercurrent operations that have stepwise separation as, for example, in the plate-type towers. The operating cycle consists of two parts: (1) a vapor-flow period when the vapor flows upward through the column while the liquid remains stationary on each plate and (2) a liquid flow period when no vapor flows and liquid drains to the next lower plate. An absorber that is cycled at a rate so as to permit all the liquid on a tower plate to be transferred before restarting the cycle gives the highest absorber efficiency for a given set of conditions [59]. Under these conditions, the cycle times vary from 20 seconds to 2 minutes and vapor flows about 80% of the time. Mathematical studies show that stage efficiencies of 200% and even higher are theoretically possible.

5.7.4 RELATIVE PRESSURE DROP OF PACKING

Relative pressure drops for various sizes and types of commercial packing are listed in Table 5.14. These pressure drop data are for countercurrent

TABLE 5.13. Typical Absorption Tower (Column) Conditions.

	Type			
	Bubble Cap	Sieve Plate	Packed	Cycled Tower, Sieve
Low liquid throughput, relative	1	1.2	0.7	1.2
Medium liquid through-put, relative	60–80	60–80		80–96
Absorption stage efficiency, %	40.6	20.32		(variable)
Overall pressure drop, cm H_2O[a]	41–81	20–81		(variable)
Plate spacing, cm	1	2/3	1 1/2	2/3
Approximate relative initial cost[b]			(with ceramic rings)	

[a]For 5-plate or 20-ft packed column.
[b]Same materials of construction.

TABLE 5.14. Relative Pressure Drops of Packing Material for Countercurrent Operation.

Packing Size and Type	Available Surface Area, Per Unit Volume, ft^2/ft^3	Pressure Drop	
		inches Water foot	cm Water meter
Munters 12060	68	0.05	0.42
1″ Koch Flexirings	65	0.90	7.50
1″ Glitsch ballast saddles	65	0.80	6.67
1″ Glitsch ballast rings	65	1.30	10.83
1″ Intalox saddles	63	0.75	6.25
1″ Norton Pall rings	63	0.90	7.50
1″ Ceilcote Tellerettes	55	0.65	5.41
2″ Maspak FN-200	43	0.75	6.25
1-1/2″ Rashig rings	40	1.60	13.32
1-1/2″ Glitsch ballast rings	40	0.84	7.00
1-1/2″ Koch Flexirings	40	0.75	6.25
1-1/2″ Norton Pall ring	39	0.75	6.25
2-1/2″ Protak P-251	39	1.00	8.33
2″ Ceilcote Tellerettes	38	0.30	2.50
2″ Koch Flexirings	35	0.45	3.75
2″ Protak P-252	34	0.82	6.83
2″ Croll Reynolds Spiral-Pak	34	0.24	2.00
2″ Glitsch ballast saddles	34	0.55	4.58
2″ Intalox saddles	33	0.50	4.16
2″ Glitsch ballast rings	32	0.55	4.58
2″ Norton Pall rings	31	0.45	3.75
2″ Heilex 200	30	0.45	3.75
2″ Rashig rings	30	1.40	11.66
3″ Ceilcote Tellerettes	30	0.24	2.00
3-1/2″ Koch Flexirings	28	0.22	1.83
3″ Glitsch ballast saddles	28	0.32	2.66
3″ Intalox saddles	27	0.30	2.50
3-1/2″ Norton Pall rings	26	0.22	1.83
3-1/2″ Glitsch ballast rings	26	0.22	1.83
3-3/4″ Maspak FN-90	25	0.36	3.00
3″ Heilex 300	23	0.27	2.25

liquid gas flow and are for a gas velocity of 8.33 ft/sec and a liquid-to-gas molar ratio of 2.0.

5.8 ADSORBERS

5.8.1 ADSORBER ARRANGEMENTS

Adsorbers range in size from small disposable units, such as on the end of cigarettes and in pill bottles, to industrial units capable of handling 300,000 Nm³/hr. The two major designs are the stationary beds and the moving beds. The stationary beds must be intermittently regenerated or disposed of for continued operation while the moving beds can operate continuously.

5.8.1.1 Stationary Bed Adsorbers

Large industrial units are usually constructed with the gases to be cleaned entering from the top and the cleaned gases leaving at the bottom. This is shown in Figure 5.30. In this sense, the gas flow direction is opposite the usual gas flow direction such as that used in countercurrent absorp-

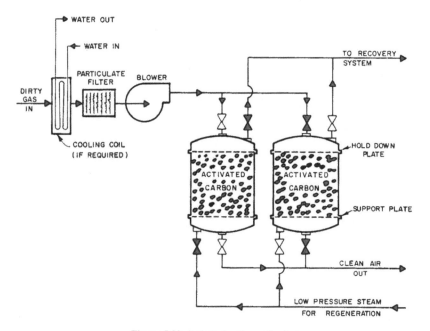

Figure 5.30 Activated carbon adsorber.

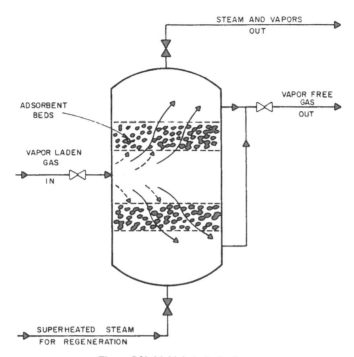

STEAM AND VAPORS
OUT

ADSORBENT
BEDS

VAPOR FREE
GAS
OUT

VAPOR LADEN
GAS
IN

SUPERHEATED STEAM
FOR REGENERATION

Figure 5.31 Multiple bed adsorber.

tion. The stationary adsorbent beds are horizontal and shallow compared to absorber bed depths with typical depths ranging from one-third to several meters. Many adsorbers appear "short" and "fat" compared to absorber towers. Very fragile adsorbents can be placed in layers using support plates below each layer. Some systems are designed with spaced multiple beds. Dirty gases enter between the beds and adsorption occurs as part of the gases pass through the bed above and part pass through the bed below the gas inlet. This provides a space savings, but regeneration is a little more difficult. This arrangement is shown in Figure 5.31.

Two adsorbers in parallel are shown in Figure 5.30. The valves in this figure are shown open for adsorption on the left adsorber and closed on the right. Regeneration valves are open on the right and closed on the left. A third adsorber in parallel could be required if extra cooldown time and/or a spare is needed.

5.8.1.2 Moving Bed Adsorbers

Partially fluidized bed adsorbers are available for continuous use. The adsorbents move slowly down through channels by gravity flow while the gases pass through in a cross flow manner as shown in Figure 5.32. Only *mild* agitation is tolerated by the fragile adsorbent. Complete fluidization would quickly destroy the adsorbent by attrition. Regeneration is accomplished by passing the adsorbent to another chamber.

Another type of moving bed adsorber is shown in Figure 5.33. This unit has a fixed bed of adsorbent that rotates. Continuous adsorption and regeneration occur simultaneously depending on location of the bed within the rotating cycle.

5.8.2 ADSORBER CAPACITY

Capacity in a regenerable adsorber depends on the type of adsorbent and adsorbate used plus amount of adsorbent, concentration of adsorbate, tem-

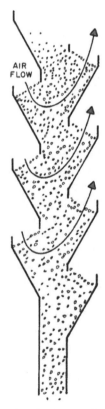

Figure 5.32 Moving-bed adsorber section showing a vertical adsorber panel with horizontal gas flow.

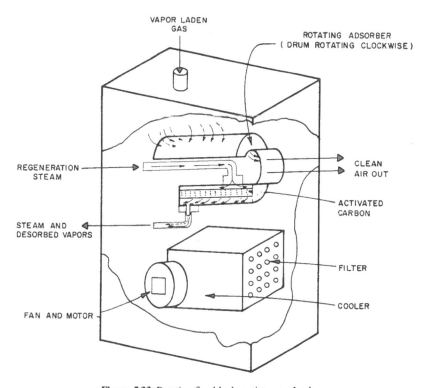

Figure 5.33 Rotating fixed-bed continuous adsorber.

perature, pressure, depositions remaining on adsorbent after regeneration, and type of carrier gas and regeneration gas. These are discussed independently, except quantity of adsorbent, which is used as the common base for all cases. Each of these parameter, should be evaluated to optimize pollutant recovery and operating cost. Figure 5.34 shows the relative difference in saturation capacities, working capacities and the independence of relation between saturation and working capacities for four different commercially available activated carbons. All beds were adsorbed and desorbed at standard conditions. Regeneration was at 10 bed volumes per minute for 20 minutes.

Typical adsorption capacity as a function of pollutant concentration is given in Figure 5.35. These data are for several odorous substances using Pittsburgh Activated Carbon, type BPL, 12 × 130 U.S. Sieve size [60] at standard conditions. Bulk density of this material is about 0.5 g/cm³. Figure 5.36 shows adsorption capacity as inversely related to temperature and directly related to concentration. These data are for BPL carbon at stan-

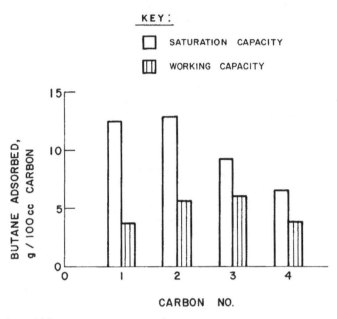

Figure 5.34 Comparative capacities of commercially available activated carbon.

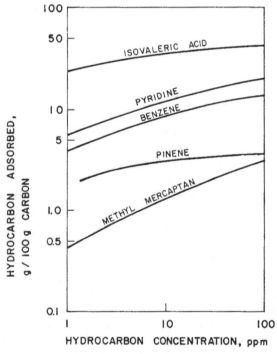

Figure 5.35 Odor adsorption as a function of concentration and species for a particular activated carbon.

370

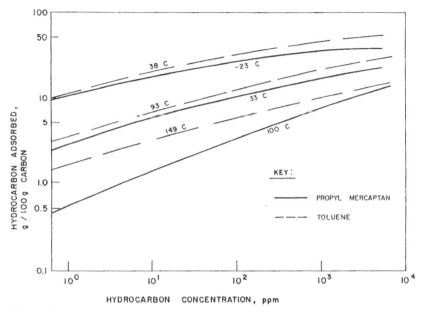

Figure 5.36 Estimated adsorption capacity as a function of temperature and concentration [61,62].

dard pressure. Adsorption should be carried out at as low a temperature as practicable. Many systems operate in the range of 25 to 55°C. Some operate as hot as 120°C. Similar curves for water vapor with a Norton Zeolon molecular sieve and silica gel are given in Figure 5.37. Water vapor concentration is 130 ppm.

Adsorption capacity increases as amount of desorption is increased. Desorption is increased as purge or as desorption temperature is increased compared to the adsorption temperature. Data for this [60] are shown in Figure 5.38 for butane and BPL carbon with a purge rate of 10 bed volumes per minute for 10 minutes. Increased desorption increases working capacity of adsorber. For example, data from the Figure 5.38 system show that increasing desorption time from 10 to 20 minutes increases working capacity by 30%. Increasing desorption from 20 to 30 minutes increases capacity by only another 10%.

Adsorbents can react chemically with the pollutants and the carrier gas stream. Chemically, the carbon in a carbon adsorbent is a reducing agent. Hydrogen sulfide, for example, is reduced and sulfur deposits at the inlet of the adsorber. This carbon must be periodically removed before adsorption capacity decreases below acceptable limits. In this system, as the concentration of oxygen in the carrier gas decreases the adsorption of hydrogen sulfide decreases.

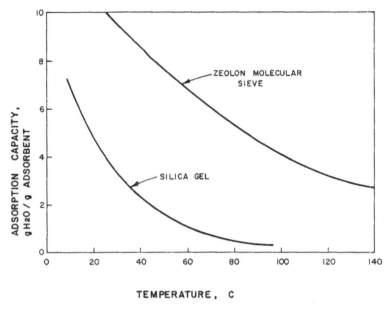

Figure 5.37 Adsorption of water vapor on different adsorbents.

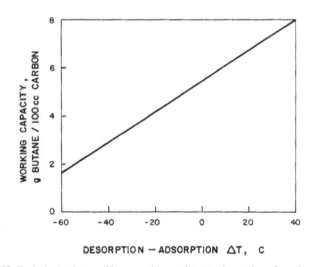

Figure 5.38 Typical adsorber working capacity as related to desorption-adsorption temperature differential.

Adequate retention time is required to remove the pollutant gases by adsorption. Typically, gas flowrates are expressed as bed volumes, e.g., 1,500 bed volumes per hour. This varies with all of the factors listed at the beginning of this section. Turk [63] presents a generalized expression which can be used to obtain an approximate adsorption time to reach adsorbent *saturation*. This is

$$t = \frac{2.9 \times 10^5 \, m}{Q y_i \overline{M}} \qquad (5.49)$$

where

t = time to adsorb, min
m = mass of adsorbent, g
Q = volumetric gas flowrate, Am³/hr
$\overline{M}$ = average molecular weight of adsorbed vapors
y = inlet vapor concentration in air, ppm

Actual working adsorption time is less than this. When several gases are adsorbed, the value for $\overline{M}$ can be found using the chain rule

$$\overline{M} = \frac{n_A M_A + n_B M_B + \ldots n_x M_x}{n_T} \qquad (5.50)$$

where

n_x = number of moles of component x
M_x = molecular weight of component x
n_T = total number of moles of gas vapor adsorbed

5.8.3 ADSORBER PRESSURE DROP

Pressure drop across the packed bed depends on size and shape of adsorbent, depth of bed and the gas flowrate. Pressure drops for several typical adsorbents are shown in Figure 5.39 for standard temperature and pressure operations. Total adsorbent bed pressure drop would be the value noted times the bed depth. Total adsorber pressure drop would include supports, valves, entrance and exit losses, plus friction losses. During normal operations, these other pressure losses would be small compared to the bed pressure loss.

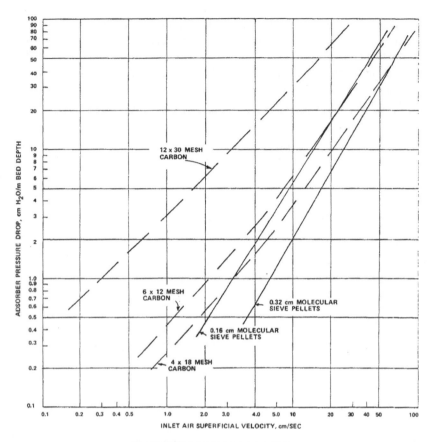

Figure 5.39 Adsorber pressure drops.

5.8.4 ADSORBER EFFICIENCIES

Adsorbers are extremely effective for removal of gaseous pollutants. Even at low inlet gas pollutant concentrations, they can be designed and operated at gas removal efficiencies of 98 to essentially 100%. The rotary bed adsorber may operate continuously at efficiencies of only 90–95%, but this may be adequate. At this time, empirical data must be used for each system to establish the needed design and operating conditions necessary to achieve the required efficiency.

5.8.5 ADSORBER DESIGN

A summary of typical design parameters is given in Table 5.15 for volatile organic compounds (VOC) in a carbon bed adsorber. These are rule-of-thumb values and are commented on as follows. The lower explosive limit, LEL, is the concentration of VOC in air necessary to produce an explosion. A concentration of <50% of the LEL is required and <10% LEL is desired. Higher VOC concentrations drive the adsorption so that a greater mass of material is removed. However, this also loads up an adsorber rapidly and, for high gas concentrations, other techniques (e.g., absorption) should probably be used for cost effectiveness. In contrast, odorous gases could have extremely low concentrations.

Gas velocities of 100 ft/minute should not be exceeded to achieve 95% adsorption efficiency. The working capacity should be established from test data. If unavailable use the Table 5.14 guide line values. In order to achieve >90% adsorption efficiencies, a deep bed (≥8″) should be used. The depth of fixed horizontal beds should not exceed 4 ft to prevent the carbon from being crushed. For best results on this, the manufacturer's data should be consulted.

Regeneration steam and condenser water are used only during the regeneration portion of the cycle. The condenser water should be clean and may be used elsewhere as hot water or it may be cooled and recycled. The con-

TABLE 5.15. Typical Carbon Adsorber Design Parameters.

Parameter	Design Range
VOC BP, °C (°F)	20–175 (70–350)
VOC molecular weight	50–200
VOC concentration, ppm	>5000
LEL, %	10–50
Gas superficial velocity, cm/sec (ft/min)	40.7–50.8 (80–100)
Working capacity, % of saturation capacity	25–30
Transfer zone depth, cm (inch)	15–46 (6–18)
Bed depth limit, m (ft)	1.22 (4)
Pressure drop, cm H_2O/m bed depth (inch H_2O/ft bed depth)	25–125 (3–15)
Carbon density, g/cm³ (lb/ft³)	0.43–0.48 (27–30)
Cycle time, minutes	60
Regeneration steam @ 15 psig, g/g solute (lb/lb solute)	4 (4)
Condenser water, ℓ/kg steam (gal/lb steam)	50 (6)

densed steam contains collected VOC which must be recovered, incinerated, or otherwise properly disposed of.

5.9 HYBRIDS

Numerous variations of standard air pollution control mechanisms have been made. These unique devices usually combine several mechanisms to achieve a specific function, such as improved fine particle collection efficiency and/or high-temperature/high-pressure operation. Several of these novel devices are discussed as examples of hybrid control devices.

5.9.1 ELECTROSTATIC FILTERS

Electrostatically augmented fabric filtration (ESFF) is a technique in which an electrostatic field is established across the fabric to cause preferential deposition of charged particles on some areas while leaving other areas open to gas flow. In normal reverse-air filters, the particles tend to follow the gas streamlines and consequently are deposited relatively evenly on the interior of the bag. Under the influence of an electrostatic field, the charged particles are selectively deposited on the lower portion of the filter bags. Therefore, the upper portions of the bags contain little particulate matter. This results in a very low filter pressure drop as the gas flows out through the clean upper sections of the bags.

Another approach consists of charging the particles then collecting them in a horizontal or vertical moving-bed filter such as that shown in Figure 5.40. This semifluidized bed is continuously recharged in an external chamber and recycled to the collector. The bed material can be nonconductive particles of solids such as sand, ceramics, and plastics. This device could be used for high temperature-pressure applications with the proper filter material.

5.9.2 MOVING BED FILTERS

Two variations of one type of moving-bed filter are discussed in the previous section. Another is the gravel bed filter collector such as the Rexnord unit shown in Figure 5.41. This semicontinuous mechanical collector consists of several units in parallel. Dirty gas enters a settling chamber then passes into a centrifugal section of the device. The partially cleaned gases pass through quartz grain filter beds, then leave the device. Cleaning is done by passing a reverse flush of air through the gravel beds while stirring them with a mechanical rake. Cut diameters of only 1.4 μm have been

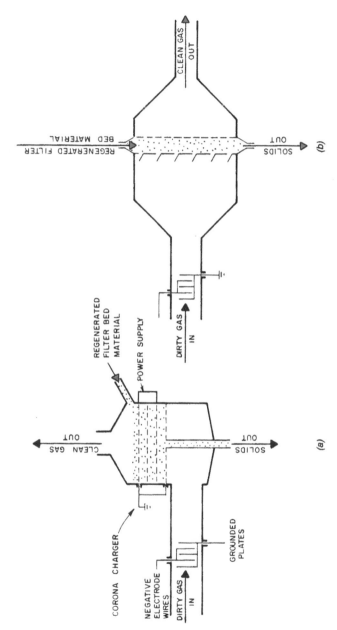

Figure 5.40 Two variations of one type electrostatic filter: (a) vertical charged filter; (b) horizontal charged filter.

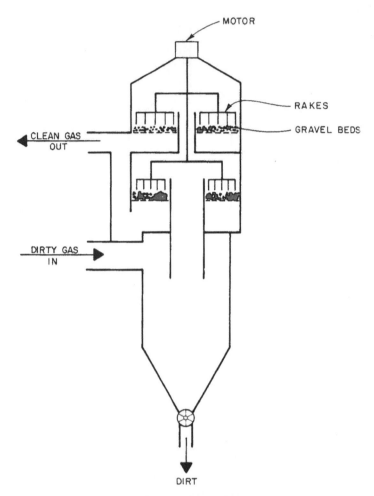

Figure 5.41 Rexnord gravel bed filter.

obtained [54]. Typical bed pressure drop is about 12 cm H_2O. This filter could be useful for high temperature-pressure particle collection.

5.9.3 WET FILTERS

The Cleanable High-Efficiency Air Filter (CHEAF) is one type of wet filter. This system consists of a rotating perforated drum wrapped with a mat of filter fibers and is cleaned by liquid sprays as shown in Figure 5.42. The liquid is present during filtration and aids to close up the fiber pores and help prevent particles from passing through the filter. Extremely high filtration velocities of about 1,000 cm/sec are used. At these velocities, the impaction collection mechanism is dominant for particles > 1 μm. Interception and impaction are dominant for particles from 0.3 to 1 μm and diffusion dominates for the < 0.1-μm particles.

This device appears promising as a fine particle collector. The cut diameter reported [64] is 0.53 μm, which is lower than that for a venturi scrubber at the same pressure drop. Operating pressure drops are about 80 cm H_2O.

5.9.4 CHARGED SCRUBBERS

Particulate wet scrubbers following ESPs are slightly more efficient because of the residual particle charges. Numerous variations of charged wet scrubbing devices are available. Note that the IWS is a charged scrubber and is discussed in Section 5.5.11.2. The procedure consists of charging either or both the particles and the collecting liquid/droplets. The charges may be positive or negative and if both are charged they may be alike or opposite. Some systems use a corona discharge section before the scrubber

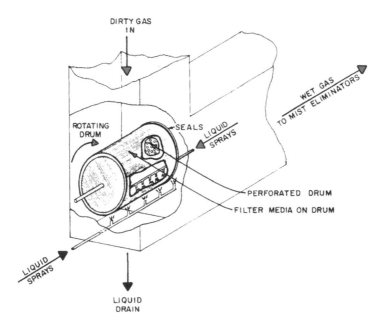

Figure 5.42 CHEAF-type wet filter.

and some use high-intensity probes at the scrubber inlet to charge the entering particles.

5.9.5 EJECTOR SCRUBBERS

The ejector scrubbers are those that utilize waste steam or hot water expended through nozzles as the gas-pumping energy. No blowers are required. This fluid also serves as the collection droplets for particle removal. The hot water (200°C) is kept under pressure to prevent flashing until it expands through the ejector nozzle. At this point, some of the water converts to vapor, but most of the water is in the form of finely atomized collection droplets. These devices are very efficient, producing cut diameters of < 1 μm [64]. Performance is directly related to scrubber energy input so they are high-energy users. If waste heat is available, these devices should be considered, otherwise conventional scrubbers may be more effective energywise.

5.10 GAS CONDITIONING

Gas conditioning is important to the success of control devices. It affects collection in mechanical collectors, changes the particle resistivity of an ESP, protects the bags of filter systems, and provides a positive diffusiophoretic force for scrubbers. Flue gas conditioning normally consists of quenching the gas with a fluid, usually water. Flue gases can also be cooled by dilution air and by radiant heat losses, as described in Section 6.1.3. Cooling by water sprays adds moisture to the gas. Conditioning flue gases with water is the most practical technique and is the only one discussed here.

5.10.1 QUENCHING

Water sprayed into a hot, unsaturated flue gas results in essentially an adiabatic cooling. The gases can be cooled until they reach the adiabatic saturation temperature by a single liquid–gas interaction. Generalized equations are developed by the author and presented here for evaporative cooling of dry air, moist air or combustion gases with up to 12% CO_2 (wet), and 40% inlet gas water vapor using water spray at 70°F. These equations are good for cooling gases down to about 220°F. For $\Delta T \geq 600°F$

$$g = 0.024(\Delta T)^{1.2}0.9993^t\{1 + [CO_2(5.83 \times 10^{-3})$$

$$+ H_2O(2.33 \times 10^{-3})]\} \tag{5.51}$$

where

g = gal H_2O/hr per 1,000 acfm of inlet gas
t = inlet gas temperature, °F
ΔT = gas temperature drop, °F
CO_2 = CO_2 in inlet gas, % (wet)
H_2O = water in inlet gas, %

For $\Delta T < 600°F$

$$g = (3.12 \times 10^{-2} - 1.2 \times 10^{-5}\Delta T)(\Delta T)^{1.2}0.9993^t$$

$$\times \{1 + [CO_2(5.83 \times 10^{-3}) + H_2O(2.33 \times 10^{-3})]\}$$

$$(5.52)$$

Note: Actual gas quench systems may require more (e.g., 10%) spray water due to unevaporated spray.

Normal fossil fuel combustion gases leaving a boiler at about 350°F would reach adiabatic saturation at almost 130°F. Hot incinerator gases would reach adiabatic saturation at higher temperatures depending on inlet gas temperature and moisture content. Equations (5.51) and (5.52) do not take into account the fact that adiabatic saturation may have been reached. Adiabatic saturation temperature and gas moisture content can be found by performing a mass/energy balance or more quickly may be estimated using Figure 5.43. This chart assumes adiabatic saturation using 70°F water and air at 1 atm pressure. Accuracy of the chart can be compared to flue gas using Example Problem 5.10. Note that as the inlet gases contain more water, adiabatic saturation temperature increases for specific inlet gas temperatures. Combustion of wastes usually result in higher moisture content flue gases.

Gas quenching requires the use of very finely atomized liquid droplets plus about 1.5 seconds of gas/spray droplet contact time. Some quenching sprays are directed into the gas as it enters the quench chamber, then the gas and droplets travel cocurrently through the chamber. By using small droplets and adequate time, the gas can become fully saturated.

Use the following for designing quench tanks for *dry scrubbers*:

- Design tank so all liquid is evaporated.
- Provide adequate detention time: 2 seconds may be theoretically adequate, but 6 second minimum should be used.
- Provide large enough diameter to prevent spray from impinging on walls and causing erosion and corrosion.
- Use downflow design (i.e., flue gas enters at top and leaves at bot-

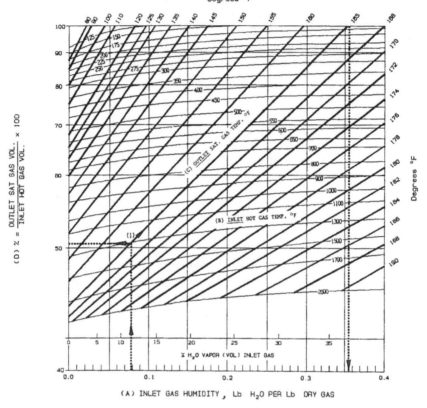

Figure 5.43 Chart for predicting adiabatic gas saturation conditions. (Courtesy Western Precipitation Div. of Joy Technologies, Inc.)

Use of figure:

1. To obtain adiabatic saturation temperature, find intersection of moisture content (A) and Inlet Hot Gas Temp (B). This is point (1). Read upward to the right along Outlet Sat Gas Temp (C) to find answer in °F at 100% on scale (D).

2. To obtain outlet gas moisture content, move vertically downward from point just located to scale (A). Note that no moisture values greater than 0.4 lb water per lb dry gas can be obtained from this figure.

3. To obtain saturated gas volume, move horizontally from point (1) to scale (D) and multiply value times hot gas volume divided by 100.

tom): this allows liquid to drain at bottom and keeps quench bottom cool.
- Use fresh water, rather than recycled bleed slurry.
- Flue gas flow distribution entering quench tank must be uniform: existing tanks with 90° elbows or T's require turning vanes (these only last 3-6 months); on new tanks install straight run inlet of at least 4 duct diameters.
- Spray nozzles:
  —Use Hastelloy C-276.
  —Try Wheelabrator Air Pollution Control Co. proprietary nozzles.
- Spray manifolds:
  —May interfere with gas distribution in small tanks.
  —Try using pipe lance parallel to gas flow.
  —Consider installing nozzles in conical inlet of tank, out of direct gas flow contact.
- Consider "wallpapering" a thin layer of corrosion-resistant metal to protect existing tanks.
- Material of construction: Hastelloy C-276 or alloy 59 are good.
- If the dry scrubbers are spray dryer reactors:
  —Rotary atomizers use reactor height to diameter of 0.8 to 1.
  —Pneumatic atomizers use reactor height to diameter of 1.5 to 1.

Saturated gases have many advantages, noted in Section 5.10. Of special importance is the elimination of negative diffusiophoretic forces in wet scrubbers. This is achieved because the saturated gases do not cause droplet evaporation (without simultaneous condensation) when they pass into a wet scrubber that depends upon droplets for collection and/or absorption.

Example Problem 5.10

Flue gases with a dry analysis of 11% CO_2 and 8% O_2 containing 12% water enter an adiabatic saturator at 1300°F. Quench water at 70°F is used. Find:

a. Adiabatic saturation temperature
b. Saturated gas water content
c. Volume of saturated gas relative to inlet gas

Solution

Basis: *100 lb moles inlet gas*

Use Figure 5.43 to estimate a saturation temperature of 165°F.

From steam table data and Equation (4.37) (or Figure 5.43), the saturation water content of the gases would be = (5.335 psi/14.7 psi) × 100 = 36.3 %. Assuming no leaks and that all water is evaporated, the mass balance is:

$$\text{lb wet gas in} + \text{lb water added} = \text{lb wet gas out}$$

Further, assuming adiabatic saturation, the energy balance relative to 70°F and using thermodynamic data from Section 2.11 and steam tables is

$$m_w(h_g - h_\ell) = \left[m_d \left(\frac{0.27 \text{ Btu}}{\text{lb °F}} \right) + m_v(0.52) \right](1300 - t)$$

where

h_g = enthalpy of water vapor at assumed saturation temp., Btu/lb
h_ℓ = enthalpy of liquid water at inlet temp., Btu/lb
m_w = mass of water added, lb
m_d = mass of dry gas in and out, lb
m_v = mass of water vapor in, lb
t = calculated adiabatic saturation temp., °F

$$\text{MW of dry gas in} = (0.11)(44) + (0.08)(32) + (0.81)(28)$$

$$= 30.08 \frac{\text{lb}}{\text{lb mole}}$$

$$m_d = (100 - 12) \text{ moles } \frac{30.08 \text{ lb}}{\text{lb moles}} = 2{,}647 \text{ lb}$$

$$m_v = (12 \text{ moles}) \frac{18 \text{ lb}}{\text{mole}} = 216 \text{ lb}$$

$$m_w = (88 \text{ moles dry gas}) \left(\frac{36.3 \text{ moles water}}{100 - 36.3} \right)$$

$$\times \left(\frac{18 \text{ lb}}{\text{mole}} \right) - 216 = 686.7 \text{ lb}$$

Substituting into the mass and energy balance and solving, t = 930°F when assumed saturation temperature is 165°F. This is a trial and error solution which converges very rapidly for assumed temperature.

At 169.5°F: saturation H_2O = 40.30%, m_w = 853.16, and calculated t = 169.4°F

Answers are:

a. 169.5°F

b. 40.3%

Part *c.* is found:

$$\text{moles gas out} = (88)\left(\frac{100}{100 - 40.3}\right) = 147.4 \text{ per } 100 \text{ moles in}$$

$$\text{Gas Vol ratio out to in} = \frac{147.4}{100}\left(\frac{460 + 169.5}{460 + 1300}\right) = 0.53$$

5.10.2 USING CONDENSATION FORCE

When saturated gases enter a wet scrubber, they may be further cooled. This results in the formation of condensing droplets. These droplets may form on very fine particles, causing them to become larger and more easily captured, or they may condense on the scrubber collector droplets, sweeping very fine particles with them. Both mechanisms are beneficial to wet scrubber operation. Use of these positive diffusiophoretic forces is called "flux-force condensation" (FFC) or "condensing scrubbers." Schifftner describes FFC as the opposite of spray-drying evaporation. Condensing forces work best when the saturated gases are 165°F or hotter. Example Problem 5.11 shows how just a relatively small temperature drop can condense a large amount of water vapor from a saturated gas stream. This makes wet scrubbers very attractive for control of toxic metals from incineration processes, due to both the condensation of the metals and the capture of the fine particles.

Example Problem 5.11

(a) The same gas as initially used in Example Problem 5.10 is to be adiabatically quenched with a hot water spray. The hot water is at 148°F. Find the conditioned gas saturation temperature and moisture content.

Solution

By the same procedures as used in Example Problem 5.10, find $t = 171°F$ and water content = 41.71%.

(b) This conditioned gas is now cooled 16°F (to 155°F) using cold water sprays. Show how much water can be condensed from the gas.

Solution

On the same basis (i.e., 88 moles dry gas):

$$\text{Water vapor before condensation} = (88)\,\frac{41.71}{100-41.71}\left(\frac{18\text{ lb}}{\text{mole}}\right)$$

$$= 1{,}133.5\text{ lb}$$

$$\text{Water vapor content after cooling} = \frac{4.203}{14.7}\,100 = 28.59\%$$

$$\text{Water vapor not condensed from gas} = (88)\,\frac{28.59}{100-28.59}\,(18)$$

Therefore,

$$\text{Water condensed} = 1{,}133.5 - 634.2$$

$$= 499.3\text{ lb (which is 44\% of}$$
$$\text{water vapor)}$$

5.11 SPECIAL CASE—PARTICLE COLLECTION IN SPRAY TOWERS

Spray towers are used for control of both particulate and gaseous pollutants. Theoretical and empirical data are used by Hesketh [66] to develop particle collection efficiency equations for cocurrent and countercurrent spray towers. The range of parameters studied are typical of those used in the practical operation of conventional spray towers. This work shows that the most important parameter in cocurrent spray scrubbers is inlet dust particle size. In decreasing order of importance are gas velocity, collector droplet size, liquid-to-gas ratio and length of scrubber. Dust collection parameters in countercurrent scrubbers in order of decreasing importance are collector droplet size, liquid-to-gas ratio, length of scrubber and gas velocity. Some of the parameters are directly related to collection and some are indirectly related as shown below.

5.11.1 PARTICLE COLLECTION FORCES

At least a dozen forces can be considered for cocurrent and countercurrent spray towers. These include:

Drag Forces—Important in both types of spray towers for both the large spray droplets and the particles to be collected. Stokes drag relationships can be used for particle (droplet) Reynolds numbers (Re_p) up to about 0.1. Above this, Newton's equations are applied. Drag forces result in particles traveling essentially at gas velocities.

Gravitational Force—Important for the large droplets, but negligible for the small particles traveling with the gas.

Electrostatic Force—Charged inlet particles could be attracted to neutral droplets in an electrically conducting liquid. The effects would only be significant at the initial point of gas-liquid contacting.

Inertial Impaction—Significant in the collection of particles on droplets and other targets in the system.

Scavaging or Sweeping Impaction—This sequential inertial collection by droplets falling through a gas is a very significant force.

Condensation/Flux Force/Diffusiophoresis—These are considered to be the same force and may or may not be significant, depending on operating parameters.

Centrifugal Force—Effective for removal of spray droplets from the gas, but negligible for dust particles.

Other Forces—Interception, thermal diffusivity (Brownian motion), thermal phoresis, photophoresis and sonic phoresis are considered to be negligible, based on current spray tower design and operating practices.

Droplets used in spray towers are usually large. Diameters of 700 to 1200 μm are common. These droplets have an initial downward vertical velocity as they leave the spray nozzle dependent on the design and operating parameters. They fall under the force of gravity balanced by drag forces due to the gas movement, which either opposes (countercurrent towers) or aids (cocurrent towers) this movement.

Using this procedure, the behavior of the particles or doplets can be determined as a function of pressure, viscous and other forces. Other forces include gravity, but electrostatic forces are considered to be very small. Combining this method for determining drag forces with a spray tower design and operating parameters makes it possible to apply the various collection mechanisms. The significant collection mechanisms modeled from these data are inertial impaction, sweeping impaction and diffusiophoresis.

Sweeping impaction is a little-discussed collection procedure and is introduced here. In spray towers, impaction between dust particles and liquid collector droplets occurs repeatedly throughout the length of the contacting zone. The procedure used here to account for this is proposed by Cheng of the U.S. Bureau of Mines [67]:

$$E_i = 1 - \exp - \frac{3\epsilon_i L Q_l}{2 D Q_g} \qquad (5.53)$$

where

E_i = overall collection efficiency for collecting droplets by sweeping impaction over the collecting length
ϵ_i = droplet-particle single impaction collection efficiency, fraction
L = collecting length over which the capture process is effective
D = average droplet diameter
Q_l = volumetric flowrate of liquid
Q_g = volumetric flowrate of gas

Units for these must be consistent so as to produce a dimensionless value for E_i.

5.11.2 APPLICABILITY

The particle collection efficiency equations developed are applicable to many industrial spray tower systems. The operating parameter ranges are representative of systems that use a "cool" or cooled inlet gas stream. The initial model is for a large power station flue gas desulfurization scrubbing system that requires particulate removal, in addition to SO_2 absorption.

Note that effective wet scrubbing consists of cooling (conditioning) hot gases before they enter the scrubber to minimize the adverse effects of negative diffusiophoresis and, if possible, to develop a positive flux force. If this is not carried out, the actual spray tower particle collection will be less than that predicted by the following equations.

5.11.2.1 Cocurrent Spray Tower Particle Collection Efficiency Model

Cocurrent spray tower particle collection efficiency is essentially only by impaction. Generalized particle collection in a 2 stage cocurrent spray tower is:

$$Pt_{o,i} = \frac{(2.50) D_c^{0.574} V^{0.69}}{\bar{d}_M^{1.60} [(L/G)L]^{0.393}} \qquad (5.54)$$

where

$Pt_{o,i}$ = penetration due to impaction, percent
      = $100 - E_{o,i}$
$E_{o,i}$ = impaction collection efficiency, percent
$\bar{d}_M$ = inlet dust mass mean diameter, μm
D_c = spray droplet mean diameter, μm
L/G = liquid-to-gas ratio, l/m^3
   L = effective collecting length below 2nd spray, ft
   V = cocurrent gas velocity, ft/sec

The conditions and range of variables examined for the cocurrent spray tower are:

Variable	Range
$\bar{d}_M$, μm	2.5–4.0
D_c, μm	700–900
L/G, l/m^3	2.5–3.5
L, ft	35–45
V, ft/sec	18–22

The model also shows that the most significant variable is inlet dust particle size. The second is gas velocity, the third is collector droplet size and the least significant are liquid-to-gas ratio and collecting length. Collection efficiency varies directly with particle size, L/G, and L and varies indirectly with collector size and cocurrent gas velocity.

5.11.2.2 Countercurrent Spray Tower Particle Collection Efficiency Model

In a 5-stage spray tower with countercurrent gas flow and following a quencher or prescrubber that releases only saturated particles, the particle collection is mainly by sweeping impaction of the falling droplets on the rising particles. A diffusiophoretic enhancement occurs due to condensation and growth of the smaller particles (i.e., those ≤ 1.5 μm). The resulting general equation for sweeping impaction including condensation growth is then:

$$Pt_o = \frac{(0.0568)D_c^{1.155}}{(L/G)^{0.606}L^{0.237}V^{0.126}} \qquad (5.55)$$

where

Pt_o = overall penetration for the 5 spray levels, %
   = 100 − E_o
E_o = overall efficiency, %
D_c = spray droplet mean diameter, μm
(L/G) = total scrubber liquid-to-gas ratio, l/m³
  L = effective collecting length below lowest spray nozzles, ft
  V = countercurrent gas velocity, ft/sec

The conditions and range of the variables studied for this countercurrent collection efficiency model are:

Variable	Range
D_c, μm	700–1420
L/G, l/m³	7.6–9.4
L, ft	8–16
V, ft/sec	9–11

The most important variable is spray droplet size. Second is L/G and third is L. Gas velocity change has little effect. Efficiency is directly related to L/G, L and V and indirectly to D_c. Note that in a countercurrent unit as L is increased, the ΔV decreases so the effect of increasing L to improve efficiency is not as significant as in the cocurrent spray tower.

Example Problem 5.12—Cocurrent Spray Tower

Cooled gases containing particulate matter with a mass mean diameter of 3.2 μm and a geometric deviation of 2.3 pass through a two-stage cocurrent spray tower at 20 ft/sec inlet velocity. The spray nozzles and recycle liquid nozzle spray bar pressure result in a spray droplet with a mean diameter of 750 μm. The effective particle collecting length in the tower is 35 ft, and the recycle liquid-to-gas ratio is 3 l/m³. Predict the particle mass collection efficiency.

Solution [Use Equation (5.54)]

$$Pt = \frac{(2.50)(750)^{0.574}(20)^{0.69}}{(3.2)^{1.60}[(3)(35)]^{0.393}} = 22.05\%$$

Therefore: efficiency = 100 − 22.05 = 77.95%

Example Problem 5.13—Countercurrent Spray Tower

Cooled gases containing particulate matter pass through a five-stage countercurrent spray tower at 10 ft/sec inlet velocity. The spray droplets have a mean diameter of 750 μm. The effective collecting length is 15 ft, and the total recycle liquid-to-gas rate is 8 l/m³. Predict the particle mass collection efficiency.

Solution [Use Equation (5.55)]

$$Pt = \frac{(0.0568)(750)^{1.155}}{(8)^{0.606}(15)^{0.237}(10)^{0.126}} = 13.28\%$$

Therefore: efficiency $= 100 - 13.28 = 86.72\%$

Example Problem 5.14—Combination System

Predict the overall particle mass collection efficiency for a series system consisting of the Example 5.12 cocurrent scrubber followed by the Example 5.13 countercurrent scrubber.

Solution

$$\text{Overall Efficiency} = \left[1 - \left(\frac{Pt_1}{100}\right)\left(\frac{Pt_2}{100}\right)\right]100$$

$$= [1 - (0.2205)(0.1328)]100 = 97.07\%$$

5.12 DEVICE SIZING

Data are presented in specific sections that permit the extrapolation of some collection information to other similar devices. In addition to these data, it is necessary to keep the following in mind. Direct extrapolation of data would require the same gaseous and particulate components and concentrations, the same operating conditions, the same temperature, humidity, pressure, pressure drop, holdup time, velocity, and system configuration. It is also necessary that there be similarity of motion. Similarity of motion in the model and the full-scale device is achieved by having equal values for several gas flow and particulate dimensionless quantities. This includes the Reynolds flow number, Re_f [Equation (3.1)], Stokes number, St [Equations (3.6), (3.49), or (3.50)], the Mach number, Ma [Equation

(1.15)], the Euler number, *Eu*, and the Froude number, *Fr*. The Reynolds number is the ratio of inertial to viscous forces. Stokes number is the ratio of stopping distance to device characteristic dimension. The Mach number is the ratio of the square root of the inertial to elastic forces. The Euler number is the ratio of the square root of pressure to inertial forces and is

$$Eu = \frac{\Delta P}{1/2\, \varrho_g v_g^2} \qquad\qquad (5.56)$$

Fr is the ratio of inertial to gravitational forces and is important only if gravitational forces are significant in the unit. Froude number is

$$Fr = \frac{v_g^2}{D_g} \qquad\qquad (5.57)$$

where *D* is the device characteristic dimension.

If the inertial forces as given by Re_f and St are constant for two different systems and all other physical properties of the substances are constant, then the ratio $(St/Re_f)^{1/2}$ would be a constant. When collector diameter equals device diameter this gives inertial force as proportional to d/D. The system with the larger diameter would have a lower impaction effectiveness and a design compensation would be required to account for this if the same effectiveness is desired.

5.13 CHAPTER PROBLEMS

5.13.1 HIGH-EFFICIENCY CYCLONE OPTIMIZATION

A high-efficiency, Stairmand-type cyclone is to be used to remove particles from a 4,000 Nm³/hr gas stream. Assume the particles have a mass mean diameter of 2.6 μm, a standard geometric deviation of 2.0, and a density of 1.4 g/cm³. The gas is air at 75°C and contains 25 g/m³ dust.

a. Establish the cyclone dimensions based on an inlet velocity of 18 m/sec (60 ft/sec).
b. Determine pressure drop.
c. Calculate the cut diameter using Lapple equation and N_e of 8.
d. Calculate overall collection efficiency using Theodore method.
e. How much dust is removed from the cyclone per hour?

5.13.2 ESP AND FILTER

(a) Assume that the dust in Problem 5.13.1 is fly ash at 130°C and that it is introduced into an ESP operating at 10 nanoamps/cm² with an SCA of 11 m² per 1000 m³/hr.

a. Give your suggestion as to the approximate ESP collection efficiency.

b. Discuss how this can be increased to 99.5%.

(b) The same dust is introduced into a reverse air-plus-shake filter containing a total air-to-cloth ratio of 33.4 m³/hr per m² and an active air-to-cloth ratio of 36.6. The cleaned cloth pressure drop is 10 cm H₂O.

a. Determine the pressure after 5 minutes and . . .

b. . . . after 10 minutes of operation.

5.13.3 VENTURI SCRUBBER

Dust in air enters a venturi scrubber at 180°C. The dust has a number mean diameter of 0.95 μm, a geometric standard deviation of 2.2, a density of 1.9 g/cm³, and an inlet concentration of 10 g/m³. The inlet gas contains 5.5% water vapor. Assume that the adiabatic saturation temperature is 55°C. Gas flow rate in is 100,000 Nm³/hr. Good particulate efficiency is desired (no gas absorption). This part of the venturi is under 50 cm H₂O vacuum. The scrubber aerodynamic cut diameter is 0.8 μm.

a. Determine the approximate throat diameter for a conventional open throat designed venturi.

b. Estimate the overall efficiency that should be possible for a well-designed and operated venturi with this dust.

c. Determine the dust outlet concentration in g/m³.

5.13.4 VENTURI SCRUBBER

Use the data from Problem 5.13.3.

a. Estimate the scrubber pressure drop needed.

b. Under these conditions, how much liquid would be required in 1 min?

c. Determine the contacting power for this system if the nozzle pressure drop is 1 atm.

d. Calculate the dimensionless pressure ratio for this unit.

e. Where the scrubbing liquid is water at SC injected tangentially at the throat, determine the optimum length of the venturi throat.

5.13.5 PACKED TOWER ABSORBER

A packed tower is to handle 100,000 Nm³/hr of 45°C gas consisting of air with 5.1% ammonia. The concentration of ammonia leaving is not to exceed 0.1%. Use a packing operating pressure drop of 2 cm H₂O per meter of packing. The packing is to be ceramic Raschig rings. The scrubbing liquid is a weak acid containing no NH_3. Assume its properties are those of water at SC and that it enters at the rate of 1 mole of liquid per mole of inlet gas. Adiabatic saturation occurs at 30°C.

a. Determine the tower diameter.
b. Calculate superficial gas velocity in cm/sec.
c. If the number of overall gas transfer units equals 12.1, approximate the height of the tower packing.
d. Determine the tower, ΔP.
e. Find a $K_G a$ for this system if the partial pressure of NH_3 in the outlet liquid is 0.030 atm.
f. Accounting for gas absorbed and water of saturation, what is the minimum size mist eliminator section needed for vertical baffles with 30°C gas?

5.13.6 WET SCRUBBING

A "mini scrubber" is a bench scale test scrubber. It has an orifice throat, but otherwise operates similar to a venturi scrubber. Consider that a "mini scrubber" operates with inlet gas at 95°C. This gas has a humidity of 0.01 g H_2O per gram dry air. The scrubbing water is at 20°C. The flow rate of the outlet gas is 60 cfh and the scrubber ΔP is 20 inches water. Throat orifice is 5/32" in diameter.

a. Determine adiabatic saturation temperature.
b. Calculate throat velocity.
c. Determine the liquid-to-gas ratio. (Note that the venturi equation is applicable when half the orifice scrubber ΔP is used as the venturi ΔP.)

REFERENCES

1 Stairmand, C. J. 1951. *Trans. Inst. Chem. Eng.*, 29:356.
2 Swift, P. 1969. *Steam Heating Eng.*, 38:453.
3 Lapple, C. E. 1967. *Air Pollution Engineer Manual.* J. A. Danielson, ed. U.S. Department of Health, Education and Welfare, PHS No. 999-AP-40, p. 95.
4 Leith, D. and W. Licht. 1972. *AIChE Symposium Series*, 68(126):126.
5 Koch, W. H. and W. Licht. 1977. "New Design Approach Boosts Cyclone Efficiency," *Chem. Eng.*, 34:80–88.
6 Shepherd, C. B. and C. E. Lapple. 1939. *Ind. Eng. Chem.*, 31:972.
7 Briggs, L. W. 1946. "Effect of Dust Concentration of Cyclone Performance," *Trans. Am. Inst. Chem. Eng.*, 42(30):511–526.

8 Theodore, L. and V. DePaola. 1980. "Predicting Cyclone Efficiency," *JAPCA*, 30(10):1132–1133.

9 Lapple, C. E. 1951. "Processes Use Many Collection Types," *Chem. Eng.*, 58(8): 145–151.

10 Parker, R., R. Jain, S. Calvert, D. Drehmel, and J. Abbott. 1981. "Particle Collection in Cyclones at High Temperatures and High Pressure," *ES&T*, 15(4):451–458.

11 Frisch, N. W. and T. P. Dorchak. 1978. "Impact of Fuel on Precipitator Performance," *Poll. Eng.*, 10(5):63–65.

12 Hesketh, H. E. 1985. "A Simplified Procedure for Approximating Electrostatic Precipitator Size/Collection Efficiency Relationships for Coal-Fired Boilers," *JAPCA*, 35(11):1188–1189.

13 Billings, C. E. et al. 1970. *Handbook of Fabric Filter Technology Vol. 1, Fabric Filter Systems Study*, National Technical Information Service No. PB 200 648.

14 Cooper, D. W. and V. Hampl. 1976. "Fabric Filtration Performance Model," in *Conference on Particulate Collection Problems in Converting to Low Sulfur Coals*, EPA-600/7-76-016, pp. 149–185.

15 Carr, R. C. 1984. "Fabric Filter Technology," *JAPCA*, 31(4):399.

16 Bush, P. V., T. R. Snyder, and R. L. Chang. 1989. "Determination of Baghouse Performance from Coal and Ash Properties," *JAPCA*, 39(2):228–237.

17 Abbott, J. H. and D. C. Drehmel. 1976. "Control of Fine Particulate Emissions," *Chem. Eng. Prog.*, 72(12):47.

18 Bergmann, L. 1974. "New Fabrics and Their Potential Application," *J. Air Poll. Control Assoc.*, 24(12):1187–1192.

19 Hesketh, H. E. 1974. *Understanding and Controlling Air Pollution, 2nd ed.* Ann Arbor, MI: Ann Arbor Science Publishers, Inc.

20 Hesketh, H. E., A. J. Engel, and S. Calvert. 1970. "Atomization—A New Type for Better Gas Scrubbing," *Atmos. Environ.*, 4(7):639–690.

21 Behie, S. W. 1974. "Aerosol Capture, Jet Dispersion and Pressure Drop Studies of a Large and Small Venturi Scrubber, Ph.D. Thesis, The University of Western Ontario.

22 Nukiyama, S. and T. Tanasawa. 1938. "An Experiment on the Atomization of Liquid by Means of an Air Stream," *Trans. Soc. Mech. Eng.* (Japan), 4(14)86.

23 Hesketh, H. E. 1972. "Gas Cleaning Using Venturi Scrubbing," paper 72-111, presented at 65th National Air Pollution Control Association Meeting, Miami.

24 Marshall, W. R., Jr. 1954. "Atomization and Spray Drying," *Chem. Eng.*, Progress Monograph Series, 50(2).

25 Green, H. L. and W. R. Lane. 1964. *Particulate Clouds—Dusts, Smokes and Mists, 2nd ed.* London: E. & F. N. Spon, Ltd., Chapter 2.

26 Calvert, S. et al. 1976. "APS Electrostatic Scrubber Evaluation," EPA-600/2-76-154a.

27 Semrau, K. T. 1977. "Practical Process Design of Particulate Scrubbers," *Chem. Eng.*, 84(20):87–91.

28 Calvert, S. 1977. "How to Choose a Particulate Scrubber," *Chem. Eng.*, 84(18): 54–68.

29 Ekman, F. O. and H. F. Johnstone. 1951. "Collection of Aerosols in a Venturi Scrubber," *Ind. Eng. Chem.*, 43:1358.

30 Hesketh, H. E. 1974. "Atomization and Cloud Behavior in Wet Scrubbers," U.S.-U.S.S.R. Symp. on Control of Fine Particulate Emissions, San Francisco, CA, January 15–18.

31 Yung, S. C., H. F. Barbarika, and S. Calvert. 1977. "Pressure Loss in Venturi Scrubbers," *J. Air Poll. Control Assoc.*, 27(4):348–351.

32 Boll, R. H. 1973. "Particle Collection and Pressure Drop in Venturi Scrubbers," *Ind. Eng. Chem. Fund.*, 12(40).

33 Theodore, L. 1977. "Pressure Loss in Venturi Scrubbers," *J. Air Poll. Control Assoc.*, 27(10):938.

34 Hesketh, H. E. 1986. *Fine Particles in Gaseous Media, 2nd ed.* Lewis Publishers, Inc.

35 Chilton, T. H. and A. P. Colburn. 1934. "Mass Transfer Coefficients," *Ind. Eng. Chem.*, 26:1183.

36 Galeano, S. F. 1966. "Removal and Recovery of Sulfur Dioxide in the Pulp Mill Industry," Ph.D. Dissertation, University of Florida.

37 Koehler, G. R. and E. J. Dober. 1974. "New England SO_2 Control Project Final Results," EPA FGD Symposium, Atlanta.

38 Gleason, R. J. 1971. "Venturi-Type Contactor for Removal of SO_2 by the Limestone Wet-Scrubbing System Process," Cottrell Environmental Systems Inc., EPA Project No. CES-116.

39 1974. *The McIlvaine Scrubber Manual.* Northbrook, IL: Th. McIlvaine Company.

40 Yung, S. C., S. Calvert, H. F. Barbarika, and L. E. Sparks. 1978. "Venturi Scrubber Performance Model," *Environ. Sci. Technol.*, 12(4):456–459.

41 Hesketh, H. E. and K. Mohan. 1983. "Specifying Venturi Scrubber Throat Length for Effective Particle Capture at Minimum Pressure Loss Penalty," *JAPCA*, 33(9):854–857.

42 McNulty, K. J., J. P. Monat, and O. V. Hansen. 1987. "Performance of Commercial Chevron Mist Eliminator," *CEP*, 83(5):48–55.

43 Calvert, S., S. C. Yung, and J. J. Leung. 1975. "Entrainment Separators for Scrubbers—Final Report," EPA-650/2-74-119b.

44 Bell, C. G. and W. Strauss. 1973. "Effectiveness of Vertical Mist Eliminators in a Cross Flow Scrubber," *J. Air Poll. Control Assoc.*, 23(11):967.

45 Ostroff, N. and T. D. Rahmlow, Jr. 1976. "Demister Studies in a TCA Scrubber," MASS-APCA Meeting, Drexel University, Philadelphia.

46 York, O. H. 1954. "Performance of Wire Mesh Demisters," *Chem. Eng. Prog.*, 50(8):412.

47 Calvert, S. 1978. "Guidelines for Selecting Mist Eliminators," *Chem. Eng.*, 85(5):109–112.

48 Spink, D. R. 1987. "Gas Solid Separation and Mass Transfer Using a Unique Scrubbing Concept," *Proceedings of the AIChE National Meeting*, Houston, Mar. 29–Apr. 2.

49 Johnson, J. M. et al. 1976. "Scrubber Experience at Mohave," Paper 11, EPA-600/7-76-106, p. 208.

50 Edwards, W. M. and P. Huang. 1977. "Air Pollution Control: The Kellog-Weir Air Quality Control System," *Chem. Eng. Prog.*, 73(8):64.

51 Holzman, M. I. and R. S. Atkins. 1988. "Retrofitting Acid Gas Controls: A Comparison of Technologies," *Solid Waste & Power*, 2(5):28–32.

52 Rhudy, Richard. 1988. "Spray Drying for High-Sulfur Coal," *EPRI Journal*, 13(6): 43–44.

53 Sandell, M. A. et al. 1989. "Fabric Filters or Electrostatic Precipitators—Which Works Best with a Spray Dryer?" *Solid Waste and Power*, III(1):34–41.

54 Jozewics, W. and G. T. Rochelle. 1986. *J. Env. Progress*, 5:218–223.

55 Peterson, J. R. and G. T. Rochelle. 1988. "Production of Lime/Fly Ash Absorbents for Flue Gas Desulfurization," *Proceedings of the EPA/EPRI/APCA 1st Combined FGD and Dry SO$_2$ Symposium*, St. Louis, Paper 9A-1.

56 Eckert, J.S. 1970. "No Mystery in Packed Bed Design," *Oil Gas J.*

57 Perry, J. H. 1950. *Chemical Engineer's Handbook, 3rd ed.* New York: McGraw-Hill Book Company.

58 Eckert, J. S. 1975. "How Tower Packings Behave," *Chem. Eng.*, 82(4):70–76.

59 Morgan, W. D. 1970. "The Sorption of Sulfur Dioxide in a Cycled Column: A Comparison Study," M.S. Thesis, The Pennsylvania State University, CAES Publication No. 116-69.

60 Clapham, T. M., T. J., Junker and G. S. Tobias. 1970. *ASHRAE Trans.*, 76(II):75.

61 Faulkner, W. D., W. G. Schuliger, and J. E. Urbanic. 1973. "Odor Control Methods Using Granular Activated Carbon," 74th National AIChE Meeting, New Orleans.

62 Lovett, W. D. and R. L. Poltorak. 1974. "The Use of Activated Carbon for Controlling Odorous Air Pollutants," APCA Specialty Conference, Pittsburgh, March 7–8.

63 Turk, A. 1968. "Source Control by Gas-Solid Adsorption," in *Air Pollution*. A. Stern, ed. New York: Academic Press, Inc.

64 Drehmel, D. C. 1977. "Fine Particle Control Technology," *J. Air Poll. Control Assoc.*, 27(2):138.

65 Schifftner, K. 1988. "Condensing Flue-Gas Scrubbers Vie for Gas-Cleanup Duties," *Power* (May):41–42.

66 Hesketh, H. E. 1995. "Predict Particle Collection in Spray Towers," *Chem. Eng. Progress*, 91(10):98–100.

67 Cheng, L. 1973. "Collection of Airborne Dust by Water Sprayer," *IEC Process Design Develop.*, 12(3):221–225.

CHAPTER 6

Control Systems

CHAPTER 2 discussed the control of air emissions and of hazardous wastes by incineration, and stressed the need for tail-end emissions controls when dealing with combustion, incineration, and other emissions sources. Chapter 3 presented the theoretical mechanisms for the control of particulates, while Chapter 4 presented the control mechanisms for gaseous pollutants. Chapter 5 goes on to discuss the application of these mechanisms to the design and operation of emissions-control devices.

This chapter now breaks away from theory, and discusses the practical aspects of emissions-control systems. The collection mechanisms and devices are treated here as parts of entire systems that may be used for the control of particulate and/or gaseous pollutants. In general, the systems are considered as installed "flange-to-flange" facilities. This means that the required auxiliaries and components such as gas ducts, stacks, enclosures, and land resources are not included as part of the system and system costs. What is included are the basic hardware of the device, the blowers, integral pipes and ducts, structural supports, and utility connections. Several important upstream and downstream factors relevant to the successful operation of the systems are presented as separate sections in this chapter.

Operation, maintenance, and cost information are discussed as appropriate with each specific system. In addition, several sections of general information on these subjects are included near the end of the chapter. Good operation and maintenance are absolutely necessary for effective economical operation of control systems.

A relative cost perspective for some of the various control systems can be seen in Figure 6.1. These data show the sales profile for the mid 1980s. This information is extracted from the Industrial Gas Cleaning Institute, Inc. (IGCI) member sales reports. Costs of shipping, erection, and auxil-

399

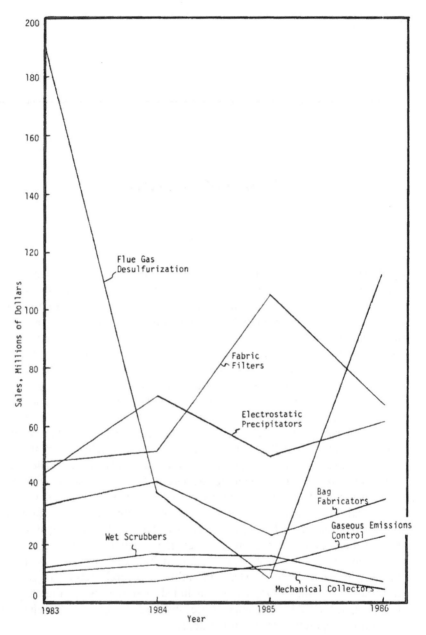

Figure 6.1 Recent trends in control equipment sales.

iaries are not included in these data and the data simply show sales reported in a given year. About 31 member companies reported these sales data.

If the data are smoothed, one can see that the sales of most systems are relatively consistent except for FGD systems which are decreasing. Again, with smoothed data, the magnitude of sales of ESPs, filters, and FGD systems are about equal. During the period from 1966 through the 1970s, sales were increasing at an exponential rate for ESPs, fabric filters, and FGD systems. This boom is over, but could be repeated as more incineration system controls are required. Currently, the level of sales dollars generated from the installation of replacement filter bags is about half that generated from the construction of new baghouses.

Wet scrubber sales, other than for FGD and gaseous emission control (GEC), run very close to and parallel with those of mechanical collectors. GEC systems do not include wet scrubbers but do include absorbers, dry scrubbers, and adsorbers. Incineration controls could spur sales of both wet scrubbers and GEC systems.

6.1 UPSTREAM OF THE CONTROL SYSTEM

Process modifications, mechanical cleaners, and gas conditioning can be considered possible modifications to help reduce emissions either with or without any other control system. Mechanical cleaners are included in this category rather than as a control system even though they do control emissions. Mechanical cleaners are often used as precleaners before most efficient control systems. They also may be integral parts of control systems, e.g., dewatering sections of scrubbing systems. The relatively constant sales of mechanical cleaners, as shown in Figure 6.1, have been consistent through the 1960s and 1970s.

6.1.1 PROCESS MODIFICATIONS

Process modifications have been mentioned previously as methods of reducing emissions. Chapter 2 presents detailed discussions related to emission products and methods of reducing these emissions. Continuous monitoring of emissions, either before or after the control system or at both places, may be valuable in determining how the process is being operated to minimize emissions. Monitoring after the control system also shows how the control system is performing.

Chapter 2 shows that combustion operation has a critical effect on how much NO_x and pollutants are emitted. Continuous monitoring of the flue gases for either excess oxygen or carbon monoxide, fuel analyses, fuel preparation, pressure and temperature measurements, flame appearance, and sound and stack tests are all important. However, the most important factor in minimizing emissions and maximizing efficiency is the continuous monitoring of excess oxygen and/or carbon monoxide. Guidelines to help improve combustion operations have been published by the U.S. EPA for industrial and commercial boilers, and hazardous waste incineration procedures are available from government and other sources [1-3].

6.1.2 MECHANICAL COLLECTORS

Whenever the emissions consist of significant amounts of > 10 μm material, mechanical collectors should be considered before the collection system. Settling chambers are the cheapest to install and operate. Cyclonic collectors are more expensive but are more effective. Typically, up to about 80% of fly ash emissions can be reduced by cyclonic precleaners. These devices should be constructed so as to last at least 5 years, with 10–15 year lives preferred. Abrasion and corrosion must be considered in designing these collectors. For example, if corrosion is no problem, the relative life of a white iron cyclone is twice that of grey iron. Chromehard cyclones would last 1.5 times longer than grey iron units.

Costs of cyclones to handle various gas flows are based on the procedure by Horzella [4]. These costs, given in Table 6.1, are adjusted to 1988 dollars and are for the cyclone plus dust discharge air lock. However, actual costs should be evaluated for the cyclone plus the control system required as a result of having the cyclone upstream. For example, use of a mechanical precleaner could make it desirable to use fans before instead of after the collection system.

Annual costs for operating a cyclone can be determined knowing the hours-per-year of operation, cost of power, interest on the capital, capital

TABLE 6.1. Costs of Cyclones in 1988 Dollars.

Item	Gas Volumetric Flowrate (Am³/hr)		
	17,000	85,000	170,000
Cyclone diameter, m	1.32	2.03	2.03
Number in parallel	1	2	4
Cost in dollars for:			
Carbon steel, 0.5 cm (3/16 in) thick	8,410	33,640	67,270
Carbon steel, 0.5 cm (3/16 in) thick plus 10 cm refractory	11,770	50,460	100,910
304 stainless, 0.5 cm thick	20,180	84,090	168,200

costs, operating, maintenance, and supervision costs, economic life of the cyclone, and pressure drop. Cyclones are normally operated at 7.5 to 15 cm water pressure drop.

Example Problem 6.1

Determine the annual cost of operating an 85,000 Am³/hr cyclone for 6,000 hr/yr. Consider that the cyclone is made of 0.5 cm carbon steel, economic life is 10 years, capital charges are 8%, electrical costs are $0.06/ kWh, and pressure drop is 15 cm water. Assume that operation, maintenance, supervision, and product disposal costs are negligible.

Solution

Capital cost from Table 6.1 is $33,640

Gas pumping costs using
$$\text{Equation (5.25)} = (6 \times 10^{-5})(15 \text{ cm})(85,000)(0.06/\text{kWh})(6,000)$$

$$= \$27,540/\text{yr}$$

Use capital recovery factor (CRF, .08,10) of 0.14903, then:

Operating Costs

$$\text{depreciation} = (\$33,640)(0.14903) = \quad \$5,013$$

$$\text{interest} = (\$33,640)(0.08) = \quad 2,691$$

$$\text{power} = \quad 27,540$$

$$\text{Total} = \$35,244$$

or

$$\left(\frac{\$35,244}{\text{yr}}\right)\left(\frac{\text{hr}}{85,000 \text{ Am}^3}\right)\left(\frac{\text{yr}}{6,000 \text{ hr}}\right)\left(\frac{100\cent}{\$}\right)(1,000) = 6.9\cent/1,000 \text{ Am}^3$$

6.1.3 GAS CONDITIONING

There are various methods of gas conditioning that can improve the operation of a collection system. Gases can be cooled by heat exchangers, humidification, or dilution to obtain a desirable temperature for the control system or to reduce volume of gas pumped (and therefore reduce operating costs). They can be humidified to improve scrubber performance. Chemicals can be added to change dust resistivity and improve ESP efficiency. For example, about 5–20 ppm of SO_3 addition will often produce fly ash with a resistivity in the desired 10^{12} ohm-cm range. (This may be used on some high-resistivity western coal fly ash but other options are also available.) More common forms of gas conditioning would be cooling and cooling plus humidification.

Cooling of gases by humidification is the most common technique and is discussed in Section 5.10. If the cooled gases are to be sent to a filter, it is important that the gases are not saturated with water vapor, that no entrained liquid droplets are present, and that the acid dew point is not reached. This may require good fogging nozzles, adequate time for evaporation, and dilution to avoid saturation. Figure 6.2 shows how air volume changes with various forms of cooling. When 540°C air is cooled to the

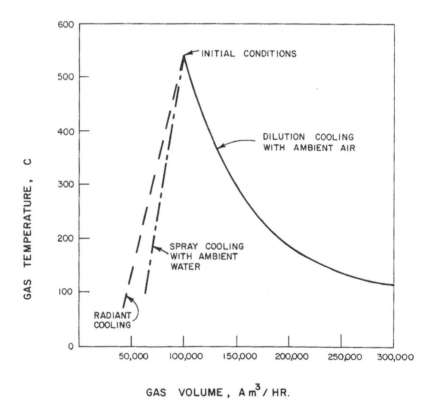

Figure 6.2 Effect on volume when cooling 100,000 Am^3 of air initially at 540°C.

maximum filtration temperature of 200°C by dilution, it requires nearly 200% additional air. When cooled to a normal filtration temperature of 120°C, it requires about 470% additional ambient air. This increases gas-pumping costs and volume of gas to be cleaned, and reduces particulate concentration. These are very undesirable factors in collection systems.

Figure 6.2 indicates that radiant cooling is most effective in reducing gas volume, but the danger of plugging the heat exchangers with particulates and corrosion, erosion, and pressure drop are disadvantages of this procedure. Spray cooling is nearly as effective as radiant cooling and can produce additional benefits of coagulation and positive phoretic forces. As an alternate to the flue gas cooling equation in Section 5.10, the amount of water required to cool a gas, assuming complete evaporation, can be approximated by

$$M_w = \frac{M_g C_p (t_h - t_f)}{(t_f - t_l) + 550} \tag{6.1}$$

where

M_w = mass of water, g
M_g = mass of gas, g
C_p = specific heat of gas, cal/(g K) [for air this is $0.2269 + 5.111 \times 10^{-5}T - 7.433 \times 10^{-9}T^2$ where T is K; this equals about 0.26 cal/(g K) at SC)]
t_h = initial hot gas temperature, °C
t_f = final temperature, °C
t_l = initial low water temperature, °C

Pressure spray nozzles, e.g., fogjet-type nozzles, work well for cooling gases in the temperature range of 1,000–250°C. Operating pressures for the nozzles are about 5.5 atmospheres and higher. Finer sprays are needed to cool below this temperature and pneumatic atomizing nozzles are recommended to help keep the gas dry.

6.2 DOWNSTREAM OF THE CONTROL SYSTEM

The exhaust gases from control systems are discharged to the atmosphere through associated duct work and chimneys. It is necessary to remove these gases from the proximity of human activity because the final pollutant concentrations are higher than can be tolerated before atmospheric dispersion occurs. This is true even though essentially 90 to 100%

of the pollutants have been removed. Depending on the control system, the pollutants remaining in these exhaust gases may be abrasive, corrosive, and wet.

Scrubbing system exhaust gases are saturated and water condensation forms downstream of the system as the gases cool. On cool days, the unheated exhaust from a wet scrubbing system appears as a brilliant white plume of water droplets that dissipates quickly as the droplets evaporate. This can be eliminated, but to do so is expensive and not entirely necessary.

6.2.1 STACK GAS REHEAT

Wet stack gases can be reheated if necessary to evaporate the liquid droplets. This can consist of essentially reversing the inlet gas cooling procedures discussed previously. Many of the systems using and being designed for reheat add from 8 to 35°C reheat. Possible reheat techniques include bypass gas reheat and direct and indirect heat. Several reheat procedures are presented here in order of decreasing preference.

If a scrubbing system is effective enough to remove more than the required amount of pollutants, some of the hot inlet gases can be bypassed to provide reheat for saturated gases while still achieving emission concentrations within established limits. This is a preferred method from economic and energy conservation aspects. In this situation, it may be necessary to remove some of the particulates from the bypass gases, especially if no mechanical precleaner is used, so the blended gases are within the particulate emission limits.

Heat exchangers (economizers) could be used to provide radiant heat. One procedure would be to cool hot gases entering the control system while heating the leaving gases. This could serve the dual function of improving the operation of the control system. A disadvantage is that plugging, erosion, and pressure drops may be significant.

Indirect heating can also be supplied by heating ambient air and mixing it with the wet gases. This eliminates the plugging, erosion and corrosion that could be present in the hot polluted gases and also does not add pollutants to the cleaned gases. This is costly as more gases are pumped.

The least desirable reheat technique is direct combustion of the wet gases or of the air to be mixed. This requires use of low pollution fuels such as gas or possibly high-quality oil.

6.2.2 CHIMNEYS

Chimneys or stacks must be provided to elevate the exhausted pollutants to a height that permits acceptable dilution by atmospheric dispersion. Dispersion is a function of effective plume height as discussed in Chapter 1. Plume height depends on stack height and plume rise. Plume rise, as shown by Equation (1.28) depends on gas exit velocity and temperature, ambient temperature, stack diameter, wind speed, and atomspheric stability. Chimneys must be mechanically stable to withstand the normal weight and wind loadings as well as to resist physical abrasion and chemical attack. Because of this, cost of tall stacks alone (not including foundations) range from $6,500 to $24,000 per meter of height depending on height, diameter, construction material, and location. Abrasion due to particles is minimal following effective control systems. Chemical resistance, especially in wet gas systems, can be severe so the chimney must be designed to resist this.

The condensed liquid in a wet gas usually lacks mineral and buffer content. If gases, such as SO_2 and HCl, are absorbed, a solution of low pH is formed. It is desirable to prevent this liquid from corroding the stack and to keep these liquid droplets from leaving the stack. Possible materials of construction are plastic stacks or liners, rubber liners, steel liners with acid-resistant epoxy or ceramic coatings, and other chemically resistant masonry brick and mortar linings. Some of these would depend on the size, height, and operating temperature of the stack, and the types of gases and liquids present.

Liquid in the stack can fall out by gravitational settling or by running down the walls. Lane [5] showed that stacks lined with nonwetting surfaces resulted in drop liquid flow on the walls rather than liquid film flow. This helps reduce reentrainment of liquids. He also reports that below about 15-m/sec gas velocity few drops were entrained from the stack. However, entrainment increases at lower altitudes and for denser gases.

Chimneys should be designed for about 15-m/sec maximum gas velocities to reduce entrainment, yet about 18-m/sec stack exit velocity is desired for plume rise. This can be accomplished by constructing a truncated cone at the top of the chimney. Turning vanes are necessary at the chimney inlet to reduce turbulence and thereby reduce the resulting liquid entrainment. These vanes also lower chimney draft losses. A liquid collection gutter should be installed on the chimney wall just above the gas inlet to prevent liquid drainage from entering this inlet.

All chimneys should include provisions for testing such as test ports, platforms, and, depending on the size of the facility, test enclosures, elevators, monorails, air conditioning, heat, and other services. Opacity and other continuous or intermittent monitoring systems must be installed as required.

The cost of 5/16″ thick carbon steel stacks up to about 100 ft tall, unlined but with flange, surface coating, and guy cables and clamps in 1988 dollars is approximately

$$P = 36.2D_1^{1.06} + 14.1D_1^{0.6}H_1 \qquad (6.2)$$

where

P = purchase cost in dollars
D_1 = stack diameter, inches
H_1 = stack height, feet

Cost of heavier-walled stacks could be estimated by factoring Equation (6.2) for the wall thickness. Stack installed cost would be about 1.92 times purchase cost. Liners, insulation, foundations, and taller stacks would escalate costs considerably.

6.3 ELECTROSTATIC PRECIPITATORS

Even though much theoretical design information has been presented in Chapter 5 and in other chapters, and even though more dollars have been spent for ESPs in the past than for other types of control equipment, there are still many factors that make it difficult to predict accurate ESP design information. Therefore, new facilities are usually modeled before construction begins. Systems that have not been modeled are likely to be over-designed to account for these uncertainties if emission guarantees are required. This tends to discourage some potential ESP customers.

The electric utility use of ESPs exceeds that of all other industries combined, so most of the following discussions are related to utility ESPs. Other important ESP use occurs in the cement, iron and steel, pulp and paper, and chemical industries.

6.3.1 MODEL TESTING

Complete model testing is usually recommended before any new ESP system is designed. These models are 1/16 to 1/50 actual size and may include up to 45 different tests. Flow patterns are thoroughly checked according to IGCI and customer requirements. The model test costs range from about $15,000 to $150,000 but are considered worthwhile investments.

Tests on actual flue gas and dust, for example, from the specific boiler and specific coal under dynamic conditions, are often important in determining amount of collection surface required. This is because the dust resistivity may vary between an in-situ test and a test of synthetically produced dust cake. An in-situ dust resistivity test using the point-to-plane probe as discussed by Smith [6] and Nichols [7] is the most reliable procedure.

6.3.2 SYSTEM SIZING

The amount of collection area based on the gas flowrate and composition, moisture content, particulate concentration, type and size, operating temperature, dust resistivity, and collection efficiency required can be translated into an ESP model arrangement based on available space at the facility. The model, of course, would be designed first and necessary modifications to obtain proper gas flow and particulate distribution would be made. Allowances must be made for dust cake resistivity and thickness, gas sneakage around the effective ESP collecting areas, electrical field strength and permissible gas velocity.

Precipitators generally consist of various combinations of chambers and fields. Each chamber is an independent compartment, isolated so that gas does not pass from one compartment to another if more than one exists. In Figure 6.3, two compartments are shown in this European-style ESP, one on the right and one on the left. Each chamber contains a number of

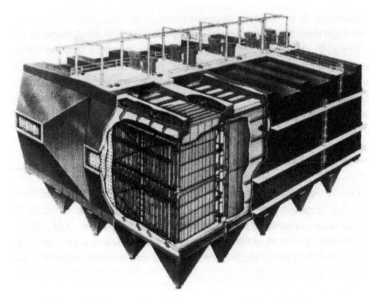

Figure 6.3 Electrostatic precipitator (courtesy Babcock & Wilcox).

fields in series. Each field is mechanically and structurally independent from the other and preferably is electrically independent. Each chamber in Figure 6.3 contains five fields as evidenced by the number of discharge hoppers shown. Note that each field in Figure 6.3 has two dust discharge hoppers. For this system of 10 fields there are 20 hoppers.

The wire and plate precipitator of Figure 6.3 has rappers for the upper and lower wire frames at the front of each field and plate rappers at the rear of each field. Three gas distribution baffles are placed at the inlet to each chamber. A separate electrical section is shown on top for each of the 10 fields and the rapper drives are shown on the sides.

6.3.3 OPERATION PARAMETERS

Average operating voltage for cold-side precipitators varies from 25 to 45 kV for a 23-cm plate-to-plate spacing. Hot-side precipitators, because of the lower dielectric (breakdown) strength of hot air, operate at 20 to 35 kV for the same plate spacing. For this system arrangement, the electrical potential at any point between the corona wire and the plate varies with location and size of wire. This is shown in Figure 6.4. Current density in an ESP in units of amp/m² is the power density in watt/cm² of collecting area divided by the applied voltage. Values of design power density as a function of dust resistivity are given in Table 6.2. Experimental average current densities agree almost exactly with the theoretically determined values [8].

Actual particle charging depends on particle size, field strength, contact time available, and whether particles are able to sneak around charging sections. Typical cold-side ESP systems as installed may be expected to have collection efficiencies based on specific collection area and coal sulfur content as shown in Figure 6.5 for pulverized subbituminous coal. Equation (5.14) also correlates SCA, efficiency, and sulfur content for coal combustion.

6.3.4 INSTALLATION

There are many complex factors to consider when installing ESPs. This section is intended to present a few simple suggestions that one can use to check an installed precipitator before startup to help insure successful operation of an otherwise properly designed and installed ESP. Many ESPs are constructed in modular form and assembled on the site. Errors at this stage result in substandard operation and can be discovered by a careful walk-through inspection.

ESPs usually operate with a warm-to-hot gas stream, so many problems

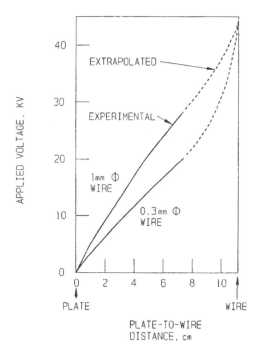

Figure 6.4 ESP electrical potential as a function of corona wire size and location with respect to the plate [8].

are related to heat. Thermal insulation is usually required to prevent undesired condensation, so a check for complete coverage should be made. The center of the ESP will be the hottest part during operation. As the hot gases enter, thermal expansion occurs in proportion to the temperature, so provisions for this expansion must be available. Free-hanging wires will expand and must not touch bottom or any other conductor as they will short out.

A check of the electrical components should be made with particular attention to broken or cracked insulators and poor electrical connections. Poor electrical connections that get hot during operation may be sensed by thermal indicators. A check for correct power line connections should be made, i.e., 230 V wires must be on 230 V connections and the same for 460 V, 575 V, and so on. An electrical-mechanical problem exists if plates are not plumb. Discharge wires may hang by gravity and must hang centered between the plates—wires cannot hang centered between plates that are not plumb. Loose or poorly centered wires should be replaced. If nothing else, they should be cut out.

TABLE 6.2. ESP Design Power Density [9].

| Fly Ash Resistivity (ohm-cm) | Approximately Equivalent for Pulverized Coal | | Power Density, watt/m² of Collecting Plate Area |
	Eastern Bituminous with Average Sulfur Content, %	Lignite with Average Na₂O Content, %	
10^4 to 10^7	4.8		40
10^7 to 10^8	4.0		30
10^9 to 10^{10}	3.3		25
10^{11}	2.8	5.0	20
10^{12}	2.1	2.5	15
$>10^{12}$	1.5	<1	10

A check should be made to make certain the system is airtight and that all mechanical connections are secure. All openings, access doors, inlets, exits, gaskets, ducts, gas distribution plates, and collector plates should be checked for proper installation, welding, positioning, and fit. Hopper discharge valves, screws, or other conveyors should be checked for closure and alignment. Rapping devices should be in line so hammers strike the

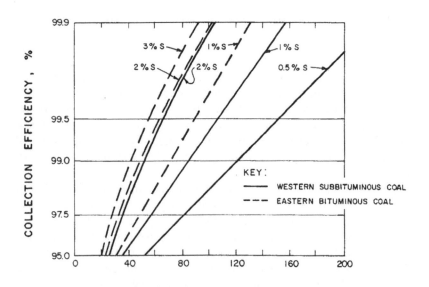

Figure 6.5 Approximate cold-side electrostatic precipitator efficiency for pulverized coal boilers [9].

anvils properly. The gas sneakage baffles and plate antiswing devices must be checked for proper installation. Vibrating cleaners must be properly tuned to provide adequate resonating impulse for cleaning the wires.

6.3.5 MAINTENANCE

ESPs have relatively low maintenance requirements specific to the precipitator itself. Most maintenance relates to the electrical portion of the system, and the electronic controls and instruments. Whenever efficiency drops, the instruments indicate a problem, or the system is shut down, the ESP should be purged and, after activating the interlock system to cut off all high voltage, entered for a physical inspection. The discharge electrodes should be inspected for hanger support and gasket failures, for wire breaks, for dust buildup, and for proper rapping operation. Collector plates should be inspected for evidence of warping, looseness, locations of excessive arcing, dust buildup, and proper rapping. The inside of the outer walls on insulated precipitators should be checked for evidence of cold spots. At least once a year, the dust should be cleaned from all plates, wires, frames, distribution plates, and guides.

The electrical supply compartments for the top and side insulators and the rapping devices should be inspected to see that they are physically intact and clean. Some may be enclosed in a penthouse (small closed building) or other cover. If a blower is provided to supply filtered air, the blower operation should be checked and the filters kept clean. At least once every six months, the insulators should be inspected to ensure that they are kept heated above the dew point and are clean.

The mechanical arms, gears, and drives must be periodically serviced. At least once every three months the rapping gears and drives should be greased and the oil levels in the motors checked. Hopper discharge mechanisms should be checked also.

6.3.6 COSTS

6.3.6.1 Capital Costs

It has been said that the basic difference between European- and American-designed precipitators is the cost. Whether or not this is true, there certainly can be a difference in ESP costs based on component quality. Precipitator lifetime ranges from 5 to 40 years depending on quality and use; average is about 20 years. Insulated precipitators also cost more than uninsulated. The costs given here are specific to the ESP system, general costs are discussed in Section 6.9.

Figure 6.6 gives typical capital cost values of ESPs for pulverized coal combustion. These values can vary depending on the quality of the unit, but they are the average reported in 1988 dollars. These costs are for large, dry, single-stage ESPs and include casing, hoppers, internals, insulation, transformers-rectifiers, rappers, and enclosures. Costs of uninsulated ESPs would be about 2/3 of the values given in Figure 6.6. Capital cost for ESPs on a $/KW basis can be estimated for specific types of coal using Figures 6.5 and 6.6. Note that ESP installation costs equal 1.24 times the ESP and accessories purchase cost. Thus the ratio of an ESP system installed cost to purchase cost equals 2.24. Capital cost is installed cost of an ESP.

6.3.6.2 Operating Costs

Operating costs for ESPs consist of direct (or variable) annual costs, indirect (or fixed) annual costs, and recovery credits if appropriate. These are itemized in Table 6.3 for typical dry ESPs using average industry values and 1988 year dollars following the procedure of Turner et al. [10].

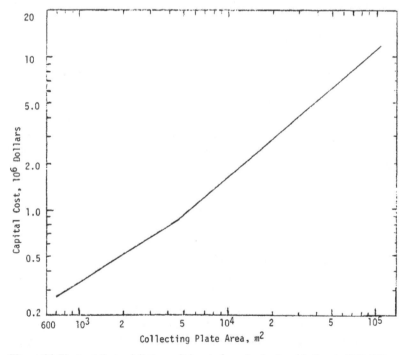

Figure 6.6 Electrostatic precipitator capital costs for pulverized coal boilers in 1988 dollars.

TABLE 6.3. Typical ESP Operating Costs Based on 1988 Changes.

Direct Costs

Chemical for gas conditioning: varies from 0 upwards
Operating Labor: (3 hr/day)($13.00/hr)(no. days operation/yr)
Supervision: 15% of operating labor cost
Utilities:
  Water: 0 for dry ESPs
  Electricity:

$$\text{Fan} = (KW_B)(\text{hr/yr})(\$0.06/\text{KWh})$$
$$\text{Corona \& Rappers} = (1.94 \times 10^{-3})(\text{ft}^2 \text{ plate area})(\text{hr/yr})(0.06/\text{KWh})$$
$$\text{or} = (2.09 \times 10^{-2})(\text{m}^2 \text{ plate area})(\text{hr/yr})(0.06/\text{KWh})$$
$$\text{Hopper heaters} = (4 \times 10^{-5})(\text{acfm})(\text{hr/yr})(\$0.06/\text{KWh})$$
$$\text{or} = (2.35 \times 10^{-5})(\text{m}^3/\text{hr})(\text{hr/yr})(\$0.06/\text{KWh})$$

Maintenance, labor, and materials:
  For >50,000 ft² plate area = (0.01)(purchase cost) + (0.084)(ft² plate area)
      or, for >4650 m² = (0.01)(purchase cost) + (0.90)(m² plate area)
  For <50,000 ft² or <4650 m² = (0.01)(purchase cost) + 4,200

Indirect Costs

Capital recovery: (A/P, 10%, 20 yrs)(capital cost) = (0.1175)(capital cost)
Plant overhead: (60%)(oper. labor + supervision + maintenance)
Administration: (2%)(capital cost)
Property taxes and insurance: (2%)(capital cost)

Total Annual Cost

  = Direct costs + indirect costs + accounting for waste charges or credits

---

No water pumping costs are shown as the system is dry. Compressed air changes are included in the fan electricity costs.

Utility fly ash is quite valuable in the production of lightweight building materials and, if this is sold, a credit should be shown. If disposal is required, costs may be $30/ton for nonhazardous waste disposal up to $200/ton for hazardous waste disposal. Transportation costs plus unloading (tipping) fees would need to be added to these costs.

The direct costs shown are typical for ESPs, but can vary from one location to another. Capital recovery is based on a 20-year lifetime for the ESP and a 10% interest rate. (This is longer than normal scrubber lifetimes.) Values given in Table 6.3 for overhead, administration, taxes and insurance are appropriate for ESP systems.

Average U.S. utility load factor is about 65% [11] with a 73% availability factor [12]. A 70% load factor is *equivalent* to operating at full load for 6,132 hours of an 8,760 hr/year.

6.3.6.3 Energy Costs

ESP energy requirements are mainly for gas pumping and corona discharge. Hopper heaters, rappers, discharge conveyor drives, instruments, and other power requirements would be small in comparison, but can be calculated or outlined in Table 6.3. In utility fly ash precipitators, corona power requirements increase as dust resistivity increases. White's data [13] show that for 98.5 to 99.7% collection efficiency of fly ash from burning 3% sulfur coal, the total corona power equals about 0.06% of the utility station power and, for an average of 0.2% sulfur coal, the corona power is about 0.10% of station power.

Gas pumping energy costs can be determined for a specific system by Equation (5.25). A well-designed new facility would have minimum ductwork from the boiler and to the stack, so little additional pressure drop would result. Retrofit precipitator systems on older boilers could result in additional ductwork and, therefore, additional pressure drop.

Example Problem 6.2

Estimate the capital and operating costs in 1988 year dollars for a 500 MW cold-side, dry ESP to collect fly ash at a 99.5% efficiency from the combustion of pulverized western coal with a 0.8% sulfur content. The system operates at 70% load for 300 days per year. About what percentage of station power would this system require at a 4-cm H_2O pressure drop?

Solution

Interpolating from Figure 6.5 shows that the required specific collecting area is about 120 m^2/Am^3 per sec for western coal. Using the approximation that there are about 520,000 Am^3/hr of hot flue gases at full load per 100 MW, the collection area required is

$$A = \left(\frac{120 \text{ m}^2}{Am^3/\text{sec}}\right)\left(\frac{520,000 \text{ Am}^3}{\text{hr 100 MW}}\right)(500 \text{ MW})\left(\frac{\text{hr}}{3,600 \text{ sec}}\right)$$

$$= 87,000 \text{ m}^2 \text{ or } 8.7 \times 10^4 \text{ m}^2$$

Capital cost from Figure 6.6 is equal to about 10.0×10^6. This is $20/KW.

This gives a purchased cost for ESP and components of

$$\frac{\$10 \times 10^6}{2.24} = \$4,464,290.$$

Annual costs are found using Table 6.3 as a guide:

Direct Costs

Operating Labor $= (3)(13)(300)$ $\qquad = \$ \ 11,700$

Supervision $= (0.15)(11,700)$ $\qquad = \$ \ 1,760$

Utilities

Electricity

$$KW_B = \left[6 \times 10^{-5}(4)\, \frac{520,000}{100}(500 \text{ MW}) \right] = 624 \text{ KW}$$

fan $= [(624)(0.70)][(300)(24)](\$0.06/\text{KWH})$ $\qquad = \$ \ 188,700$

Corona and Rappers

$= (2.09 \times 10^{-2})(87,000 \text{ m}^2)(300)(24)(0.06) = \$ \ 785,500$

Hopper Heaters

$$= (2.35 \times 10^{-5})\left[520,000 \left(\frac{500}{100}\right) 0.70 \right](300)(24)(.06) = \$ \ 18,480$$

Maintenance, Labor, and Materials

$$= (0.01)(4,464,290) + (0.90)(87,000) = \underline{\$ \ 122,940}$$

Subtotal Direct Costs $= \$1,129,080$

Indirect Costs

Capital Recovery $= (0.1175)(10 \times 10^6)$ $\qquad = \$1,175,000$

Plant Overhead $= (0.60)(11,700 + 1760 + 122,940)$ $\qquad = \$ \ 81,840$

Administration $= (0.02)(10 \times 10^6)$ $\qquad = \$ \ 200,000$

Property Taxes and Insurance $= (0.02)(10 \times 10^6)$ $\qquad = \underline{\$ \ 200,000}$

Subtotal Indirect Costs $= \$1,636,840$

Total Annual Costs $= \$2,785,920$

This is equal to

$$\frac{\$2,785,920}{\text{yr}} \ \frac{\text{yr}}{(300)(24)\text{hr}} \ \frac{1}{(500,000)(.70)\text{KW}} \ \frac{1,000 \text{ mills}}{\$} = 1.11 \ \frac{\text{mills}}{\text{KWH}}$$

Electricity used $= 188,700 + 785,500 + 18,480 = \$992,680/\text{yr}$

and Total KW used $= \dfrac{\$992,680}{\text{yr}} \dfrac{\text{KWH}}{\$0.06} \dfrac{\text{yr}}{(300)(24)\text{hr}} = 2,300 \text{ KW}$

$\therefore$ % of station $= \dfrac{2,300}{500,000(.70)} \ 100 = 0.66\%$

6.4 FILTERS

As shown in Figure 6.1, fabric filters are a popular type of control device. Based on sales dollars, they are currently slightly more popular than ESPs. The future is bright for filter vendors, as dry scrubbing systems will incorporate more filters so that high efficiency gas and particle removal can be obtained.

Operational data can be very useful in determining specifications and operating parameters for filter systems. Pulse jet filter use on 346 utility and industrial coal-fired boilers was summarized by Dean and Cushing [14]. This includes equipment from 16 large domestic and nondomestic vendors, plus as many small vendors. The two largest vendors, Standard Havens and Flakt, shared about 20% each of the total number of systems. Of the systems surveyed, most (58.8%) used air-to-cloth (A/C) velocities of 4–5 ft/minute (2–2.5 cm/sec). A/C values of 3–4 ft/minute (1.5–2 cm/sec) were used by 16.5% of the systems and only 12.2% used 5–6 ft/minute (2.5–3 cm/sec). The survey showed that of the domestic vendor systems, woven glass was the most popular filter media. Woven glass and glass felt accounted for 79% of the bags. In the non-domestic systems, Teflon felt accounted for 39.8% of the bags, nomex for 27.6%, and glass for only 15.4%.

The second largest U.S. filter system is the TVA Shawnee station which consists of 10 175-MW boilers. The facility has been operational since 1956 and burns <1% sulfur coal with no SO_2 control. The 10 identical filter systems have been operational since 1982. Each system has 10 filter compartments containing 324 30-cm diameter bags with a 35:1 length-to-diameter ratio. The bags are 0.76 g/cm² (14 oz/yd²) fiberglass with a Teflon finish. Clean filter permeability required is 10–18 cm/sec per cm² and filter velocity is about 2.2 ft/min (1.13 cm/sec). A 3-yr bag life is obtained. Each compartment is cleaned about once/hr by reverse air. The cleaning is initiated when the tube sheet pressure drop reaches 4″ (10 cm) water gauge. Normal gas flow into each system is 994,000 m³/hr at 163°C with 1.7 to 5.7 g/m³ inlet dust loading. Maximum outlet dust loading is 0.01 g/m³, giving design collection efficiencies of 99.35 to 99.81%. Top bag clamps are sewn in and bottom clamps are quick release. All bag hardware is 304 stainless steel or equivalent. Each compartment has 2 inlet, 2 outlet, and 2 reverse-air dampers.

6.4.1 SYSTEM SIZING

Fabric filter system sizes depend on the type of filter system used as the air-to-cloth ratio varies with type. The system configurations can consist of either envelope design filters or cylindrical filters and normal upflow with outside or inside filtering.

In addition to the fact that average filter velocities depend on filter-system type as discussed in Sections 6.4 and 5.3, actual filter velocities also depend on where dust cake buildup occurs in the operating system. Local dust buildup and local filtration velocities are both shown in Figure 6.7 for a pulse jet fabric filter as a function of distance from the bag inlet. This shows that actual velocities vary considerably from the average. The pulse jet bags used to obtain these data were 11.4 cm diameter by 2.44 m long. Superficial filtration velocity was 10 cm/sec at a pressure drop of 14 cm water. The cleaning pulse was at 6.8 atm pressure, once a minute.

Filtration effectiveness decreases as cake thickness decreases and as filtration velocity increases. A filter system's effectiveness can be varied considerably by the alteration of numerous parameters. It is, therefore, good to size a filter system based on actual test model data. The parameters that can be varied include type of filtration system, size of bags and baghouse, bag shape, number and type of bags, bag material weight, operating pressure drop range, air-to-cloth ratio (filtration velocity), and bag-cleaning intensity and frequency.

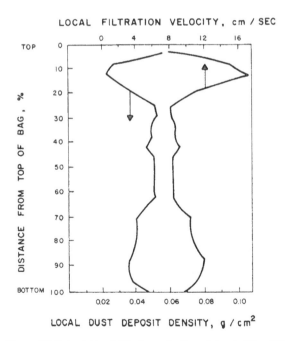

Figure 6.7 Pulse jet local filtration velocity and dust buildup [15].

It is important to size the baghouse to the system. Pulse-jet units may be on-line or off-line cleaning. If on-line is used, the pressure drop and gas flow may vary with time as the filter cake builds up and is cleaned, depending upon the number of bags and the cleaning frequency. Reverse air systems are off-line cleaning and the baghouse must be sized to account for the off-line compartment. The fan must be capable of moving the gases through the filter system at maximum pressure drop, and also must be able to overcome the pressure drops due to ducts, boilers, stacks, and other components.

6.4.2 OPERATION PARAMETERS

Average baghouse life is 20 years, though some systems have lives of up to 40 years. Bags have typical life periods of 3–5 years and, as shown by Figure 6.1, replacement bag fabrication is a big expense. Once the system is installed, it is worthwhile to adjust the operating parameters for optimum cleaning efficiency and cost effectiveness. A study should be made to determine the effects of varying the filtration time (usually by changing the maximum pressure drop to initiate cleaning), the cleaning time, the cleaning intensity and, if the system is continuous, the effects of varying the cleaning cycle sequence. Air-to-cloth ratio may be changed if the system is overdesigned and/or if some compartments can be isolated and process flowrates decreased for the study period. If the data provide suggestions to improve the system operation, other types of fabrics and fabric treatments could be tried when replacement bags are needed. Gas temperature, composition, dew point, and particle cohesiveness and adhesiveness are critical to successful bag life.

Filter pressure drop is the most important operating parameter. Each compartment should have a pressure-drop manometer. Together with gas flowrate, emissions rate, and dust collection rate this will provide data to indicate how the system is functioning. Increased pressure drop may mean increased air flow, cloth blinding, full hoppers, condensation, or an inoperative cleaning mechanism. Low pressure drop may mean fan problems, loose or broken bags, or leakage between filter compartments. Temperatures and humidities should also be recorded to prevent bag damage due to overheating or cooling below the dew point.

6.4.3 INSTALLATION

A careful physical observation before startup will point out installation problems that should be corrected. Many baghouses are constructed from modular compartments which should reduce installation faults. Checks should be made for the obvious items such as bag placement, proper bag tension (squeeze each bag), clamps tightly installed but not so much that they cut the fabric, bag tears or indication of abuse in any other way, tools left loose inside the baghouse, door gasket placement and door closure, damper and valve operation, cleaning mechanism operation, baffles in place and secure, discharge mechanism working, and insulation complete if needed. Filter bags are usually constructed well and new bag defects are the least-frequent source of filter problems.

6.4.4 MAINTENANCE

Bags are the greatest maintenance concern yet most of the maintenance labor costs are expended looking for and repairing leaks. Bag replacement requires relatively little time. Some bags will last only a few months while others last several years. Fabrics can be treated to increase bag life by reducing abrasion. This treatment consists of adding special lubricants such as silicone, graphite, or Teflon. Bag clamping procedures and clamp types should be studied to devise the least abrasive method. Cleaning frequency and intensity should also be examined if bag failure is high, to determine the most effective but gentle cleaning arrangement. High and low temperatures and corrosive gases all contribute to reduced bag life.

Recorded information pertinent to the filter fabric should be kept. This should include number, size and shape of bags, type and weight of fabric, number of bags per compartment, bag placement and a history of bag failures relative to placement within the system, location on the bag, and reason for failure, e.g., tear, seam failure, abrasion, chemical attack, moisture heat, clamping or support device, or other.

Dust-removal systems are second in maintenance problem frequency. They must be kept clean and in good working shape. Procedures applicable to other material-handling facilities would be used here. Other significant filter system maintenance items include control system, dampers and closures, reverse-air blowers, pressure taps, and heaters.

6.4.5 COSTS

6.4.5.1 Capital Costs

Capital costs of filter systems can vary with collection efficiency required, type of dust, air-to-cloth ratio, system type, and system size. However, a generalized procedure [17] has been developed based on filter cloth area. Net cloth area, A_{Net}, is found first for the filter based on desired air-to-cloth ratio and the volumetric flow rate of gas entering the filter. Gross cloth area, A_{GR}, is then found by

$$A_{GR} = 3.5A_{Net}^{0.9} \tag{6.3}$$

The purchase cost of insulated pulse-jet filters, P_{PJ}, in 1988 dollars for 4,000 to 16,000 ft² of gross cloth area units, not including bags, is

$$P_{PJ} = 11,860 + 6.92A_{GR} \tag{6.4a}$$

where A_{GR} is in ft². In metric units for 370 to 1,500 m² gross cloth area units with A'_{GR} in m², the purchase cost is

$$P_{PJ} = 11,860 + 74.39A'_{GR} \tag{6.4b}$$

If stainless steel material is required, add an additional 70%.

The purchase cost of an insulated reverse-air filter, P_{RA}, in 1988 dollars for 100,000 to 400,000 ft² gross cloth area units not including bags is

$$P_{RA} = 304,370 + 6.57A_{GR} \tag{6.5a}$$

In metric units for 9,300 to 37,200 m², it is

$$P_{RA} = 304,370 + 70.63A'_{GR} \tag{6.5b}$$

Add an extra 40% for stainless steel material of construction.

Bag costs vary from about 15% to 100% of the base baghouse cost depending on the fabric. Some common filter bag costs are listed in Table 6.4.

To obtain total equipment cost, multiply cost of baghouse with bags times 18% to get costs of instruments, controls, taxes, and freight. In other words, total equipment cost is 1.18 times baghouse cost with bags. Cost of

TABLE 6.4. Typical Filter Bag Costs in 1988 Dollars.

Material	Use	Cost	
		$/ft²	$/m²
16 oz polyester or polypropylene	pulse jet	0.59	6.34
	reverse air w/ rings	0.48	5.16
16 oz fiberglass	pulse jet	1.30	13.97
	reverse air w/ rings	0.91	9.78
14 oz nomex	"both"	1.80	19.35

installing the system is 1.17 times the total equipment cost, giving a ratio of installed cost to equipment cost of 2.17. Capital cost is installed cost. Historically, baghouse capital costs have been about $20/KW for large utility systems and the size factoring exponent is 0.72.

6.4.5.2 Operating Costs

Total annual costs for a typical fabric filter system and procedures for estimating these are outlined in Table 6.5. This follows the procedures suggested by Turner [17]. Baghouse operating costs at large utilities have historically been about 0.8 mills/KWH. Several fan electricity charges are listed in Table 6.5. The first is the main gas pumping cost, which can be estimated from Equation (5.25) using gas flow rate entering the filter and remembering that, e.g., tube sheet pressure drop in inches of water for a reverse-air unit is approximately 3 times air-to-cloth ratio in ft/min. To this pressure drop must be added all entrance, duct, damper, and other system losses. The bag-cleaning costs would be for pulse-jet or for reverse-air cleaning.

Bag-replacement costs in Table 6.5 are based on a capital recovery period of 2 years at a 10% interest rate, i.e., the capital recovery factor = 0.5762. The indirect cost capital recovery assumes a 20-year equipment life and a 10% interest rate. Use Table 6.4 as a guide to bag cost. The bag recovery costs are why bag purchase costs are excluded from the direct cost capital recovery statement.

As with ESPs, filter system waste disposal charges may run from $30 to $200/ton plus transportation and fees. There may also be a cash credit if the material is sold—after all, baghouses were invented as a means of recovering a product.

6.4.5.3 Energy Costs

The main energy cost in a filter system is the gas-pumping cost. Material handling requirements to remove the dry cake from the hoppers and energy for cleaning the bags are relatively low compared to the gas-pumping costs. On a monetary basis, the electric costs for a filter system typically range from 5 to 13% of the total operating costs. When the bags are new, this cost would be lower and when the bags are old, the cost increases.

6.4.6 GAS REMOVAL

Although filters, like ESPs, are particulate-removal devices they can

TABLE 6.5. Typical Fabric Filter Operating
Costs Based on 1988 Charges.

Direct Costs

Operating labor: (6 hr/day)($13.00/hr)(no. days operation/yr)

Supervision: 15% of operating labor cost

Utilities:

      Water: (may be required for quenching—see Section 5.10; then use $1.25/
          1,000 gal)

   Electricity:

          gas pumping fan $= (KW_B)$(hr/yr)($0.06/KWH)

       reverse-air bag cleaning $= (KW_B)$(hr/yr)($0.06/KWH)[a]

       pulse-jet bag cleaning by

  compressed air @ 100 psig $=$ (0.002)(actual filter acfm)

$$\times \left(\frac{\$0.16}{1000 \text{ cfm}} \right) \left(\frac{60 \text{ min}}{\text{hr}} \right) \left(\frac{\text{hr operated}}{\text{yr}} \right)$$

      Fuel: (may be required for heatup; obtain Btu, then use $0.43/100,000 Btu
          nat. gas)

Maintenance:

   Labor $=$ (3 hr/day)($15.00/hr)(no. days operated/yr)

   Material $=$ maintenance labor cost

Bag replacement: [(0.10)(gross ft² bag area) + total bag cost](0.5762)

Indirect Costs

Capital recovery: (0.1175)[(capital cost − (0.10)(gross ft² bag area) − total bag cost)]

Plant overhead: (60%)(operating labor + supervision + maintenance)

Administration: (2%)(capital cost)

Property taxes and insurance: (2%)(capital cost)

Total Annual Cost

    $=$ Direct Costs + Indirect Costs + Accounting for waste charges or credits

[a]Find bag cleaning blower KW_B using Equation (5.25) but figure ΔP at 15 cm (6″) H_2O and Q at 1/10 filter system actual flow rate.

remove gases by converting the gas to a particulate which can then be captured. Particulates can be formed by gas-phase reactions, as discussed for precipitators, or the reactions can occur on the surface of an alkali chemical precoat on the filter. Possible precoat chemicals are sodium ores, such as nahcolite ($NaHCO_3$), soda ash, lime, and even alkali fly. Systems with a minimum of 90% SO_2 removal guarantees are offered for full-scale utility systems consisting of lime spray dryers and fabric filters. Cost estimates for these types of systems have been made at $70/KW capital costs and 3 mills/KWh operating costs [18]. A possible return from this type of system using a sodium alkali reactant could be from the sale of sodium sulfate. This has a value of about $66/ton to the paper industry.

6.5 SCRUBBERS

Figure 6.1 shows separate listings for wet scrubbers, FGD systems, and gaseous emissions control. In the future, as incineration emission control devices become more prevalent, it could be expected that separate listings could be shown to distinguish gaseous emissions control devices as dry scrubbers, wet absorbers, and adsorbers (adsorbers are only a very small portion of the group). The following subsections will consider separately wet particulate scrubbers (other than FGD), wet FGD and acid gas controls, dry scrubbers (spray dry and dry injection), and absorbers. However, many of the comments in Section 6.5.1 are applicable to all scrubbers.

6.5.1 WET PARTICULATE SCRUBBERS

Wet particulate scrubbers serve many useful emission-control functions as discussed in detail by Schifftner and Hesketh [19]. Figure 6.1 shows that they average only about 1/3 the total sales dollars of either ESPs or fabric filters. Gas atomized spray venturi scrubbers have been the industry workhorses for particle removal, and this section will cover that type.

Wet scrubbing systems may include all of gas conditioning, the scrubber, entrainment separator and associated ducts, recirculation tanks, pipes, fan, pumps, and controls. These components will vary with the system and must be estimated individually.

Purchase cost, P_v, of a 1/4″ carbon steel, fixed-throat venturi scrubber in 1988 dollars can be estimated using

$$P_v = 63.6(acfm)^{0.57} \tag{6.6a}$$

where *acfm* is actual scrubber inlet volumetric flow rate. This equation is good from 6,000 to 200,000 *acfm*. Note that the sizing exponent is 0.57. Cost of different thickness or different type material of construction must be factored in as discussed later in this chapter. A manually adjusted venturi throat would add $4,060 to the cost and an automatic throat would add $7,480. Instruments and controls would add another 10%.

If a *quencher* is added preceding the scrubber, purchase cost of a carbon steel quencher in 1988 year dollars is

$$P_Q = 0.284(acfm) + 10,400 \qquad (6.6b)$$

where *acfm* is actual quencher inlet volumetric flowrate. Note that scrubber costs given by Equation (6.6a) must be recalculated based on a saturated *acfm* leaving the quencher and now entering the scrubber.

Following the procedure of Vatavuk and Neveril [20], a fiberglass liner would add 15% to the purchase cost and a 3/16" rubber liner would cost about $5.50/ft^2 of internal surface area additional. Costs of all the auxiliary equipment would also be extra.

Installation costs would be about 91% of the venturi system purchase cost. Therefore, the ratio of total installed capital cost to purchase cost equals 1.91. Annual costs can be estimated following the procedure of Table 6.6. Capital recovery is based on a 10-year life and 10% rate of interest. Stack reheat and waste disposal charges must be added to these costs as appropriate. Note that the water rate is make-up and washing water which may be very different from recycle-rate water.

6.5.1.1 Operation Parameters

Pressure drop is the most important parameter to assist in the operation of a wet scrubber. Liquid-to-gas ratio and gas flowrate are the next most important key operational factors. These three quantities make it possible to determine scrubber particulate and gas removal performance. Scrubber pressure drop is a function of liquid and gas rates as well as unit size as pointed out in Section 5.4. Note that if gas composition changes are possible, increased gas density can increase pressure drop. If liquid and gas flowrates are held constant and scrubber size is not varied (if variable throat area is maintained constant), a decreasing pressure drop could indicate fan failure, erosion of the unit, and leaks. More likely, low pressure drop indicates a decrease in gas and/or liquid flowrates.

Increased pressure drop would indicate an increased liquid and/or gas flowrate, but if these are held constant it could be caused by a buildup of solids within the scrubber or mist eliminators or a dislodged section of the

TABLE 6.6. Venturi Scrubber System Operating Costs in 1988 Dollars.

Direct Costs

Operating Labor: (6 hr/day)($13.00/hr)(no. days operation/yr)
Supervision: 15% of operating labor cost
Utilities:
 Electricity $= (KW_B + KW_p)$(hr/yr)($0.06/KWH)

$$\text{Water} = (gpm)\left(\frac{60 \text{ min}}{\text{hr}}\right)\left(\frac{\$1.25}{1,000 \text{ gal}}\right)(\text{hr oper/yr})$$

Maintenance:
 Labor = (3 hr/day)($15.00/hr)(no. days operated/yr)
 Material = maintenance cost labor

Indirect Costs

Capital recovery: (0.1628)(capital cost)
Plant overhead: (60%)(operating labor + supervision + maintenance)
Administration: (2%)(capital cost)
Property taxes & insurance: (2%)(capital cost)

Total Annual Cost

 = Direct costs + indirect costs + waste disposal & stack gas reheat costs

unit which is blocking flow. High-pressure drop scrubbers operating under negative pressure are especially vulnerable to leaks. This reduces efficiency and wastes energy. Leaks are frequently due to improper construction or construction materials, and poor maintenance and operation. Negative pressure drops cannot be increased beyond the point of drawing in air through the liquid drain legs or the point of evacuating and collapsing the equipment.

When SO_2 is scrubbed or other gases are absorbed, wet scrubbing becomes a chemical process. It is necessary to maintain data on system pH, mist eliminator wash rate, wash frequency and pressure drop, recycle slurry solids concentration, feed chemical rate, and inlet and outlet SO_2 concentrations. In both particulate and gas scrubbers, data should be recorded for bleed slurry solids concentration and rate, entering flue gas temperature, and O_2 or CO concentration, leaving gas temperature, recycle slurry temperatures and levels, pH if necessary, makeup water rate, and, on occasion, the gas stream inlet and outlet particulate size and concentration.

In FGD systems, any or all of thickeners, rotary vacuum filters, and centrifuges may be used to dewater the solids leaving the system. Operating data should be recorded pertinent to these facilities. If system reheat is used, data on reheat ΔT and reheat system pressure drop and/or fuel consumption should be recorded as necessary. Systems that produce an oxidized sulfate product may require retention-time and oxidant-rate data. Spray nozzle pressures are needed to show whether nozzles are plugged or worn and how much liquid is being delivered.

Proper operating velocities must be maintained in the particulate scrubber, gas adsorber, and mist eliminators for effective operation. Low velocities in the scrubber reduce particulate collection and high velocities reduce gas absorption. Mist eliminator carryover occurs at high velocities yet mist elimination is ineffective if the separators are operated at <30% of design capacity.

The scrubbing system must be operated within the limitations of the materials from which the system is constructed. Temperatures should not be excessive. Examples of reasonable upper limits are 80°C for rubber, 90°C for fiberglass reinforced plastic (FRP), and 120°C for lead. Metals and ceramics can be used above this, but they too have temperature and chemical composition limits. Rate of localized attack on stainless steel increases as slurry pH and chloride ion concentration increase and as temperature increases. If such a condition develops in the scrubbing system, corrosion rate studies should be made to determine operating conditions for maximum temperature-concentration-pH combinations with the materials of construction used in the system. Otherwise, the system must be redesigned with new materials of construction. For example, one scrubbing system designed for acidic 1,100°C inlet gas temperatures uses a series of linings. the shell is carbon steel. The layers in order from the shell inward are rubber lining, ceramic insulation, foamed glass insulation, acid-resistant ceramic insulation, acid brick, and carborundum facing.

6.5.1.2 Installation

Scrubbing systems should be considered as chemical processing systems. As such, numerous checks must be made of the liquid-, gas-, and solids-handling systems. For the liquid-handling portion of the system, it is necessary to physically allow all lines to determine whether the connections are correct and that they are properly made. Pumps must rotate correctly, usually as shown by the arrow, pump mechanical seals must be polished and parallel, packed seals must be packed correctly, all seals must be adjusted before and after break-in, seal flushes should be connected and turned on during operation, strainer baskets and valves must be in place and positioned properly, and all lines must be sloped to a low point where they can be drained. Heat tracing of liquid lines may be necessary for cold weather operation and shutdowns. Flow meters and pressure gauges should be installed appropriately with bypass lines where needed. Agitation, recycle and clean out provisions may be required for the slurry lines to prevent plug up.

The gas ducts, connections, and controls should be inspected to be certain the installation is as specified. Scrubbing system dampers and gas line closures are subject to more corrosion than dry systems and this should be taken into account.

The scrubbing system should be supported so as to withhold not only the weight of the assembled scrubber, but also to hold the added weight of all the liquid that will be withheld by the lines, tanks, scrubber, and other parts of the system. The corrosive erosive gases and abrasive slurries sometimes present in scrubbers make special linings necessary. These should be low-friction as well as corrosion-resistant materials. A check of all linings should be made if possible to note for completeness, good bonding to the metal walls, tears, scratches, and other imperfections. Linings of multiple layers could be checked as each layer is installed. Temperature-sensitive linings must have provisions for one and preferably two emergency quench systems to keep them from burning up in systems where the inlet gases are hot. These are usually water quenches and they should be checked periodically. The two systems could be one process water and one city water system.

Check all material of construction to be sure that it conforms to specifications. This is especially important for protective shields. Welds should be checked for type of material and completeness. All openings should have gaskets and closures that secure "vacuum" tight to minimize leaks, especially in high-pressure drop systems.

6.5.1.3 Maintenance

Wet scrubbing systems require more and better maintenance than control systems that operate dry. This should not be surprising in that scrubbing systems not only operate wet but they have wet-dry interface areas and can handle corrosive and erosive gases and liquids. Partial downtime should be scheduled each week so that the scrubbing system can be inspected, cleaned, and maintained in the best practicable condition. Plugging historically has been a problem in scrubbing systems, but this usually originates from poor design and/or poor operation and maintenance. Nozzles are installed at critical locations to wash surfaces that are more subject to deposit buildup. These nozzles must be in service and require frequent checking.

Figure 6.8 is a simplified wet scrubbing system schematic showing the "dirty dozen" areas of high maintenance. Area number 1 is the first stage of the scrubbing system, which often is a venturi for particulate scrubbing and terminal gas conditioning. The wet–dry interface is subject to buildup of soft deposits. Preconditioned gases are less likely to produce these deposits, so the conditioning portion of the system and the wet–dry interface and associated sprays must be checked often. Venturi throat areas must be checked to see that erosion due to high velocities has not destroyed the protective lining or the shell metal and that deposits are not present. Scrubber linings must be inspected to see that they are not deteriorating. Sprays and liquid inlets are subject to wear, plugging, and misalignment.

The flooded elbow is the second area noted. It is subject to rapid wear if not kept flooded with liquid. The absorber tower is area number 3. Pack-

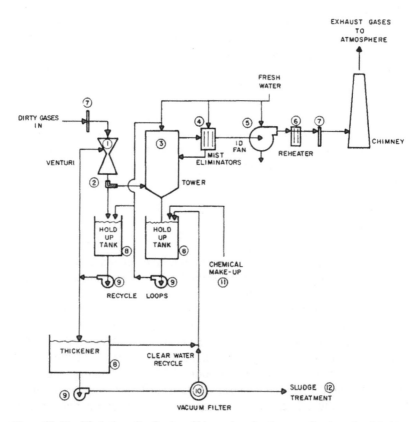

Figure 6.8 Simplified schematic of wet scrubbing system showing areas for planned maintenance.

ing or plate condition is important to be sure that breaking, wear, and gas and liquid distribution are acceptable. Strainers must be cleaned to keep liquid draining freely. Spray nozzles and liquid distributors should be checked. Any plastic packing or system parts must not be steam-cleaned.

Mist eliminators, number 4, may operate with vertical gas flow and be inside the tower or with horizontal gas flow as shown. Horizontal is better. The wash spray nozzles must be properly sized, oriented, pressurized, and in condition so as to cover the entire eliminator area. Some wash may consist of recycle liquor which would plug and wear nozzles more rapidly than freshwater wash. Sprays must be adequate for washing and not so fine that the droplets are carried over. Nozzle pressures should operate at no more than 2 atmospheres. Pressure drop in chevron-type eliminators should be about 1–2.5 cm H_2O. Physical checks are needed to see that chevrons, wires and other parts of the eliminators are not distorted or moved out of place. All types should be wired or otherwise secured in place.

Fans on these systems are large and subject to stress fatigue. They must be kept in balance and rotating freely. Blade wear, cracking and deposits can all result in imbalance and destruction of the unit. Sprays are used to wash deposits from the fan tips during operation. Soot blowers can be used to keep the bearings clean. The fan shown as number 5 operates "wet." Fan energy heats the gases by compression about 0.1°C for every centimeter water pressure drop developed assuming no liquid droplets are present. For a 50-cm pressure drop scrubbing system, this would result in a 5°C gas temperature rise. If slurry droplets were present, this would cause evaporation and depositing of dissolved and suspended solids on the blades.

Reheaters, if present, operate at high temperatures and are subject to pluggage due to evaporation of liquids, corrosion, and stress corrosion. Checks must be made for leaks and wear. Reheaters may be in-line after the fan, as shown by number 6, or they may be before the fan or indirect. Isolation valves and/or control dampers, shown as number 7, must be worked periodically and kept available for use.

The chimney is not a part of a flange-to-flange scrubbing system, but special maintenance may be required, especially if the chimney follows a wet scrubbing system. Notes on this are given in Section 6.2.2.

All holdup, storage, and processing tanks, such as those shown as number 8, need to be checked for deterioration of linings or metal walls, wear, plugging, agitator condition, and strainers. Pumps, number 9, are a source of high maintenance. The packing and seals must be adjusted continuously (usually by the operator) and replaced as necessary. Lubrication as frequently as daily or per-shift may be required, but note that over-lubrication is as bad as underlubrication. All oil reservoirs need to be checked weekly.

Filter cloth, scraper blades, vacuum pumps, and other parts of the vacuum filter, centrifuge, or sludge dewatering facility are noted as number 10. They must be checked continuously and adjusted or replaced as necessary. Chemical makeup and sludge processing material handling systems, numbers 11 and 12, need to be maintained for continuous service. This includes maintenance of associated chemical and control instruments used in the process.

6.5.2 WET FGD AND ACID GAS CONTROL SYSTEMS

Wet flue gas desulfurization and acid gas control systems such as those used on incinerators to control SO_2, HCl, and HF can be treated together, as the systems are basically similar. At the present time, the FDG systems are larger, but even so they are usually comprised of a number of smaller units in parallel. A *maximum* capacity equivalent to about 125 MW per scrubber is suggested (i.e., about 380,000 acfm or 650,000 Am³/hr).

A 10-year life is typical for abrasive/corrosive cost scrubbers which is about half that of ESPs and fabric filters. However, modern wet FGD systems are now expected to last up to 30 years at a 90% or greater reliability. Test scrubbers are usually used to design these systems. Test units as small as 170 m³/hr (100 cfm) and 1.7 m³/hr (1 cfm) are used.

Some wet FGD and acid gas control systems are not effective for gas removal. They are predominantly particle-control units, with chemicals added to control pH. Others, for example, can include both a venturi for particles and an absorber for gases. This is shown in Table 6.7 for five different utility FGD systems. The only true gas-control system is the Cholla unit. With no ESP present, FGD wet scrubbers can reduce particulates to 0.054 g/10⁶ cal (0.03 lb/10⁶ Btu) and, when properly designed and operated, remove up to 98% of the SO_2 on a short-term basis. For sustained operation, values of SO_2 removal up to 95% are used.

Costs for FGD wet scrubbing systems are available and are presented here as the acid gas control example. System costs for 90% or better SO_2 removal, including associated chemical feed, sludge dewatering, mist eliminators, fan, pumps, ducts, pipes, and controls, are considerable. The utility industry reports that capital costs of installed FGD systems of the non-regenerable type in 1988 year dollars are about $120/KW for low-sulfur systems and about $180/KW for high-sulfur systems when using limestone or lime-base systems. This is equivalent to about $40/acfm ($23/Am³ per hr) and $60/acfm ($34.6/Am³ per hr), respectively. These costs are for 500 MW size units. Other non-regenerable FGD systems costs fall within this range. Costs of regenerable FGD units of the same size are about twice the cost of non-regenerable systems. Note that vendors

TABLE 6.7. Sample Listing of Operating Wet Scrubbers [9,18]

Utility company[a]	APS	PP&L	PSC	MP&L	APS
Station	Four Corners	Dave Johnston	Cherokee	Clay Boswell	Cholla
Startup date	12/71	4/72	11/72–7/74	5/73	10/73
Application	Particulate	Particulate	Particulate	Particulate	SO$_2$ + Particulate
Design	Venturi	Venturi	3-Stage TCA	Hi-Press Spray	Venturi + Packed Tower
Scrubber capacity, MW	575	330	660	350	115
Number scrubber modules	6	3	9	1	2
Reheated gas	Yes	No	Yes	No	Yes
Liquid-to-gas ratio, ℓ/m^3	1.2	1.7	6.7	1.1	{ 2 to venturi 6 to tower
ΔP, cm H$_2$O	71	38	25–38	10	51
Electricity required, % of station power	3–4	2.3	4.0	0.86	2.43
Manpower, total operators	8			4	4
SO$_2$ removal, %	30	40	20	20	92
Particulate removal, %	99.2	99.0	95–97.5	99	99

[a]APS = Arizona Public Service; PP&L = Pacific Power and Light; PSC = Public Service Co. of Colorado; MP&L = Minnesota Power and Light.

433

reported FGD system installed cost for non-regenerable type units to be
$75 to $120/KW [21]. A breakdown of typical scrubbing system capital
costs is given in Table 6.8.

Operating costs for FGD systems range from 10–20 mills/KWh for large
non-regenerable systems based on a 65% capacity or load factor (i.e., system operates 65% of the time or 5,698 hr/yr). These costs also assume that
the systems last for 30 years, and may not cover the total cost of waste disposal. Limestone costs $5–10/ton plus delivery and fine powder limestone
is $10–12/ton plus delivery. Lime costs $50–100/ton plus delivery.

The following generalized example is given to estimate FGD wet scrubbing system costs. This system consists of a new 500 MW limestone scrubbing system with the assumptions listed below. The system can remove
consistently 90% of the SO_2. A high-efficiency ESP would precede the
scrubbing system to keep overall particulate levels to 0.054 g/10^6 cal. Waste
disposal is by ponding untreated sludge at 2.5 mill/KWh. No escalation of
costs is considered. Costs are in 1988 year dollars. Assumptions are:

- 3,400 Nm³/hr per MW, 2.5 × 10^9 cal/MWh
- 65% load factor (the system operates 5,698 hrs/yr)
- 3 operators/shift

TABLE 6.8. Breakdown Typical of Scrubbing
System Capital Costs.

Item	Percent of Installed Cost
Excavation	0.8
Foundations	3.3
Structural	9.0
Process equipment	25.1
Piping	5.4
Electrical	9.5
Painting	0.5
Instruments	4.6
Insulation	2.2
Indirect field construction	11.0
Overhead	2.6
Engineering	7.2
Revisions	1.4
Startup	3.0
Contractor fee	6.4
Interest during construction	5.0
Contingency	3.0
Total	100.0

- 6 maintenance workers/shift
- Maintenance material equal to maintenance labor
- 1.2 times $CaCO_3$ stoichiometry for limestone needs
- Annual household use is 12,637 KWh of electricity
- Scrubbers operate all times boilers operate
- 95% conversion of S to SO_2
- Coal has 2.7% S, 7,200 cal/g and 10% ash
- FGD capital costs are $120/KW
- Gases are ideal
- System requires 2.27×10^{11} g steam/yr (5×10^8 lb/yr)
- Process water needs are 950 ac ft/yr (3.1×10^8 gal/yr)
- Scrubbing system requires 2% of station power

Other assumptions are used as noted.

From these data, it can be calculated that 77,400 tons (English) of SO_2 would be produced per year. Then the amount of limestone needed is

$$(77,400)\left(\frac{100 \text{ ton } CaCO_3}{64 \text{ ton } SO_2}\right)(1.2)(0.90) = 130,600 \text{ ton/yr}$$

Electricity needs are

$$(500,000 \text{ KW})(5,698 \text{ hr/yr})(0.02) = 5.7 \times 10^7 \text{ KWh/yr}$$

Operator and maintenance man-hours are

$$(3)(5,698) = 17,094 \text{ hr/yr}$$

Capital recovery factor for 30 year life at 10% interest rate = 0.1061. These data are assembled in Table 6.9 to show typical breakdown of an FGD scrubbing system operating costs.

The equivalent unit operating costs can be calculated as

$$\left(\frac{\$17,452,670/\text{yr}}{500,000 \text{ KW}}\right)\left(\frac{\text{yr}}{5,698 \text{ hr}}\right)\left(\frac{1,000 \text{ mills}}{\$}\right) = 6.13 \text{ mills/KWh}$$

Adding the cost of sludge disposal, the total is 8.63 mills/KWh. The cost to a householder equals

$$\left(\frac{\$8.63 \times 10^{-3}}{\text{KWh}}\right)(12,637 \text{ KWh/yr}) = \$109/\text{yr}$$

TABLE 6.9. FGD Scrubbing System Costs in 1988 Dollars.

Direct Costs	Annual Cost, $	Percent of Operating Costs
Limestone delivered = (130,600 ton)($14/ton)	1,828,400	10.48
Operating labor = (17,094 hr)($13.00 hr)	205,130	1.18
Supervision = (0.15)(205,128)	30,770	0.18
Utilities:		
Steam = (5 × 10⁸ lb)($2.50/1000 lb)	1,250,000	7.16
Water = (3.1 × 10⁸ gal)($1.25/1000 gal)	387,500	2.22
Electricity = (5.7 × 10⁷ KWh)($0.06/KWh)	3,420,000	19.60
Maintenance:		
Labor = (2)(17,094 hr)($15.00/hr)	512,820	2.94
Materials = same as labor	512,820	2.94
Analyses, assumed at	90,000	0.52
Direct Cost Subtotal	8,237,440	47.22
Indirect Costs		
Capital recovery = (0.1061)(120)(500,000)	6,366,000	36.48
Plant overhead = (0.60)(205,128 + 30,769 + 512,820)	449,230	2.56
Administration = (0.02)(120)(500,000)	1,200,000	6.88
Property taxes and insurance = (.02)(120)(500,000)	1,200,000	6.88
Indirect Cost Subtotal	9,215,230	52.80
Total annual operating cost (not incl. waste disposal)	17,452,670	100.00

Note from Table 6.9 that the two largest direct costs are for electrical energy and for chemicals. However, waste disposal in this example accounts for 29% of the total annual costs. Treatment and/or disposal of hazardous wastes would make these costs even greater.

Energy costs reported for dual-alkali scrubbing to remove 95% of the SO_2 [22] are less than 1.1% of the station power for a large system of 300 MW operating at a total maximum pressure drop of 24 cm H_2O. Of this, 60% is used for booster fans, 10% for reheat fans, and 30% for pumps, thickeners, vacuum filters, instruments, lights, feeders, and all other requirements. If less SO_2 were to be removed, the 30% value would decrease proportionately.

Water losses with the sludge depend on the amount of SO_2 removed and the solids content of the sludge. For a 60% solids sludge, the water losses could equal about 6×10^3 g/hr per MW per percent sulfur scrubbed. Evaporation from open thickeners also contributes to water losses. With typical ambient temperatures, humidity, and wind speeds, this could equal about 10^4 g/hr per MW.

Chemical additives are required for SO_2 removal, whether by wet scrubbing or by collection in a filter or ESP. Scrubbers would use less chemicals because stoichiometry requirements are less, being about 1:1 for lime, sodium, and magnesium reactants and about 1.2:1 for limestone. In terms of lime, this equals about 45 ton/yr per MW per percent of sulfur removed.

6.5.2.1 Wet FGD Systems

FGD systems have been used since about 1850 [23]. There are currently at least seven significant FGD wet scrubbing processes. These are the lime, limestone, alkali fly ash, Wellmann-Lord/Allied Chemical, double-alkali, magnesium oxide, and sodium carbonate processes. Each of those could use basically similar scrubbing devices. The system variations exist mainly in the chemicals used, the chemical sequence, and the type and disposition of the product. A very brief description of some of these systems is presented here.

Lime and Limestone Scrubbing

These processes are basically similar, and both produce a similar nonregenerable sludge. The overall chemical processes can be characterized for limestone as

$$SO_2 + 1/2H_2O + CaCO_3 \rightarrow CaSO_3 \cdot 1/2H_2O + CO_2$$

and for lime (slaked)

$$SO_2 + Ca(OH)_2 \rightarrow CaSO_3 \cdot 1/2H_2O + 1/2H_2O$$

Some of the produce from both processes would go on reacting

$$CaSO_3 \cdot 1/2H_2O + 1\ 1/2H_2O + 1/2O_2 \rightarrow CaSO_4 \cdot 2H_2O$$

Systems with lower SO_2 concentrations and higher oxidation conditions produce more of the gypsum product ($CaSO_4 \cdot 2H_2O$).

Older FGD scrubbing systems frequently consist of a venturi scrubber followed by an absorption tower. The newer systems use spray and/or packed absorbers only if particulate removal requirements are low. This reduces pressure drops (and therefore operating costs), reduces scaling and improves SO_2 removal. With proper gas conditioning, much of the particulates are also removed. Typical SO_2 removals are 95% on both low- and high-sulfur coal.

Operation of these systems is critical and many parameters must be monitored and controlled to assure proper operation. The chemical feed rate must be adjusted to maintain the desired pH as SO_2 concentration and/or boiler load changes. Lime systems pH is about 7, and for limestone systems, it is about 5.7. Hold tank residence time is critical in that crystal- lization on nuclei in suspension within the hold tank is desired instead of crystallization on scrubbing system internals. The concentration of sus- pended solids must be high enough to provide crystallization sites, yet not so heavy that pluggage due to settling occurs. For these reasons relatively high liquid-to-gas ratios are used to maintain desired pH as absorption occurs, and to provide adequate crystal sites. Locations for adding the makeup chemical are critical. Chemicals can be added at the hold tanks or at various stages within the absorber to maintain desired pH to prevent or encourage crystallization.

Wellman-Lord/Allied Chemical Scrubbing System

The Wellman-Lord process is a regenerable system. SO_2 is absorbed and reacted with sodium sulfite solution to produce sodium bisulfite as follows

$$SO_2 + Na_2SO_3 + H_2O \rightarrow 2NaHSO_3$$

Some of this product oxidizes to sulfate and is lost. Thermal decomposi- tion causes the reaction shown to reverse producing a rich SO_2 stream which is treated in an Allied Chemical Claus process to produce a high- purity (99.9%) elemental sulfur. Other sulfur recovery systems could be used in place of the Allied Chemical Claus systems.

The scrubbing system consists of a venturi followed by an absorber. Over 90% of the SO_2 is removed when burning high-sulfur coal. The small amount of oxidized sodium sulfate is purged from the system, evapo- rated and sold to the paper industry. Soda ash is added to make up the lost sodium content. This system is unique in that no sludge is produced. The product, elemental sulfur, is of high (reagent-grade) purity. A variation of this process would be to produce sulfuric acid from the recovered SO_2.

Double-Alkali Scrubbing

This process consists of scrubbing the SO_2 with a sodium solution, then contacting this solution in an external reactor with a calcium solution to produce a calcium sludge. The sodium solution has the advantage of being more reactive and the products formed in the scrubber portion of the system are not soluble so plugging is less likely to occur. The Cane Run #6 scrubbing system of Louisville Gas and Electric is guaranteed to remove 95% of the SO_2 when 5% sulfur coal is burned. Tests have shown that 99% SO_2 removal can be achieved. The lime stoichiometry in the reactor is 1.05 or less per mole SO_2 removed from the flue gas. This 300-MW system consists of two absorber modules and two reactor trains in parallel and can operate from 20 to 100% of peak capacity. Each reactor train has two reactors in series. One thickener and three filters handle the sludge product. The overall system gas pressure drop is 24 cm H_2O including a 10–15 cm H_2O pressure drop across the absorber trays. Absorber liquid-to-gas ratio is 0.76 ℓ/Am^3. This system is guaranteed to have a one year 90% availability.

General System Notes

Older systems may be able to increase SO_2 removal by various procedures. Use of 2–8% MgO additive in lime scrubbers, i.e., thiosorbic lime, can increase SO_2 removal efficiency at no system operating penalty. This can also reduce scaling potential. Operating changes such as increasing the liquid-to-gas ratio and decreasing gas velocity may also increase SO_2 removal. However, increased pressure drops and/or decreased system capacity could result. System changes such as the addition of more effective absorber packing, spray arrangement modification, or addition of more sprays would increase SO_2 removal. Adjustment of pH, chemical addition amounts and locations, holdup time, and slurry concentrations are other possibilities. Some of these could require system modifications. Tank, pipe, and pump sizes may have to be increased and additional chemical facilities installed. Increased SO_2 removal means that more sludge would be handled, so these facilities may have to be modified also.

6.5.2.2 Wet FGD Sludge

The nonregenerable scrubbing systems produce a sludge by-product. As the lime/limestone systems comprise about 87% of the scrubbing facilities, only the use of this type of sludge is discussed. These sludges are thixotropic in nature and may contain about 30–70% calcium sulfite, 20–65% calcium sulfate, plus calcium carbonate, fly ash, and inerts. Wet density is about 1.28 g/cm^3. More sulfate is formed when low-sulfur coal flue gas is scrubbed and more sulfite is formed when high-sulfur coal is scrubbed. Amount of sulfate can be increased by oxidation of the solution. Sludge mixed with fly ash results in a composite volume less than the sum of volumes of the original components. Production rate of dry sludge from these systems is about 10^6 g/hr per MW per percent of sulfur removed. This value can be corrected to account for mass of wet sludge produced using the percent of water. High-sulfate sludges dewater easier and contain more solids. Properly dewatered sludge contains 55–70% solids.

Most sludge is disposed in the dewatered condition by ponding. There are many potential uses of this material. When dried it is hard and brittle, like dry clay. It must be fixed by a chemical additive, otherwise it reverts to the thixotropic state when wetted. Simple oxidation converts this sludge to gypsum. Gypsum is currently consumed in the U.S. at a rate of 16 million ton/yr for wallboard and Portland cement.

There are a number of commercial fixation processes being sold to convert scrubbing system sludge into a stable material. Most of these processes are proprietary. They convert the sludge into landfill with a compressive strength of from 5×10^7 to 45×10^7 g/m². For comparison, the allowable bearing strength of clay, sand, and gravel ranges from $5–10 \times 10^7$ g/m² and for hardpan, shale, and rock it is $10–35 \times 10^7$ g/m². This shows that fixed sludge can be a good construction base material.

Treated sludges have a low permeability of about 10^{-6} to 10^{-7} which makes it essentially impervious to water penetration. Water penetrates 0.3 m (1 ft) per year at 10^{-6} and 0.03 m (0.1 ft) at 10^{-7} permeability. Sludge treated in this manner can be used as fill to improve the value of land. FGD sludge in the dewatered form is being studied as a method of sealing up mine shafts as the mining operation withdraws to reduce water pollution and subsidence problems.

6.5.2.3 Material of Construction

Wet FGD systems are subject to high degrees of corrosion and erosion because of low liquor pH, high chloride ion concentration, abrasive solids, and other chemicals present. Linings and use of high chromium, molybdenum, and nickel alloy steels are required to provide acceptable equipment life. In general, rubber linings are highly suitable for lining many wet scrubbing system vessels, pipes, pumps, and fans. Rubber linings in the late 60s and early 70s were not prepared and applied properly, and many failed. However, the state of the art has progressed so that these linings now give good life. Fiberglass, epoxy, plastic, and ceramic liners can also give good service in wet scrubbing systems as noted in Section 6.5.2.

Plain-carbon and high-carbon stainless steels are usually not satisfactory without liners in FGD scrubbing systems. Low-carbon alloys are not always acceptable in low-pH high-chloride environments because of the resulting localized attack due to pitting and crevice corrosion. Stress corrosion is usually not a problem in scrubbers at temperatures below 60°C (140°F).

Table 6.10 lists several alloys of interest for SO_2 scrubbing systems and the results of an FGD system corrosion test [25]. These data were obtained from a coal-fired boiler scrubbing system at average conditions reported as pH of 2, temperature of 66°C, velocity of 1,350 cm/sec, and chloride ion concentration of 200 ppm. Composition of 317 L stainless steel differs from the composition of 317 S/S shown in the table in that the "L" signifies low-carbon content (maximum is 0.03%).

6.5.3 LIME SPRAY DRYER

This system is used for FGD and acid gas control from incinerators. It consists of a lime shaker which produces $1-3$-μm particles which are slurried in water and atomized as 75-μm liquid droplets into a large reaction chamber. The droplets contact high velocity (about 100 ft/sec) hot flue gases where they dry (in 3–10 milliseconds) and react with the acids. The particles remaining and reaction products formed are usually removed in a fabric filter. Reactions occur as described in the Lime FGD Wet Scrubber section. Notes relative to reactant costs and other factors are given in the Wet FGD section.

The installed cost of a 500 MW (2.6×10^6 Am³/hr) lime spray dry FGD system in 1988 year dollars is the same as a utility-evaluated low-sulfur lime wet scrubber, i.e., \$120/KW. This is \$40/acfm or \$23/Am³ per hr. A spray dry system with fabric filter [25] on a 278,000 Am³/hr MSW incineration system has an installed capital cost of \$58.4/acfm (33.6/Am³ per hr). This suggests a sizing exponent of 0.73. An installed cost-to-purchase cost ratio of 2.17 is used.

Spray dryers *may* have a small capital cost advantage for acid gas control over wet scrubbing systems with thickeners and liquid and sludge handling facilities, but the extra alkali required increases operating costs. Typical stoichiometric ratios of alkali to acid gas are 2.5. Operating costs can be estimated using these data and by following procedures outlined in the preceding scrubber and filter sections.

6.5.4 DRY INJECTION

Fine lime injection has been tried for both FGD and incineration acid gas control. This procedure is adequate for control of low concentrations of HCl and SO_2, but usually is not adequate for most FGD needs. A more

TABLE 6.10. Some Alloys of Commercial Interest in FGD Scrubbers [24].

| | Nominal Composition (%) | | | | | | | FGD Corrosion Testing | | |
	Cr	Mo	Ni	Fe	Cu	C	Other	Average Corrosion Rate (mm/yr)	Maximum Pit Depth (mm)	Maximum Crevice Corrosion
Wrought alloys										
Inconel[a] 625	22	9	Bal	3		0.05	3.6 Cd + Ta	0.07	none	incipient
Hastelloy[b] C-276	16	16	Bal	5		0.02[d]	2.5 Co[d], 4W	0.08	none	none
Hastelloy[b] C-4	16	16	Bal	3[d]		0.015[d]	2 Co[d], 0.7 Ti	0.08	none	incipient
Incoloy[a] 825	22	3	Bal	30	2	0.03		0.11	0.05	incipient
317 S/S	19	3.5	13	Bal		0.08[d]		0.11	0.25	incipient
Carpenter[c] 20 Cb-3	20	2.5	35	Bal	3.5	0.07		0.11	0.05	incipient
316 S/S	17	2.5	12	Bal		0.08[d]		0.11	0.10	incipient
Cast alloys										
IN-862	21	5	24	Bal		0.07[d]		0.08	none	none
CD-4M Cu	26	2	5	Bal	3	0.04[d]		0.10	0.08	none

[a]Trademark of The International Nickel Co., Inc.
[b]Trademark of Cabot Corporation.
[c]Trademark of Eastern Stainless Steel Corporation.
[d]Maximum.

reactive alkali such as trona or nacholite ore is better for this purpose. The process consists of conditioning the gases by water spray cooling to about 120°C (250°F), but with no liquid water present. Then, finely ground alkali is injected into the gas. A fabric filter is required to collect the particulates and to provide a reaction bed.

Capital cost for dry injection on a 278,000 Am^3/hr incineration system is reported [26] to be $54.3/acfm ($31.2/Am^3 per hr). This is slightly less expensive than a spray dry system but can require even more alkali with less efficient acid gas control. A capital cost sizing exponent of 0.73 should be used. Operating costs of dry injection systems is essentially the same as for spray dry systems [27].

6.5.5 ABSORBERS

The two major types of absorbers used for the control of gaseous emission are countercurrent spray towers and countercurrent and crosscurrent packed towers. Purchase cost of absorber towers, P_A, in 1988 year dollars can be estimated by

$$P_A = 1100D_A^{0.75} \tag{6.7}$$

where D_A is column diameter in inches from 10 to 200 inches and up to 40 ft high. This is based on data of Hall et al. [28]. This cost is for a system that can accommodate double packed beds, each 5 ft. high. Equation (6.7) shows a sizing exponent of 0.75.

The packing cost must be added to the tower cost for packed beds. This purchase cost in 1988 year dollars can be estimated for polyethylene packing, P_p, in dollars per ft^3 packing by

$$P_p = 26.5S_p^{-1.05} \tag{6.8}$$

where S_p = packing size in inches from 1 to 3 inches. This is applicable to Pall and Raschig Rings and Intalox Saddles types of packing. These packings in stainless steel material of construction would cost about 5 times more.

Costs of tanks, controls, instruments, platforms, fans, pumps, and mist eliminators must be included in equipment costs. Installed capital costs would then be 2.20 times equipment costs. As a rule of thumb, installed absorber costs in 1988 year dollars run from $7.5–15/scfm for large systems > 10,000 scfm.

Operating costs could be calculated using the procedures in the previous sections (e.g., as in fabric filters). A 5-year packing life and a 20-year columns life can be assumed.

6.5.6 MIST ELIMINATORS

The final stage of all wet scrubbers is the mist eliminators. These costs are included in some presentations, but may need to be estimated for initial or replacement purposes. Installed cost of a high quality, two-stage mist eliminator in 1988 year dollars is about $100/ft^2 of cross-sectional area. This is a relatively modest cost (e.g., equivalent to $0.50/KW), but very necessary to the success of the wet scrubber or absorber.

6.6 ADSORBERS

6.6.1 SIZING AND OPERATING

Adsorbers are among the simplest air pollution control systems, yet their operation is nothing less than an art. Working capacity data are obtained by trial and error for each specific absorbent, adsorbate, and system configuration as discussed in Section 5.8. This should aid in establishing and documenting initial operating sequences. It then is a simple matter of repeating these established sequences.

Automatic sequencing and control enables adsorbers to operate with little attention for long periods of time. Usually, an odorous emission signals that something is wrong with the system. There are certain operating parameters that should be observed. These include pressure drop measurements across the beds, temperatures of gas and beds, gas flowrates, regeneration fluid flowrates, relative humidity, and concentrations in and out. Gas adsorption flowrates should be set to provide an optimal superficial velocity, e.g., 30 cm/sec. At constant flowrates, an increase in pressure drop could indicate buildup of solids or deposits in the bed or deterioration of the bed. If solids are depositing, precleaners should be used. This includes filtering out solids and/or condensing materials that could be more easily removed by condensation. Pressure decreases on a negative system could indicate blower problems or other downstream problems such as air leaks. Temperature measurements are needed to prevent (or indicate) thermal decomposition of the beds.

Recovery of the desorbed gas is an important part of the control system and often the part that provides an attractive return on the adsorbing system investment. This includes the simple physical process of decantation plus the chemical processes of distillation, stripping, extraction and others. Included in the "others" category is incineration. However, unless the heat is recovered, this would be wasteful and should be used only if the other techniques are not suitable.

Recovery by decantation is convenient if the desorbed vapors can be condensed and if they are insoluble in water. Many organic vapors meet these criteria and can be recovered by steam desorption and decantation. In a continuous decanter, the siphon leg must have an air break. The leg height fixes the location of the interface where the inlet discharge should be located. The light "oils" are removed from a top overflow and the water is recovered from the siphon leg.

The recovery system must be designed and operated with safety in mind as many recovered materials are flammable vapors. Operation must be either above or below the explosive limit and an inert gas purge should be used if it is necessary to cycle the equipment from one side of the explosive limit to the other. Sparks must be eliminated, so grounding of all discharge lines (especially liquid lines), receivers, and other equipment is required.

6.6.2 INSTALLATION AND MAINTENANCE

Adsorbent bed material is fragile and must be handled with care. Checks should be made to determine that the beds are packed carefully and uniformly to the specified depths. All gas and liquid lines and control systems should be inspected for proper connections, grounding, adequate valves, and good working order.

Most adsorption systems require relatively little maintenance. Bed replacement is the most demanding item. Controls, pumps, and fans require periodic routine inspection and lubrication.

6.6.3 COSTS

Completely assembled, small adsorbers that can handle up to 170 m³/hr (100 cfm) of odorous gas can be purchased for about $500. These systems have no pumps or controls and are simply coated canisters (barrels) prepacked with adsorbent. They can be discarded when the adsorbent is expended. Regeneration costs would be extra.

Capital costs of regenerable carbon adsorbers based on vendor data and DuPont installation data [29] and adjusted to 1988 year dollars can be found by

$$CC_A = 110Q^{0.72} \tag{6.9}$$

where

CC_A = adsorber capital cost, dollars
Q = inlet gas flow rate, Nm³/hr

A minimum Q of 1,000 Nm³/hr applied to Equation (6.9). Below this, base unit cost is fixed at $16,000. These costs include the initial carbon bed and all costs except regeneration system costs for low inlet gas concentration adsorbers. These should include cost of continuous multiple-bed systems operating on an adsorption-desorption-cooldown cycle.

Operating costs vary greatly with method of operation and regeneration. Another cost is that of replacement carbon, which runs about $1.50/lb. The life of the carbon adsorbent is typically 3 years. One procedure for estimating operating cost of an adsorber is given in the following example.

Some assumptions are

condenser water = 6 gal/lb steam used
steam consumption: for carbon regeneration = 4 g steam/g organic
                   for product recovery processes = 1 g steam/g organic per separation stage (i.e., by steam stripping)
      depreciation: 10-yr straight line
      maintenance: 6% of investment/yr
   operating labor: 1 hr per shift per control device at $22/hr to include all supervision and overhead
carbon replacement
   per 3-yr life = $0.027/yr per g organic/hr

A credit for recovered product should be made based on purchase price or production cost.

Example Problem 6.3

Estimate the operating costs and operating conditions of a 50,000 Am³/hr carbon adsorber used to recover benzene (C_6H_6) from a process stream. Inlet concentration is 0.2% C_6H_6 at 35°C and outlet is to be 20 ppm. Carbon steel construction is to be used with 6 × 12 mesh activated carbon operating at a superficial velocity of 45 cm/sec. The carbon bulk density is 0.4 g/cm³. Bed depths are 50 cm. Pilot tests indicate a working solvent loading at the 2% breakthrough desired is 0.15 g solvent/g carbon. The system is to operate 8,000 hr/yr on a time cycle of one-third adsorption, one-third regeneration, and one-third cooldown and standby. Benzene is sold for $0.40/kg ($1.32/gal).

Solution

$$\text{Adsorber cross-section area} = \left(\frac{50,000 \text{ Am}^3/\text{hr}}{45 \text{ cm/sec}}\right)\left(\frac{\text{hr}}{3,600 \text{ sec}}\right)\left(\frac{100 \text{ cm}}{\text{m}}\right)^3$$

$$= 3.09 \times 10^5 \text{ cm}^2$$

Therefore, diameter $= 6.27$ m.

$$\text{Carbon mass per bed} = (3.09 \times 10^5 \text{ cm}^2)(50 \text{ cm})\left(\frac{0.4 \text{ g}}{\text{cm}^3}\right)$$

$$= 6.17 \times 10^6 \text{ g}$$

$$\text{Benzene adsorption rate} = \left(\frac{2,000 - 20 \text{ moles C}_6\text{H}_6}{10^6 \text{ moles gas}}\right)(50,000 \text{ Am}^3/\text{hr})$$

$$\times \left(\frac{273}{273 + 35}\right)\left(\frac{78 \text{ g}}{22.4 \text{ l}}\right)\left(\frac{1000 \text{ l}}{\text{m}^3}\right)$$

$$= 3.06 \times 10^5 \text{ g C}_6\text{H}_6/\text{hr}$$

$$\text{Time to load one bed with C}_6\text{H}_6 = \left(\frac{0.15 \text{ g solvent}}{\text{g carbon}}\right)(9.26 \times 10^6 \text{ g})$$

$$\times \left(\frac{\text{hr}}{3.06 \times 10^5 \text{ g solvent}}\right) = 4.5 \text{ hr}$$

$$\text{Gas flow rate} = (50,000 \text{ Am}^3/\text{hr})\left(\frac{273 + 20}{273 + 35}\right) = 47,560 \text{ Nm}^3/\text{hr}$$

Capital costs from Equation (6.9): $CC_A = 110(47,560)^{0.72} = \$256,500$

(assume this includes a decanter regeneration system cost).

Figure 5.39 gives bed pressure drop of about (20 cm H_2O)/m. Estimate twice this for total system pressure drop $= (20 \text{ cm/m})(0.5 \text{ m})(2) = 20$ cm H_2O. Electrical consumption using Equation (5.25) $= (6 \times 10^{-5}) \times (20)(50,000) = 60$ KW, which is acceptable for this system.

$$\text{Steam consumption} = \left(\frac{4 \text{ g steam}}{\text{g organic}}\right)(3.06 \times 10^5 \text{ g/hr})$$

$$= 1.22 \times 10^6 \text{ g/hr}$$

Annual costs of adsorption system using the DuPont [29] technique are

$$\text{Depreciation} = \frac{\$256,500}{10} \qquad\qquad\qquad = \$ \ 25,650$$

$$\text{Maintenance} = (0.06)(256,500) \qquad\qquad = \quad 15,390$$

$$\text{Operating Labor} = \left(\frac{1 \text{ hr}}{\text{shift}}\right)\left(\frac{3 \text{ shift}}{24 \text{ hr}}\right)\left(\frac{8,000 \text{ hr}}{\text{yr}}\right)\left(\frac{\$22}{\text{hr}}\right) \quad = \quad 22,000$$

$$\text{Carbon replacement} = \left(\frac{\$0.027/\text{yr}}{\text{g organic/hr}}\right)\left(\frac{3.06 \times 10^5 \text{ g } C_6H_6}{\text{hr}}\right)$$

$$= \quad 8,280$$

Utilities

$$\text{Electric} = (60 \text{ KW})(8,000 \text{ hr/yr})(\$0.06/\text{KWh}) \qquad = \quad 28,800$$

$$\text{Steam} = \left(\frac{1.22 \times 10^6 \text{ g}}{\text{hr}}\right)(8,000 \text{ hr})\left(\frac{\text{lb}}{454 \text{ g}}\right)\left(\frac{\$3.70}{1,000 \text{ lb}}\right) = \quad 79,540$$

$$\text{Water} = \left(\frac{1.22 \times 10^6 \text{ g}}{454}\right)(8,000)\left(\frac{6 \text{ gal water}}{\text{lb steam}}\right)\left(\frac{\$1.25}{1,000 \text{ gal}}\right)$$

$$= \quad 161,230$$

$$\text{Capital charges at } 15\% = (0.15)(256,500) \qquad = \quad 38,480$$

$$\textit{Total Annual Costs} = \$359,370$$

$$\text{Annual income from sale of product} = \left(\frac{\$0.40}{\text{kg}}\right)\left(\frac{3.06 \times 10^5 \text{ g}}{\text{hr}}\right)$$

$$\times \left(\frac{\text{kg}}{1,000 \text{ g}}\right)\left(\frac{8,000 \text{ hr}}{\text{yr}}\right)$$

$$= \$979,200$$

Net annual *profits* are \$979,200 − 359,370 = \$619,830 per year

6.7 INCINERATORS

The two basic groups of incinerators are the fixed and the transportable. Both systems have relatively short lives when operations are such that they are frequently heated up and cooled down. The transportable have shortest lives (e.g., 5 years) because of the mechanical shock during movement. Well-cared-for fixed incinerators can operate for 10 to 20 years.

Three types of incinerators are discussed here. The constraint for volumetric heat release for these are:

Hearth incinerator   15,000 to 25,000 Btu/(hr ft³)
Rotary kiln incinerator   25,000 to 40,000 Btu/(hr ft³)
Liquid injection incinerator   25,000 to 60,000 Btu/(hr ft³)

where ft³ is the volume of the incinerator primary. The secondary, if any, is sized to provide about 2 seconds of gas residence time.

6.7.1 CAPITAL COST

Heat content and approximate composition of the waste and auxiliary fuel should be known. Assuming complete volatilization and combustion, the volumetric flow rate of flue gases from both the waste and auxiliary fuel (if required) can be determined. The gas flowrate can then be used, following the procedure in Reference [30], to establish purchase cost of the basic incinerator system. This includes refractory and shell, forced draft (FD) and induced draft (ID) fans, burners, feed and ash-discharge system, and necessary drive motors. Purchase cost in 1988 year dollars of rotary kiln incinerators, P_{RK}, is

$$P_{RK} = 738,500 f_1^{0.52} \tag{6.10}$$

for f_1 from 5 to 30. The size factor, f_1, can be found by

$$f_1 = Q_{sc} \left(\frac{t_e + 460}{2.66 \times 10^6} \right) \tag{6.11}$$

where

Q_{sc} = incinerator exit flue gas volumetric flow rate, scfm
t_e = incinerator secondary or exit gas temperature, °F

Incinerator installed cost is 1.64 times total purchase cost.
    Purchase cost of hearth incinerators, P_H, in 1988 year dollars is

$$P_H = 261,500 f_1^{0.69} \tag{6.12}$$

for f_1 from 4 to 14.
    Purchase cost of liquid injection incinerators, P_{LI}, in 1988 year dol-

lars is

$$P_{LI} = 280,500 f_1^{0.39} \tag{6.13}$$

for f_1 from 5 to 30.

The three purchase costs just presented are based on a system with 2 seconds gas residence time. These costs can be adjusted for shorter or longer times by multiplying Equations (6.10), (6.12), or (6.13) by the adjustment factor, f_2

$$f_2 = 1.067 - \frac{0.133}{\theta} \tag{6.14}$$

where θ is gas residence time in seconds.

Total system purchase cost would include emission controls and may also include quenching and/or a heat recovery boiler. Emission control costs are provided in the previous sections of this chapter. Quench chamber purchase costs, P_Q, adjusted to 1988 year dollars can be estimated using [31]

$$P_Q = 0.28 Q_a + 10,300 \tag{6.15}$$

where Q_a is actual chamber inlet gas volumetric flow rate, acfm. Installed cost is 1.61 times purchase cost of quencher. A spray chamber could be used also for this at a purchase cost, P_{sc}, in 1988 year dollars of

$$P_{sc} = 0.30 Q_a + 55,500 \tag{6.16}$$

Boiler purchase costs [32] escalated to 1988 year dollars, P_B, can be estimated by

$$P_B = 190 Q_{sc}^{0.79} \tag{6.17}$$

where Q_{sc} is boiler inlet volumetric flow rate, scfm. Boiler installed cost is 1.64 times boiler purchase cost.

Total purchase cost of an incineration system includes the purchase cost of all components plus 10% for instruments and controls plus 3% for taxes and 5% for freight. Total capital cost of the installed system is then 1.64 times this purchase cost total.

6.7.2 OPERATING COST

Operating costs must include costs relative to each component of the system. Fuel for startup must also be included as it takes about three hours for heat-up each time the incinerator is brought on-line from a cold start. This is equivalent to about 10,000 Btu/(hr ft³ of combustion chamber). Table 6.11 gives a procedure that can be followed. Man-hours shown are higher than for many other systems because hazardous or toxic materials may be handled.

To the incinerator costs must be added costs for waste disposal and emission control. Costs due to steam recovery must be credited to operating costs. Refractory life is assumed to be four years and interest rate is 10%. Incinerator life of 20 years is assumed.

Steam credit needs to be estimated. One way to do this is to assume that 60% of the gross incinerator heat input goes to steam. Equate this steam quantity to the cost of producing it with an oil-fired boiler operating at 80% thermal efficiency. Assuming #6 fuel oil at $.70/gal with 153,000 Btu/gal, the value is $4.58/10⁶ Btu. Then

Steam credit in $/hr = (incinerator heat input in Btu/hr)

$$\times \left(\frac{0.60}{0.80}\right)\left(\frac{\$4.58}{10^6 \text{ Btu}}\right)$$

$$= (3.44 \times 10^{-6})(\text{heat input Btu/hr}) \quad (6.18)$$

Equation (6.18) can be adjusted to account for the substitution of different fuels and for varying fuel costs.

6.8 OPERATION AND MAINTENANCE

A properly designed and installed control system is necessary to achieve the required emission limits. However, these limits cannot be achieved without good operation and maintenance. Assuming adequate equipment, instruments, and tools are available, this requires training, familiarization with the facility, and a desire to do a good job. Specific comments on operation and maintenance are given in the preceding control systems sections. The following comments are additional and apply in general to all systems.

6.8.1 OPERATION

Engineers, designers, supervisors, and others involved in air pollution control systems, as well as the actual operators, need to understand the operation of the system. This understanding should consist of formalized schooling, on-the-job training, and self instruction.

TABLE 6.11. Assumed Operating Costs for Incinerator in 1988 Year Dollars.

Direct Costs

Operating labor: (4/hr)(operating hr/yr)($13.00/hr)

Supervision: 25% of operating labor cost

Utilities:
   Electric: $(\Sigma\, KW_B)$(1.1)(hr/yr)($0.06/KWh)
     Fuel: as required for start-up + operation
     Water: (for ash quench + gas quencher in gal/hr)(hr/yr)($1.25/1000 gal)

Maintenance:
      Labor: (2/hr)(hr/yr)($15.00/hr)
     Material: equal to maintenance labor cost
     Refractory: (0.3155)(refractory installed cost)

Indirect Costs

Capital recovery: (11.75%)(capital cost − refractory installed cost)

Plant overhead: (60%)(operating labor + supervision + maintenance)

Administration: (2%)(capital cost)

Property taxes and insurance: (2%)(capital cost)

Total Annual Cost

= Direct costs + indirect costs + waste disposal

+ emission control costs − steam credit

Pressure drops across the control system have been noted as one of the most important parameters in all systems. Subtle factors such as keeping liquids out of the instrument lines and keeping the lines free from pluggage are necessary for obtaining meaningful readings. An understanding of fluid flow is important in that when there is a significant gas velocity, gas phase pressure drop measurements must be made on the same size pipe for both inlet and outlet readings for the same volumetric gas flow. Otherwise, a correction must be made to account for the inversion of kinetic and static pressure heads before meaningful system or device pressure drops can be obtained.

It is advisable to record as much meaningful data as possible on a new system at the time of startup. These data can later be referred to when trying to determine why certain types of difficulties seem to appear. Some important parameters to measure, if pertinent to the system, are: pressure drops across all components; gas static pressure at the fan and the equipment; gas and liquid flowrates; liquid pressures; fan (especially) and pump electrical current; gas and liquid temperatures and compositions; gas humidities; liquid pH, solids concentrations, densities and rates; and stack appearance and/or emission rates. At the time of startup, it is imperative to check items such as fan and pump rotation. Some of these can deliver up to 60% of rated output when turning in the wrong direction. A simple reversing of any two wires in a three-wire connection changes rotation direction.

Complete operating instructions should be formalized, recorded, and followed. The equipment supplier should provide some assistance in this. Literature, including suggested procedures and forms, is also available [33].

6.8.2 MAINTENANCE

Pollution-control systems cannot function very long without good maintenance; the better the maintenance, the better the assurance that the system will perform within the required emission limits. Preventive maintenance includes routine inspection and correction of the plant facilities. Preventive maintenance does not have to be directed merely at maintaining the status-quo level of performance—it can be used to continuously upgrade the facilities to achieve reduced emissions and conserve energy. This is no minor assignment and requires highly qualified technical assistance, training, and a desire to do the job well on the part of all involved.

6.9 COSTS

There are certainly costs to providing pollution-control facilities. However, these costs do provide certain tangible and intangible savings. The President's Council on Environmental Quality has predicted that, in the U.S., air pollution expenditures provide about a 15% return on the investments. There are many situations where incineration has saved large disposal costs and the recovery of energy from the combusted materials results in a net profit. These examples indicate that not all costs are outgoing.

6.9.1 COMPARATIVE COSTS

Comparative costs are presented here with reservations. Specific cost examples are given as appropriate with each system in the preceding sections. One cannot compare the various systems on the basis of cost alone. Each control system has specific advantages and limitations, and even systems in the same category can vary widely in effectiveness, material of construction, life, and costs.

Keeping in mind the above reservations, generalized cost data are presented for various systems in Figure 6.9. The "Example FGD" curve extrapolates the operating cost data of 6.13 mill/KWh for the 500-MW system example in Section 6.5.2. The dashed curves for low and high FGD operating costs are extrapolated reported data [34] of 0.4 and 8.1 mill/KWh. Note from the several FGD costs shown that, depending on how the costs are determined and what is included, costs of a single type of system can span nearly the entire cost range. The spray dry and dry injection values are extrapolated data [27].

6.9.2 GENERALIZED COSTS

Cost estimating can be summarized using the following procedures. The first is for capital (installed) costs, the second is for annual (operating and maintenance) costs. Capital costs can be based on costs of purchased

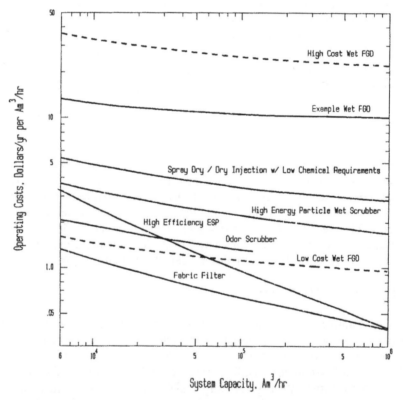

Figure 6.9 Typical air pollution control system operating costs in 1988 dollars.

TABLE 6.12. Capital Cost Factors.

System	Ratio Installed to Purchase Cost	Relative Equipment Cost	Relative Installation Costs		
			Direct	Indirect	Total
ESP	2.24	1.0	0.67	0.57	1.24
Filter	2.17	1.0	0.72	0.45	1.17
Wet scrubber	1.91	1.0	0.56	0.35	0.91
Wet FGD	1.91	1.0	0.56	0.35	0.91
Spray dry	2.17	1.0	0.72	0.45	1.17
Dry injection	2.17	1.0	0.72	0.45	1.17
Absorber	2.20	1.0	0.85	0.35	1.20
Adsorber	1.61	1.0	0.30	0.31	0.61
Quencher	1.61	1.0	0.30	0.31	0.61
Incinerator	1.64	1.0	0.30	0.34	0.64

equipment including all necessary auxiliaries. Instruments and controls for the equipment may be included. If not, add 10% extra to cover purchase cost of these items. Taxes add another 3%. Delivery freight charges can either be determined directly or can be estimated at an additional 5%. All of these costs equal the equipment total purchase cost.

Capital costs can be estimated by multiplying the purchase cost by the ratio of installed to purchase cost for each type of system. These are given in Table 6.12 for the common control system and are based on the work of Vatavuk and Neveril [35]. Note that the ratios given do not include site preparation, buildings, land, or extensive start-up or testing charges. Included are nominal installation charges for the following:

Direct costs

---

foundations/supports
handling/erection
electrical/piping
insulation/painting

Indirect costs

---

engineering/supervision
field construction and fees
startup/testing
modeling
contingencies

Annual operating costs have been listed with most systems. The commonly accepted procedure [36] for this is outlined in Table 6.13. CRF stands for capital recovery factor. Typical interest rate is 10%. Thus, the CRF of bag replacement every two years is 0.5762. For refractory replacement every four years, it is 0.3155. For equipment with a 5-year life at 10% interest, it is 0.2638; for 10-year life it is 0.1628; for 20-year life, it is 0.1175; and for 30-year life it is 0.1061. Annual operating costs equal the sum of direct costs plus indirect costs plus credit for by-product sales, if any.

Costs of utilities, chemicals, waste disposal, and wages all vary. They are influenced by such things as gross national productivity, supply and demand, location from source, and costs of processing. Use the Table 6.13 values as guides, but check locally where possible.

Capital and operating costs vary with geographical location. This includes climatic factors, local tax structures, financial conditions, interest rates, local money and labor markets, and utility costs. Factors to consider when evaluating labor expenses are costs and productivity. The nominal productivity of craftsmen in Los Angeles paid for 8 hours of work is about 6.5 hours or 81% at best. Using this as a basis of 1.00, data can be obtained to show productivity of workers in other areas. For example, in one year the average pipefitter in Houston took 1.16 times longer to do the same job. Wages the same year show that the average Los Angeles pipefitter received $11.50/hr, whereas the Houston pipefitter received $8.12/hr. Comparison of costs for work accomplished shows $11.50/base hour in Los Angeles and 8.12 × 1.16 or $9.42/base hour in Houston.

6.9.3 COST EXTRAPOLATION

Cost data may frequently be obtained for a specific system and system size. If this system is similar to the desired system, it may be possible to obtain approximate cost estimates for the desired system using cost extrapolating techniques. Data corrections can be made to account for equipment size, material of construction, and date of purchase (cost escalation).

6.9.3.1 Equipment Size

Cost of a certain piece of equipment can be estimated knowing the cost of similar equipment. If costs of two or more different size units are available, plot the data as log of cost versus equipment capacity to find the equipment cost. Otherwise, it is possible to estimate approximate costs using the exponential (logarithmic) relationship

$$I_b = I_a \left(\frac{C_b}{C_a} \right)^n \tag{6.19}$$

Direct (Variable) Costs

Chemicals: actual consumption × cost
  (e.g., limestone = \$5–10/ton + delivery
  powder limestone = \$10–12/ton + delivery
      lime = \$50–100/ton + delivery)

Operating labor: $\left(\dfrac{\text{no. operator hrs}}{\text{yr}}\right)\left(\dfrac{\$13.00}{\text{hr}}\right)$

Supervision: 15% of operating labor cost

Utilities:

Electricity: $(\text{no. KW})\left(\dfrac{\text{no. hr operated}}{\text{yr}}\right)\left(\dfrac{\$0.06}{\text{KWH}}\right)$

Water: $(\text{no. gpm})\left(\dfrac{60 \text{ min}}{\text{hr}}\right)\left(\dfrac{\text{no. hr operated}}{\text{yr}}\right)\left(\dfrac{\$1.25}{1000 \text{ gal}}\right)$

Fuel: $\dfrac{\text{gal}}{\text{yr}} \times \dfrac{\$1.00}{\text{gal}}$ for #2 oil (or: @137,000 Btu/gal = \$7.30/10⁶ Btu)

$\dfrac{\text{gal}}{\text{yr}} \times \dfrac{\$0.70}{\text{gal}}$ for #6 oil (or: @153,000 Btu/gal = \$4.58/10⁶ Btu)

$\dfrac{\text{std ft}^3}{\text{yr}} \times \dfrac{\$0.43}{100 \text{ ft}^3}$ for nat. gas (or: @1020 Btu/scf = \$4.22/10⁶ gal)

Steam: $\dfrac{\text{lb}}{\text{yr}} \times \dfrac{\$3.70}{1,000 \text{ lb}}$ for coal steam

$\dfrac{\text{lb}}{\text{yr}} \times \dfrac{\$6.60}{1,000 \text{ lb}}$ for #6 oil steam

Compressed air: $\dfrac{\text{ft}^3}{\text{yr}} \times \dfrac{\$0.16}{1,000 \text{ ft}^3}$

Maintenance:

Labor: $\left(\dfrac{\text{no. operator hrs}}{\text{yr}}\right)\left(\dfrac{\$15.00}{\text{hr}}\right)$

Material: Actual costs or assume equal to maintenance labor

Replacement
  changes: e.g. (bags + replacement costs)(CFR)
        (refractory + replacement costs)(CRF)
Waste disposal: (actual ton/yr)(\$30–200/ton) + transportation and fees

---

Indirect (Fixed) Costs

Capital recovery: (CRF)(capital cost less any capital charges for replacement items)
Plant overhead: (60%)(operating labor + supervision + maintenance)
Administration: (2%)(capital cost)
Property taxes and insurance: (2%)(capital cost)

where

C = capacity or size of equipment a or b
I = capital investment of equipment
n = sizing factor exponent

This sizing exponent should be specific to the equipment being considered if possible. Some values for this are given in Table 6.14. In the absence of other data, the six-tenths rule can be used. This suggests that n equal to 0.6 is often a suitable sizing factor. Extrapolations should only be made when the sizes are similar. Error increases as sizes differ and estimates should not be made if sizes differ by more than tenfold.

6.9.3.2 Material of Construction

Costs for different materials of construction can be obtained using relative cost factors for the specific materials. Values of several materials are carbon steel (basis = 1.0), polyvinyl-coated steel (1.4), aluminum (1.5), lead (1.6), 304 stainless (1.8), 316 stainless (2.4), 317 stainless (2.7), Monel 400 (3.6), Inconel 600 (3.9), Hastelloy (7.0), and titanium (7.7). If the cost

TABLE 6.14. Sizing Exponents (n) for Some Pollution Systems and Components.

System	Sizing Exponent	Remarks
Cyclone	0.91	
ESP	0.63	< 50,000 ft² (4650 m²) plate area
ESP	0.84	≥ 50,000 ft² (4650 m²) plate area
Fabric filter	0.72	
Wet scrubber	0.57	
Wet FGD	0.72	
Spray dryer	0.73	
Dry injection	0.73	
Absorber	0.75	
Adsorber	0.72	
Quencher	1.00	
Hearth incinerator	0.69	
Rotary kiln incinerator	0.52	
Liquid injection incinerator	0.39	
Boiler	0.78	
Ducts	1.05	
Elbows	1.66	
Dampers	1.30	Manual

of a particular device is known but the construction differs in wall thickness, use a linear extrapolation to correct for this.

6.9.3.3 Cost Escalation

The national economy escalates about 6% per year. This causes wages and supplies to rise in a similar fashion, increasing capital costs (mainly because of rising installation costs) and operating costs. It is risky to escalate equipment costs based on the general economy increases because competition and improvements can result in relatively constant or even decreasing equipment costs [37].

The Marshall & Swift (M & S) Equipment Cost Index for 8 process industries is a good index to use for adjusting capital costs to current year values. The procedure is to factor a known system cost for some given year by the ratio of M & S Index for the current year over M & S Index value for the given year. The M & S Index had a base of 100 in 1926. Table 6.15 gives annual average values for 26 years, up to 1995. These are published in *Chemical Engineering* and other reports.

6.9.3.4 Typical Costs of Utilities, Fuel, Raw Materials, and Waste Treatment

Costs in 1995 dollars in midwest U.S. are about:

Utilities
    Steam $3.00/1,000 lb
    Water $1.25/1,000 gal
    Electricity $0.06/KWh
    Brine $0.06/ton-h
    Wastewater Treatment $0.50/lb organic
    Solid Waste Disposal $50–100/yd^3
Raw Materials
    Limestone $14/ton delivered
    Lime (domestic) $55/ton
    $Mg(OH)_2$ $250/ton delivered as 60% slurry
    Sodium Bicarbonate $225/ton delivered
    Platinum Catalyst,
        ground and 98% pure $4,000/ft^3
    Activated Carbon $2.50/lb
Fuel
    Natural Gas $3.00/10^6 Btu
    Coal $20–30/ton

TABLE 6.15. Marshall and Swift Process Industries
Average Equipment Cost Index Values.

Mid Year	M & S Index, Av. Values
1970	303.3
1971	321.3
1972	332.0
1973	344.1
1974	398.1
1975	444.3
1976	472.1
1977	505.4
1978	545.3
1979	599.4
1980	659.6
1981	721.3
1982	745.6
1983	760.8
1984	780.4
1985	789.6
1986	797.7
1987	813.6
1988	852.0
1989	895.1
1990	915.1
1991	930.6
1992	943.1
1993	964.2
1994	993.4
1995	1,020 (est.)

6.9.3.5 Other Costs

Numerous one time costs are inherent to pollution control systems. A few are listed here.

Stack testing for characterizing the stream	$40 to 64,000
Compliance testing	$20 to 30,000
Permits	$3 to 6,000
Developing costs for quarterly reports and automated reporting	$2 to 4,000
Dispersion modelling	$2 to 15,000

Annual costs could include:

Permit fees for VOCs	$25/ton per year
Reports	$5 to 20,000/year
Leak detection and repairs	$2,000/year

6.10 CHAPTER PROBLEMS

6.10.1 CYCLONE COSTS AND GAS COOLING

a. Plot the data of Table 6.1 on log-log coordinates as cyclone capital costs in 1988 dollars versus gas throughput as a function of construction material.

b. Develop an equation to show capital cost as a function of gas flowrate for the 1/2 cm carbon steel cyclone.

c. If these data can be extended, predict the cost of a 1/2 cm carbon cyclone for Problem 5.13.1.

6.10.2 GAS COOLING

Flue gases containing 6% H_2O are to be cooled from 177 to 82°C by spray cooling using water. Estimate the amount of 15°C water required per 1,000 Nm^3/hr by

a. Equation (6.1) and Figure 5.43

b. Steam tables

c. Can these gases hold this much water?

6.10.3 ESP COSTS

A utility is planning to construct a 300 MW pulverized coal boiler to burn Illinois bituminous coal. The coal as received has 3.7% moisture, 3.0% sulfur, and 6% ash. The MAF heat content is 8,300 cal/g. Assume that the NSPS applies, which limits particulate emissions to 13 ng/J heat input (0.03 lb/10^6 Btu) and limits SO_2 to 85% with a maximum reduction (floor) of 86 ng/J (0.20 lb/10^6 Btu) and a maximum emission (ceiling) of 520 ng/J (1.2 lb/10^6 Btu). Assume that a cold-side ESP is planned for emission control.

Approximate:

a. Heat input

b. Coal-firing rate

fuel: 3.2% S and 10,700 Btu/lb coal
flue gas: 4,556,000 acfm at 305°F
inlet SO_2: 70,320 lb/hr
inlet particulates: 22,760 lb/hr
flue gases: reheated to 170°F
efficiency: 84.2% for SO_2 and 94% for particulates
waste slurry: 80% solids

Each boiler has 5 parallel scrubbing units.

a. Assuming that the bid was based on end-of-1978 price indices (M&S = 560), compare this scrubbing system cost to that given for 500 MW systems.

b. Compare the reported flue gas volume, SO_2 rate and particulate rate entering these scrubbers with calculated values. State any conclusions.

c. If adiabatic saturation occurs at 120°F, to what temperature does the gas leaving the first heat exchanger have to be cooled? State all assumptions.

d. Calculate exit emissions and compare with an SO_2 limit of 1.2 lb $SO_2/10^6$ Btu and a particulate limit of 0.1 lb/10^6 Btu.

REFERENCES

1 1977. "Guidelines for Industrial Boiler Performance Improvement," EPA-600/8-77-003a.

2 1976. "Guidelines for Burner Adjustments of Commercial Oil-Fired Boilers," EPA-600/2-76-088.

3 Rickman, W. S. et al. 1985. "Circulating Fluidized Bed Incineration of Hazardous Wastes," CEP, 92(5):34–38.

4 Horzella, T. I. 1978. "Selecting, Installing Cyclone Dust Collectors," CE(3):84–92.

5 Lane, W. R. and D. S. Sakata. 1978. "Design of a Chimney Downstream of a Wet Scrubber without a Flue Gas Reheater," paper 87-38.1, 71st APCA Meeting, Houston, June 23–30.

6 Smith, W. B., K. M. Cushing, and J. D. McCain. 1977. "Procedures Manual for Electrostatic Precipitator Evaluation," EPA-600/7-77-059.

7 Nichols, G. B. 1974. "Techniques for Measuring Fly Ash Resistivity," EPA-650/2-74-079.

8 McDonald, J. and L. Felix. 1977. "Particulate Control Highlights: An Electrostatic Precipitator Performance Model," EPA-600/8-77-020b.

9 Szabo, M. F. and R. W. Gerstle. 1977. "Operation and Maintenance of Particulate Control Devices on Coal-Fired Utility Boilers," EPA-600/2-77-129.

10 Turner, J. H. et al. 1988. "Sizing and Costing of Electrostatic Precipitators, Part II, Costing Considerations," JAPCA, 38(5):715–726.

11 1975. "Report on Improving the Productivity of Existing and Planned Electric Power Plants," Federal Energy Administration Interagency Task Group on Power Plant Reliability.

12 Jonke, A. A. 1980. "Fossil Plant Reliability," Argonne National Laboratory Seminar at SIUC, Carbondale (December 9).

13 White, I. J. 1977. "Electrostatic Precipitation of Fly Ash," *J. Air Poll. Control Assoc.*, 27(4):308–318.

14 Dean, A. H. and K. M. Cushing. 1988. "Survey on the Use of Pulse-Jet Fabric Filters," *JAPCA*, 38(1):90–96.

15 Ellenbecker, M. J. and D. Leith. 1978. "Pressure Drop in a High Velocity Pulse-Jet Fabric Filter," paper 78-62.6, 71st APCA Meeting, Houston, June.

16 1978. Cushing, K. M. and W. B. Smith. 1978. "Procedures Manual for Fabric Filter Evaluation," EPA-600/7-78-113.

17 Turner, J. H. et al. 1987. "Sizing and Costing of Fabric Filters, Part II, Costing Considerations," *JAPCA*, 37(9):1105–1112.

18 Laseke, B. A., Jr. 1975. "Survey of Flue Gas Desulfurization Systems: Cholla Station," Arizona Public Service Co., EPA-650/2-75-057.

19 Schifftner, K. C. and H. E. Hesketh. 1986. *Wet Scrubbers—A Practical Handbook, 2nd ed.* Lewis Publishers.

20 Vatavuk, W. M. and R. B. Neveril. 1981. "Estimating Size and Cost of Venturi Scrubbers," *Chem. Engr.*, 88:93–96.

21 Bakke, E. and O. Sarmiento. 1988. "Performance Impact of Mist Eliminators and Wet Electrostatic Precipitators on Particulate Emissions and Opacity," *Proceedings 1st Combined FGD and Dry SO₂ Control Symposium*, paper #7-5, St. Louis, MO, EPA/EPRI/APCA.

22 Van Ness, R. P. et al. 1978. "Project Manual for Full-Scale Dual-Alkali Demonstration at Louisville Gas and Electric Co.—Preliminary Design and Cost Estimate," EPA-600/7-78-010.

23 1977. "A History of Flue Gas Desulfurization Systems Since 1850," *J. Air Poll. Control Assoc.*, 27(10):948–961.

24 Delleney, R. D., J. C. Dikerman, and W. R. Menzies. 1978. "Application of Flue Gas Desulfurization to Industrial Boilers," paper 78-64.4, 71st APCA Meeting, Houston (June).

25 Michels, H. T. and E. C. Hoxie. 1977. "Some Insights into Corrosion in SO₂ Exhaust Scrubbers," ASM Conference on Materials and Reliability Problems in Fossil-Fired Power Plants.

26 Frame, G. B. 1988. "A Comparison of Air Pollution Control Systems for Municipal Solid Waste Incinerators," *JAPCA*, 38(8):1081–1087.

27 Holzman, M. I. and R. S. Atkins. 1988. "Retrofitting and Gas Controls: A Comparison of Technologies," *Solid Waste and Power*, II(5):28–32.

28 Hall, R. S., W. M. Vatavuk and J. Mately. 1988. "Estimating Process Equipment Costs," *Chem. Engr.*, 95(17):66–75.

29 Kittlemen, T. A. and R. B. Akell. 1978. The Cost of Controlling Organic Emissions, CEP 74 (4):89–91.

30 Hesketh, H. E., F. L. Cross, and J. L. Tessitore. 1989. *Incineration for Site Clean-Up and Destruction of Hazardous Wastes.* Lancaster, PA: Technomic Publishing Co., Inc.

31 Vatavuk, W. M. and R. B. Neveril. 1981. "Estimating the Size and Cost of Gas Conditioners," *CE*, 88:127–132.

32 Vogel, G. A. and E. J. Martin. 1983. "Estimating Capital Costs of Facility Components," *CE*, 90:87–90.

33 Cross, F. L., Jr. and H. E. Hesketh. 1975. *Handbook for the Operation and Maintenance of Air Pollution Control Equipment*. Lancaster, PA: Technomic Publishing Co., Inc.

34 1978. "Summary Report—Utility Flue Gas Desulfurization Systems," October–November, 1977, PEDCO Report to the EPA under Contract No. 68-01-4147, Task no. 3.

35 Vatavuk, W. M. and R. B. Neveril. 1980. "Factors for Estimating Capital and Operating Costs," *CE*, 87:157–162.

36 Neveril, R. B., J. U. Price, and K. L. Engdahl. 1978. "Capital and Operating Costs of Selected Air Pollution Control Systems," *JAPCA*, 28(12):1253–1256.

37 McCarthy, J. E. 1978. "Choosing a Flue-Gas Desulfurization System," *CE*, 85(6): 79–84.

Table of Most Commonly Used Symbols

ENGLISH SYMBOLS

A = area

A_c = area of collector, cm^2

Am^3 = actual cubic meter

a = acceleration, cm/sec^2 or interfacial area, cm^2/cm^3

atm = one atmosphere pressure

B = particle thermal mobility, sec/g

b = particle electrical mobility, $cm^2/(volt\ sec)$

C = Cunningham correction factor, dimensionless

C_a = Cunningham correction factor applied to d_a, dimensionless

$°C$ = degrees Celsius, degrees Centigrade

C_{1-9} = constants

C_7 = concentration corrected to 7% oxygen in dry flue gas

C_{12} = concentration corrected to 12% CO_2 in dry flue gas

C_{50} = concentration corrected to 50% EA

C_D = drag coefficient, dimensionless

C_p = specific heat at constant pressure

C_v = specific heat at constant volume

$C_{(x,y,z)}$ = downwind concentration from stack at location x, y, z

CC_A = capital cost of "A," dollars

D_{AB} = gas diffusion in gases, cm^2/sec

D'_{AB} = gas diffusion in liquids, cm^2/sec

D_{PM} = particle diffusivity, cm^2/sec per particle

D = diameter or equivalent of containing device

D_c = diameter of collector or characteristic dimension, cm

D_i = initial diameter, cm

D_f = final diameter, cm

D_l = diameter of streamline, cm

D_s = stack inside diameter, m

D_1 = stack diameter, inches
d = differential operator
d_{50} = aerodynamic cut diameter
d = particle diameter
d_a = aerodynamic diameter
d_c = cut diameter
d_e = equivalent diameter
d_p = particle diameter, μm
d_s = sedimentation diameter
dscf = dry standard cubic feet
dscm = dry standard cubic meter
$\bar{d}$ = mean diameter
$\bar{d}_M$ = mean diameter by mass
$\bar{d}_N$ = mean diameter by number
E = field strength, volt/cm
Eu = Euler number, Equation (5.53), dimensionless
EA = excess air
e = charge on one electron, 1.603×10^{-19} coulomb
esu = electrostatic unit
exp = natural log base e raised to the exponent following exp
F = force, constant, fraction
°F = degrees Fahrenheit
F_D = drag force
F_G = gravitational force
F_o = Fourier number, Equation (3.46), dimensionless
Fr = Froude number, Equation (5.54), dimensionless
f_1 = size factor, Equation (6.11), dimensionless
G = gas phase flowrate, g mole/(cm² hr)
g = gram or local gravitational acceleration
g_c = gravitational acceleration constant, 980.7 cm/sec²
gr = grain
H = humidity
H = plume height, m
   or Henry's law constant, atm cm³/g mole
H' = stack height, m
H_1 = stack height, ft
h = distance or height, cm
hp = horsepower
J = joule
K = degrees Kelvin, 273.16 + °C
   or rate constant
K_1 = filter cake-fabric resistance coefficient
K_G = overall mass transfer coefficient based on gas film
K_I = impaction parameter (Stokes number), dimensionless
K_L = overall mass transfer coefficient based on liquid film
Kn = Knudsen number, Equation (3.3), dimensionless

K_s = function of atmospheric stability

KW = kilowatt

KWH = kilowatt hour

k = Boltzmann's constant, 1.38×10^{-16} g cm^2/(sec^2 particle K)

k_G = gas film mass transfer coefficient, g mole/(hr cm^2 atm)

k_g = gas thermal conductivity

k_L = liquid film mass transfer coefficient, g mole/(hr cm^2 g mole/cm^3)

L = liquid phase flowrate, g mole/(cm^2 hr)

   or length, cm

L_{cr} = critical length, cm

L_G = liquid-to-gas ratio, l/m^3

M = molecular weight

Ma = Mach number, Equation (1.15), dimensionless

MMD = mass mean diameter

MPa = megapascal

N_A = mass transfer rate of moles of A, g mole/hr

N_{AG} = number of moles of component A in the gas mixture

Nm^3 = normal cubic meter, i.e., m^3 at standard conditions

N_T = number of total moles

$\overline{N}$ = number average concentration

n = number particles, etc.

ng = nanogram, 10^{-9} g

n_p = number of electron charges per particle

P_{AB} = purchase cost of "AB," dollars

P_1 = pressure in MPa

P_2 = pressure in bars

P = pressure

P_A° = vapor pressure of pure A at that temperature and pressure

Pe = Peclet number, Equation (3.76) dimensionless

Pr = Prandtl number, Equation (3.64) dimensionless

P_T = total pressure

Pt = particle penetration, fraction

p = partial pressure

p_{AL} = partial pressure of A in equilibrium over liquid solution

ppm = parts per million, by volume for a gas and by mass for a liquid or solid

Q = volumetric flowrate, m^3/hr

Q_a = actual inlet volumetric flow rate, acfm

Q_s = source strength, g/m^3

Q_{sc} = incinerator exit flue gas volumetric flow rate, scfm

R = universal gas constant, 82.05 atm cm^3/(g mole K) or 1.987 cal/(g mole K) or degrees Rankine

Re_D = collector droplet Reynolds number, i.e., use D_c in numerator of Re_p, dimensionless

Re_f = flow Reynold number, Equation (3.1), dimensionless

Re_p = particle Reynolds number, Equation (3.2), dimensionless

r = ratio of horizontal to vertical axis length

S = entropy
S_p = packing size, inches
SC = standard conditions of 20°C and 1 atm
SF = separation factor
STP = standard temperature and pressure of 0°C and 1 atm
Sc_P = particle Schmidt number, Equation (3.62), dimensionless
Sc_w = water vapor Schmidt number, Equation (3.63), dimensionless
St = Stokes number, Equation (3.6), dimensionless
T = absolute temperature, °K
T_s = stack temperature, °K
TSP = total suspended particulates
t = time, usually sec (or as specified)
    or temperature, °C
t_e = incinerator exit temperature, °F
U = overall heat transfer coefficient
$\overline{U}$ = mean wind speed, m/sec
    or bulk velocity
$\overline{U}_s$ = mean wind speed at stack exit, m/sec
V = volume
V_{AG} = volume of A in gas mixture
V_T = total volume
$\tilde{V}$ = molecular volume, cm³/g mole
v = velocity
v_d = diffusiophoretic velocity, cm/sec
v_e = stack exit velocity, cm/sec
v_g = gas velocity, cm/sec
v_i = initial velocity, cm/sec
v_n = normal centrifugal velocity, cm/sec
v_p = particle velocity, cm/sec
v_s = terminal settling velocity, cm/sec
v_T = thermophoretic velocity, cm/sec
v_t = tangential centrifugal velocity, cm/sec
v_{tf} = final tangential velocity
v_x = horizontal velocity, cm/sec
X = distance
X_A = mole fraction A in solution
x = downwind distance
x_s = stopping distance, cm
y = crosswind distance
y_A = mole fraction A in gas
z = vertical distance

GREEK SYMBOLS

γ = ratio of specific heats, C_p/C_v

ΔH = plume rise, m

ΔH_a = enthalpy of adsorption

ΔH_v = enthalpy of vaporization

ΔP = gas-phase pressure drop

Δp = liquid-phase pressure drop

ΔT_{ln} = log mean temperature difference

ΔX_B = effective boundary layer thickness over which Brownian diffusion occurs, cm

Δx_d = effective boundary-layer thickness over which diffusiophoresis occurs, cm

Δx_T = effective boundary-layer thickness over which heat is transferred, cm

$\overline{\Delta x_d}$ = mean diffusional displacement, cm

ϵ_i = individual fractional efficiency

ϵ_o = overall fractional efficiency

θ = gas residence time, sec

λ_g = gas mean free path, cm

μ_g = gas viscosity, g/(cm sec)

μm = micron, micrometer, 10^{-6} meter

π = 3.1416

ϱ_g = gas density, g/cm^3

ϱ_o = unit density, 1 g/cm^3

ϱ_p = particle density, g/cm^3

σ_g = standard geometric deviation

σ_y = horizontal standard deviation, m

σ_z = vertical standard deviation, m

τ = relaxation time, sec

Υ = surface tension, dyne/cm

χ = dynamic shape factor, dimensionless

ω = migrational velocity

Conversion Factors

LENGTH

ALL UNITS	SI UNITS
	meter = m

1 inch = 2.54 cm
1 mile = 5,280 ft = 1,609 m
1 micron = 10^{-3} mm = 10^{-6} m
1 Å = 10^{-8} cm = 10^{-10} m
1 m = 3.28 ft
1 ft = 0.305 m

MASS

ALL UNITS	SI UNITS
	kilogram = kg

1 lb = 453.6 g
1 ton = 2,000 lb
1 tonne = 2,200 lb
1 lb = 7,000 grains = 16 oz
1 grain/ft³ = 2.29 g/m³

GEOMETRY

ALL UNITS	SI UNITS

circle area = πr^2 = $\pi d^2/4$
circle circumference = πd
cylinder volume = πr^2 × height

472

sphere area $= 4\pi r^2 = \pi d^2$
sphere volume $= 4/3\pi r^3 = 1/6\pi d^3$

VISCOSITY

ALL UNITS   SI UNITS

1 poise $= 1$ g/(cm sec)
　　　$=$ absolute viscosity
1 centipoise $= 1$ cp
　　　$= 0.000672$ lb/(sec ft)
1 stoke $= $ cm²/sec $=$ poise/ϱ
　　　$=$ kinematic viscosity
air $= 1.83 \times 10^{-4}$ g/(cm sec) at SC
water $= 1$ cp $= 0.01$ g/(cm sec) at SC

ACCELERATION

ALL UNITS   SI UNITS

$g = 32.174$ ft/sec²
　$= 980.665$ cm/sec² at sea level
　$= 9.80665$ m/sec²

AREA

ALL UNITS   SI UNITS

1 km² $= 0.386$ mi²　　　　　　　　　　　　　m²
1 m² $= 1.2$ yd²
1 acre $= 43,560$ ft² $= 0.00156$ mi²
1 hectare $= 2.5$ acres

PRESSURE

ALL UNITS   SI UNITS

1 atm $= 1.01325$ bar $= 14.696$ lb$_f$/in²　　pascal $=$ Pa
　　　$= 29.92$ in Hg (32°F)
　　　　$= 760$ mm Hg (0°C)
　　　$= 33.936$ ft H$_2$O (60°F)
　　　　$= 760$ torr

$$= 407.2 \text{ in } H_2O = 1{,}034 \text{ cm } H_2O$$
$$= 0.101325 \text{ MPa} = 101{,}325 \text{ Pa}$$
1 psi $= 1 \text{ lb}_f/\text{in}^2 = 27.71'' \text{ H}_2\text{O}$
1 Pa $= \text{newton/m}^2$
1 bar $= 10^5 \text{ Pa} = 0.1 \text{ MPa}$
1 microbar $= 1 \text{ dyne/cm}^2$
$$= 1.02 \times 10^{-3} \text{ cm } H_2O$$

DENSITY

ALL UNITS

air $= 1.20 \times 10^{-3} \text{ g/cm}^3$
   $= 7.49 \times 10^{-2} \text{ lb/ft}^3$ at standard con-
       ditions
water $= 62.4 \text{ lb/ft}^3 = 1 \text{ g/cm}^3$
   $= 8.34 \text{ lb/gal at } 4°C$
mercury $= 13.6 \text{ g/cm}^3$ at $4°C$
glass $= 2.2 \text{ g/cm}^3$
earth $\cong 1.5 \text{ g/cm}^3$
limestone $= 2.1–2.9 \text{ g/cm}^3$
sandstone $= 2.0–2.6 \text{ g/cm}^3$
coal $= 1.0–1.5 \text{ g/cm}^3$
iron ore $= 3.5–5.0 \text{ g/cm}^3$
iron slag $= 2.5–3.0 \text{ g/cm}^3$
fly ash $\cong 1.0 \text{ g/cm}^3$
1 gr/scf $= 2{,}290 \text{ mg/Nm}^3$
1 g/Nm^3 $= 0.438 \text{ gr/scf}$

SI UNITS

$\text{kg/m}^3 = 10^{-3} \text{ g/cm}^3$

TIME

ALL UNITS

1 hr $= 3600 \text{ sec}$
1 work year $\cong 2{,}000$ hours per person
1 year $= 8{,}766$ total hours

SI UNITS

sec

RATE

ALL UNITS

1 scfm $= 1.7 \text{ Nm}^3/\text{hr} = 28{,}317 \text{ cm}^3/\text{min}$
   $= 472 \text{ cm}^3/\text{sec}$
1 gpm $= 0.227 \text{ m}^3/\text{hr}$
1 l/sec $= 15.85 \text{ gpm}$

SI UNITS

m^3/sec

VOLUME

ALL UNITS

SI UNITS

$1 \text{ ft}^3 = 7.481 \text{ U.S. gal} = 28.32 \text{ l}$
$\qquad = 0.028317 \text{ m}^3$
$1 \text{ gal} = 8.35 \text{ lb } H_2O @ SC = 3.785 \text{ l}$
$1 \text{ barrel} = 42 \text{ gal oil}$
$1 \text{ liter} = 1.057 \text{ quarts}$
$1 \text{ liter} = 0.2642 \text{ gal}$
$1 \text{ cm}^3 = 1 \text{ ml at } 4°C$
$1 \text{ m}^3 = 35.314 \text{ ft}^3 = 1,000 \text{ l}$
$1 \text{ gal}/1,000 \text{ ft}^3 = 0.134 \text{ l/m}^3$

FORCE

ALL UNITS

SI UNITS

$1 \text{ dyne} = \text{g cm/sec}^2$
$1 \text{ lb}_f = \text{lb}_m \text{ g/g}_c$
$1 \text{ N} = 1 \text{ kg m/sec}^2 = 10^5 \text{ dynes}$

newton = N

ENERGY

ALL UNITS

SI UNITS

$1 \text{ erg} = \text{dyne cm}$
$1 \text{ joule} = 10^7 \text{ erg}$
$1 \text{ cal} = 4.184 \text{ joule}$
$1 \text{ cal/g mole} = 1.8 \text{ Btu/lb mole}$
$1 \text{ ton refrigeration} = 200 \text{ Btu/min}$
$1 \text{ Btu} = 252.2 \text{ cal} = 778.2 \text{ ft lb}_f$
$1 \text{ KWh} = 3,412 \text{ Btu} = 1.341 \text{ hp hr}$
$\qquad = 3.6 \times 10^6 \text{ J}$
$1 \text{ hp hr} = 1.98 \times 10^6 \text{ ft lb}_f$
$1 \text{ J} = 1 \text{ Nm} = \text{Pa m}^3 = \text{kg m}^2/\text{sec}^2$

joule = J

$1 \text{ Therm} = 10^5 \text{ Btu}$
$\qquad \cong 100 \text{ ft}^3 \text{ natural gas}$

POWER

ALL UNITS

SI UNITS

$1 \text{ watt} = 1 \text{ joule/sec} = 10^7 \text{ erg/sec}$
$1 \text{ hp (mech)} \cong 746 \text{ watt}$

watt

$$= 33,000 \text{ ft lb}_f/\text{min}$$
$$= 2,545 \text{ Btu/hr}$$
$$1,000 \text{ hp (boiler)} \equiv 34,000 \text{ lb steam/hr}$$

VISCOSITY

1 slug/(ft sec) = 32.174 lb_m/(ft/sec)
1 poise = 1 g/(cm sec) = 100 cp
1 cp = 6.72 × 10^{-4} lb_m/(sec ft) = 2.42 lb_m/(hr ft)
    = 0.01 g/(cm sec) = 2.089 × 10^{-5} lb_f sec/ft^2
Kinematic viscosity = absolute viscosity/density
                    = centistoke = 0.01 cm^2/sec for H_2O @ 4°C

TEMPERATURE

ALL UNITS   SI UNITS

K = °C + 273.16   Kelvin = K
F = 1.8°C + 32   Celsius = °C
  = 2°C − 0.2°C + 32
R = °F + 459.49 = 1.8 K

SPEED

ALL UNITS   SI UNITS

1 ft/sec = 30.48 cm/sec = 0.6818 mi/hr   m/sec
1 mi/hr = 0.447 m/sec
1 ft/min = 0.508 cm/sec

CONSTANTS

ALL UNITS

g_c = 32.174 lb_m ft/(lb_f sec^2) = 980.7 cm/sec^2 = 1 kg m/(N sec^2)
natural log base e = 2.7183
π = 3.14159
ln 10 = 2.3026
Boltzmann's constant = k = 1.38 × 10^{-16} g cm^2/(sec^2 molecule K)
                          = 1.38 × 10^{-23} joule/K
Avogadro's number = 6.02 × 10^{23} atoms/g atom
gas constants = R = 1.987 cal/(g mole K) = 82.05 atm cm^3/(g mole K)
              = 1,544.6 lb_m ft/lb mole R = 83.14 × 10^6 g cm^2/(sec^2 g mole K)
              = 8.31434 Nm/(g mole K) = 8.31434 J/(g mole K)

1 lb mole $= 359$ ft^3 ideal gas at STP
1 g mole $= 22.4$ liter ideal gas at STP
$(C_p)_{H_2O} \cong 1$ Btu/(lb$_m$ R) at SC
(C_p)steel $\cong 0.3$ Btu/(lb$_m$ R) at SC
1 coulomb $= 6.24 \times 10^{18}$ electrons $= 1.038 \times 10^{-4}$ mole Ag$^+$ ion $= 1$ amp sec
$\qquad = 1$ watt sec/volt $= 10^7$ g cm^2/(volt sec^2) $= 1$ Farad volt
$\qquad = 3 \times 10^9$ statcoulomb
1 amp $= 1$ coulomb/sec $= 1$ watt/volt
1 esu $= 3.34 \times 10^{10}$ amp sec $= 3.34 \times 10^{10}$ coulombs
1 electron $= 4.8 \times 10^{-10}$ esu $= 1.602 \times 10^{-19}$ coulombs $= 9.108 \times 10^{-28}$ g
1 statvolt $= 300$ volt

AIR AT STANDARD CONDITIONS

$\mu_a = 1.83 \times 10^{-4}$ g/(cm sec)
$\quad = 3.76 \times 10^{-7}$ lb$_f$ sec/ft^2
$\quad = 1.21 \times 10^{-5}$ lb$_m$/(sec ft)
$(C_p)_{air} = 0.26$ Btu/(lb$_m$ R) $= 0.26$ Cal/(g K)
$M_{air} = 28.96 \cong 29$
$k_a = 4.02 \times 10^{-6}$ Btu/(sec ft R)
$\varrho_a = 1.20 \times 10^{-3}$ g/cm^3 $= 7.49 \times 10^{-2}$ lb$_m$/ft^3
Composition $\cong 21.0\%$ O$_2$ and 79.0% by N$_2$ by vol
$\qquad\qquad \cong 23.2\%$ O$_2$ and 76.8% N$_2$ by mass

CONVERSION FACTORS FOR AIR POLLUTION

Convert ppm to mass per volume ratio at SC

GENERAL

$$\text{(ppm)} \frac{M}{0.0241} = \mu g/m^3$$

SPECIFIC

Mult ppm at SC	By Listed Value to Obtain $\mu g/m^3$
CH$_4$	663
CO	1,160
CO$_2$	1,820
HCl	1,516
HCN	1,120
HF	828
H$_2$S	1,410
NO$_2$	1,910
O$_3$	1,990
SO$_2$	2,650

1 number/ft³ = 35.31 number/m³
1 ton/mi² = 3.125 lb/acre = 0.717 lb/ft²
1 g/m³ = 0.0283 g/ft³
1 lb/hr = 0.126 g/sec

POWER GENERATION ESTIMATING FACTORS

100 MW of gross generated output for bituminous coal
    ⇆ 9 × 10⁵ lb steam/hr @ 88.5% eff
    ≅ 1.01 × 10⁹ Btu/hr heat input
    = 40 ton/hr of 12,500 Btu/lb coal
    ≅ 520,000 Am³/hr flue gas @ 175°C, 27% excess air, and 6% H_2O
    in flue gas
    ≅ 340,000 Nm³/hr
    ≅ 306,000 acfm @ 350°F ≅ 200,000 scfm
1.444 Am³ ≅ 1 MW sec
1 lb of 12,500 Btu/lb coal ≅ 10 lb steam
100 MW of gross generated output for lignite coals
    ≅ 1.13 × 10⁹ Btu/hr heat input
    ≅ 314,000 acfm
Average pulverized coal boiler has 36.5 miles tubes/100 MW.
Power line transmission losses ≅ 3%/1,000 mi of line
Fan costs ≅ $100/inch ΔP per MW per year
Combustion Heat Release
 Stoker grate ≅ 400,000 Btu/hr per ft² grate
 Stoker furnace ≅ 30,000 Btu/hr per ft³ volume
Assumes: generator efficiency @ 38%
    boiler thermal efficiency @ 88.5%
    boiler water to steam enthalpy change = 1,000 Btu/lb
    lignite @ 14% ash, 6770 Btu/lb, 36% water as fired
Gas fired boiler at high fire
    ~4,185 cfh natural gas @ 1,000 Btu/ft³ per 100 BHP
    ~800 cfm combustion air per 100 BHP
    ~1,000 cfm ventilation plus combustion air per 100 BHP
    @ 50% efficiency, produces ~0.5 lb steam per ft³ natural gas or
     ~210 lb steam/hr per 100 hp

Toxic Chemicals Subject to Section 313 (Title III of SARA) Reporting

Chemical Abstract Service (CAS) Number	Chemical Name
75-07-0	Acetaldehyde
60-35-5	Acetamide
67-64-1	Acetone
75-05-8	Acetonitrile
53-96-3	2-Acetylaminofluorene
107-02-8	Acrolein
79-06-1	Acrylamide
79-10-7	Acrylic acid
107-13-1	Acrylonitrile
309-00-2	Aldrin
107-05-1	Allyl chloride
7429-90-5	Aluminum (fume or dust)
1344-28-1	Aluminum oxide
117-79-3	2-Aminoanthraquinone
60-09-3	4-Aminoazobenzene
92-67-1	4-Aminobiphenyl
82-28-0	1-Amino-2-methylanthraquinone
7664-41-7	Ammonia
6484-52-2	Ammonium nitrate (solution)
7783-20-2	Ammonium sulfate (solution)
62-53-3	Aminiline
90-04-0	o-Anisidine
104-94-9	p-Anisidine
134-29-2	o-Anisidine hydrochloride

(continued)

Chemical Abstract Service (CAS) Number	Chemical Name
120-12-7	Anthracene
7440-36-0	Antimony
*	Antimony compounds
7440-38-2	Arsenic
*	Arsenic compounds
1322-21-4	Asbestos (friable)
7440-39-3	Barium
*	Barium compounds
98-87-3	Benzal chloride
55-21-0	Benzamide
71-43-2	Benzene
92-87-5	Benzidine
98-07-7	Benzoic trichloride (benzotrichloride)
98-88-4	Benzoyl chloride
94-36-0	Benzoyl peroxide
100-44-7	Benzyl chloride
7400-41-7	Beryllium
*	Beryllium compounds
92-52-4	Biphenyl
111-44-4	Bis(2-chloroethyl) ether
542-88-1	Bis(chloromethyl) ether
108-60-1	Bis(2-chloro-1-methylethyl) ether
103-23-1	Bis(2-ethylhexyl) adipate
75-25-2	Bromoform (tribromomethane)
74-83-9	Bromomethane (methyl bromide)
106-99-0	1.3-Butadiene
141-32-2	Butyl acrylate
71-36-3	n-Butyl alcohol
78-92-2	sec-Butyl alcohol
75-65-0	tert-Butyl alcohol
85-68-7	Butyl benzyl phthalate
106-88-7	1.2-Butylene oxide
123-72-8	Butyraldehyde
2650-18-2	C.I. Acid Blue 9, diammonium salt
3844-45-9	C.I. Acid Blue 9, disodium salt
4680-78-8	C.I. Acid Green 3
569-64-2	C.I. Basic Green 4
989-38-8	C.I. Basic Red 1
1937-37-7	C.I. Direct Black 38
2602-46-2	C.I. Direct Blue 6
16071-86-6	C.I. Direct Brown 95
2832-40-8	C.I. Disperse Yellow 3
3761-53-3	C.I. Food Red 5
81-88-9	C.I. Food Red 15

Chemical Abstract Service (CAS) Number	Chemical Name
3118-97-6	C.I. Solvent Orange 7
97-56-3	C.I. Solvent Yellow 3
842-07-9	C.I. Solvent Yellow 14
492-80-8	C.I. Solvent Yellow 34 (Aurimine)
128-66-5	C.I. Vat Yellow 4
7440-43-9	Cadmium
*	Cadmium compounds
156-62-7	Calcium cyanamide
133-06-2	Captan
63-25-2	Carbaryl
75-15-0	Carbon disulfide
56-23-5	Carbon tetrachloride
463-58-1	Carbonyl sulfide
120-80-9	Catechol
133-90-4	Chloramben
57-74-9	Chlordane
76-13-1	Chlorinated fluorocarbon (Freon 113)
7782-50-5	Chlorine
10049-04-4	Chlorine dioxide
79-11-8	Chloroacetic acid
532-27-4	2-Chloroacetophenone
108-90-7	Chlorobenzene
510-15-6	Chlorobenzilate
75-00-3	Chloroethane (ethyl chloride)
67-66-3	Chloroform
74-87-3	Chloromethane (methyl chloride)
107-30-2	Chloromethyl methyl ether
*	Chlorophenols
126-99-8	Chloroprene
1897-45-6	Chlorothalonil
7440-47-3	Chromium
*	Chromium compounds
7400-48-4	Cobalt
*	Cobalt compounds
7440-50-8	Copper
*	Copper compounds
120-71-8	p-Cresidine
1319-77-3	Cresol (mixed isomers)
108-39-4	m-Cresol
95-48-7	o-Cresol
106-44-5	p-Cresol
98-82-8	Cumene

(continued)

Chemical Abstract Service (CAS) Number	Chemical Name
80-15-9	Cumene hydroperoxide
135-20-6	Cupferron
*	Cyanide compounds
110-82-7	Cyclohexane
94-75-7	2,4-D
1163-19-5	Decabromodiphenyl oxide
2303-16-4	Diallate
615-05-4	2,4-Diaminoanisole
39156-41-7	2,4-Diaminoanisole sulfate
101-80-4	4,4'-Diaminodiphenyl ether
25376-45-8	Diaminotoluene (mixed isomers)
95-80-7	2,4-Diaminotoluene
334-88-3	Diazomethane
132-64-9	Dibenzofuran
96-12-8	1,2-Dibromo-3-chloropropane (DBCP)
106-93-4	1,2-Dibromoethane (ethylene dibromide)
84-74-2	Dibutyl phthalate
25321-22-6	Dichlorobenzene (mixed isomers)
95-50-1	1,2-Dichlorobenzene
541-73-1	1,3-Dichlorobenzene
106-46-7	1,4-Dichlorobenzene
91-94-1	3,3'-Dichlorobenzidine
75-27-4	Dichlorobromomethane
107-06-2	1,2-Dichloroethane (ethylene dichloride)
540-59-0	1,2-Dichloroethylene
75-09-2	Dichloromethane (methylene chloride)
120-83-2	2,4-Dichlorophenol
78-87-5	1,2-Dichloropropane
542-75-6	1,3-Dichloropropylene
62-73-7	Dichlorvos
115-32-2	Dicofol
1464-53-5	Diepoxybutane
111-42-2	Diethanolamine
117-81-7	Di-(2-ethylhexyl) phthalate (DEHP)
84-66-2	Diethyl phthalate
64-67-5	Diethyl sulfate
119-90-4	3,3'-Dimethoxybenzidine
60-11-7	4-Dimethylaminoazobenzene
119-93-7	3,3'-Dimethylbenzidine (o-Tolidine)
79-44-7	Dimethylcarbamyl chloride
54-14-7	1,1-Dimethyl hydrazine
105-67-9	2,4-Dimethylphenol
131-11-3	Dimethyl phthalate
77-78-1	Dimethyl sulfate
534-52-1	4,6-Dinitro-o-cresol

Chemical Abstract Service (CAS) Number	Chemical Name
51-28-5	2,4-Dinitrophenol
121-14-2	2,4-Dinitrotoluene
606-20-2	2,6-Dinitrotoluene
117-84-0	n-Dioctyl phthalate
123-91-1	1,4-Dioxane
122-66-7	1,2-Diphenyl hydrazine (hydrazobenzene)
106-89-8	Epichlorohydrin
110-80-5	2-Ethoxyethanol
140-88-5	Ethyl acrylate
100-41-4	Ethyl benzene
541-41-3	Ethyl chloroformate
74-85-1	Ethylene
107-21-1	Ethylene glycol
151-56-4	Ethyleneimine (Aziridine)
75-21-8	Ethylene oxide
96-45-7	Ethylene thiourea
2164-17-2	Fluometuron
50-00-0	Formaldehyde
*	Glycol ethers
76-44-8	Heptachlor
118-74-1	Hexachlorobenzene
87-68-3	Hexachloro-1,3-butadiene
77-47-4	Hexachlorocyclopentadiene
67-72-1	Hexachloroethane
1335-87-1	Hexachloronaphthalene
680-31-9	Hexamethylphosphoramide
302-01-2	Hydrazine
10034-93-2	Hydrazine sulfate
7647-01-0	Hydrochloric acid
74-90-8	Hydrogen cyanide
7664-39-3	Hydrogen fluoride
123-31-9	Hydroquinone
78-84-2	Isobutyraldehyde
67-63-0	Isopropyl alcohol (only persons who manufacture by the strong acid process—no supplier notification)
80-05-7	4,4'-Isopropylidenediphenol
7439-92-1	Lead
*	Lead compounds
58-89-9	Lindane
108-31-6	Maleic anhydride
12427-38-2	Maneb
7439-96-5	Manganese
*	Manganese compounds
108-78-1	Melamine

(continued)

Chemical Abstract Service (CAS) Number	Chemical Name
7439-97-6	Mercury
*	Mercury compounds
67-56-1	Methanol
72-43-5	Methoxychlor
109-86-4	2-Methoxyethanol
96-33-3	Methyl acrylate
1634-04-4	Methyl tert-butyl ether
101-14-4	4,4'-Methylene bis(2-chloroaniline) (MBOCA)
101-61-1	4,4'-Methylene bis(N,N-dimethyl) benzenamine
101-68-8	Methylene bis(phenylisocyanate) (MBI)
74-95-3	Methylene bromide
101-77-9	4,4'-Methylene dianiline
78-93-3	Methyl ethyl ketone
60-34-4	Methyl hydrazine
74-88-4	Methyl iodide
108-10-1	Methyl isobutyl ketone
624-83-9	Methyl isocyanate
80-62-6	Methyl methacrylate
90-94-8	Michler's ketone
1313-27-5	Molybdenum trioxide
505-60-2	Mustard gas
91-20-3	Naphthalene
134-32-7	alpha-Naphthylamine
91-59-8	beta-Naphthylamine
7440-02-0	Nickel
*	Nickel compounds
7697-37-2	Nitric acid
139-13-9	Nitrilotriacetic acid
99-59-2	5-Nitro-o-anisidine
98-95-3	Nitrobenzene
92-93-3	4-Nitrobiphenyl
1836-75-5	Nitrofen
51-75-2	Nitrogen mustard
55-63-0	Nitroglycerin
88-75-5	2-Nitrophenol
100-02-7	4-Nitrophenol
79-46-9	2-Nitropropane
156-10-5	p-Nitrosodiphenylamine
121-69-7	N,N-Dimethylaniline
924-16-3	N-Nitrosodi-n-butylamine
55-18-5	N-Nitrosodiethylamine
62-75-9	N-Nitrosodimethylamine
86-30-6	N-Nitrosodiphenylamine
621-64-7	N-Nitrosodi-n-propylamine
4549-40-0	N-Nitrosomethylvinylamine

Chemical Abstract Service (CAS) Number	Chemical Name
59-89-2	N-Nitrosomorpholine
759-73-9	N-Nitroso-N-ethylurea
684-93-5	N-Nitroso-N-methylurea
16543-55-8	N-Nitrosonornicotine
100-75-4	N-Nitrosopiperidine
2234-13-1	Octachloronaphthalene
20816-12-0	Osmium tetroxide
56-38-2	Parathion
87-86-5	Pentachlorophenol (PCP)
79-21-0	Peracetic acid
108-95-2	Phenol
106-50-3	p-Phenylenediamine
90-43-7	2-Phenylphenol
75-44-5	Phosgene
7664-38-2	Phosphoric acid
7723-14-0	Phosphorus (yellow or white)
85-44-9	Phthalic anhydride
88-89-1	Picric acid
*	Polybrominated biphenyls (PBB)
1336-36-3	Polychlorinated biphenyls (PCB)
1120-71-4	Propane sultone
57-57-8	beta-Propiolactone
123-38-6	Propionaldehyde
114-26-1	Propoxur
115-07-1	Propylene (propene)
75-55-8	Propyleneimine
75-56-9	Propylene oxide
110-86-1	Pyridine
91-22-5	Quinoline
106-51-4	Quinone
82-68-8	Quintozene (pentachloronitrobenzene)
81-07-2	Saccharin (only persons who manufacture—no supplier notification)
94-59-7	Safrole
7782-49-2	Selenium
*	Selenium compounds
7440-22-4	Silver
*	Silver compounds
1310-73-2	Sodium hydroxide (solution)
7757-82-6	Sodium sulfate (solution)
100-42-5	Styrene (monomer)
96-09-3	Styrene oxide
7664-93-9	Sulfuric acid
100-21-0	Terephthalic acid

(continued)

Chemical Abstract Service (CAS) Number	Chemical Name
79-34-5	1,1,2,2-Tetrachloroethane
127-18-4	Tetrachloroethylene (perchloroethylene)
961-11-5	Tetrachlorvinphos
7440-28-0	Thallium
*	Thallium compounds
62-55-5	Thioacetamide
139-65-1	4,4'-Thiodianiline
62-56-6	Thiourea
1314-20-1	Thorium dioxide
13463-67-7	Titanium dioxide
7550-45-0	Titanium tetrachloride
108-88-3	Toluene
584-84-9	Toluene-2,4-diisocyanate
91-08-7	Toluene-2,6-diisocyanate
95-53-4	o-Toluidine
636-21-5	o-Toluidine hydrochloride
8001-35-2	Toxaphene
68-76-8	Triaziquone
52-68-6	Trichlorfon
120-82-1	1,2,4-Trichlorobenzene
71-55-6	1,1,1-Trichloroethane (methyl chloroform)
79-00-5	1,1,2-Trichlorethane
79-01-6	Trichloroethylene
95-95-4	2,4,5-Trichlorophenol
88-06-2	2,4,6-Trichlorophenol
1582-09-8	Trifluralin
95-63-6	1,2,4-Trimethyl benzene
126-72-7	Tris(2,3-dibromopropyl) phosphate
51-79-6	Urethane (ethyl carbamate)
7440-62-2	Vanadium (fume or dust)
108-05-4	Vinyl acetate
593-60-2	Vinyl bromide
75-01-4	Vinyl chloride
75-35-4	Vinylidene chloride
1330-20-7	Xylene (mixed isomers)
108-38-3	m-Xylene
95-47-6	o-Xylene
106-42-3	p-Xylene
87-62-7	2,6-Xylidine
7440-66-6	Zinc (fume or dust)
*	Zinc compounds
12122-67-7	Zineb

*See next page "Chemical Categories."

CHEMICAL CATEGORIES

Section 313 requires emissions reporting on the chemical categories listed below, in addition to specific chemicals listed above.

The compounds listed below, unless otherwise specified, are defined as including any unique chemical substance that contains the named chemical (i.e., antimony, arsenic, etc.) as part of that chemical's structure.

- Antimony compounds
- Arsenic compounds
- Barium compounds
- Beryllium compounds
- Cadmium compounds
- Chlorophenols
- Chromium compounds
- Cobalt compounds
- Copper compounds
- Cyanide compounds—X CN, where X=H, or any other group where a formal dissociation may occur. For example, KCN or Ca(CN)
- Glycol ethers—includes mono- and di-ethers of ethylene glycol, diethylene glycol, and triethylene glycol

$$R-(OCH2CH2)_n-OR'$$

where

n = 1, 2, or 3
R = alkyl or aryl groups
R' = R, H, or groups which, when removed, yield glycol ethers with the structure: $R-(OCH2CH)_n-OH$

Polymers are excluded from the glycol ether category.

- Lead compounds
- Manganese compounds
- Mercury compounds
- Nickel compounds
- Polybrominated biphenyls (PBBs)
- Selenium compounds
- Silver compounds
- Thallium compounds
- Zinc compounds

Index

Absolute Temperature, 14
Absorbers
    Bubble Cap Towers, 354
    Cost, 443
    Height, 358
    Packed Towers, 351, 357
    Packing, 351, 358
    Plate Towers, 355, 360
    Pressure Drop, 364
    Tower Capacity, 356
    Venturi, 329
Absorption, 225, 329
    Contact Stages, 245
    Driving Force, 246, 361
    Efficiency, 247, 360
    Equilibrium Curve, 234
    Gas Phase Controlling, 225
    Interfacial Area, 231
    Liquid-to-Gas Ratio, 242
    NTP, 362
    NTU, 247, 363
    Operating Lines, 240
    Stage Efficiency, 245
Acid Gas Control, 432
Activated Carbon, 258, 370
Adiabatic Gas Saturation, 382
Adsorbents, 258
Adsorbers, 350, 366
    Capacity, 264, 358, 368
    Costs, 445
    Design Parameters, 351, 374
    Hybrids, 376

    Installation, 445
    Maintenance, 445
    Operating, 444
    Pressure Drop, 364, 373
    Sizing, 444
Adsorption, 255
    Efficiency, 374
    Heat of, 257
    Saturation Capacity, 264
    Working Capacity, 264
Aerodynamic Diameter, 2, 175
Air-Fuel Ratio, 81, 112
Air Pollution, 1
Ambient Air Quality Standards (AAQS), 21
Atmospheric
    Diffusion, 51
    Dispersion, 44
    Stability, 50
Atomization, 310, 318
    Droplet Sizes, 313
    Efficiency, 316
    Gas Atomization, 318
    Gas Velocity, 315
    Power, 317
    Pressure Drop, 316
Auditing, 61

Baghouses (see also *Filters*), 294
    Design, 302
    Efficiency, 300
    Filtration Velocities
        (Air-to-Cloth Ratio), 48, 304

Baghouses *(continued)*
   Pressure Drop, 298
   Reverse Flow, 297
   Reverse Pulse (Jet Pulse), 295
   Shaker, 296
Blower (see *Fan*)
Boilers, 87
Boltzman's Constant, 177
Boyle's Law, 15

Centrifugal
   Collection, 200
   Motion, 179
Charles' Law, 15
Clapeyron Equation, 233
Cleanable High Efficiency Air Filter, 378
Coal, 84
   Ash, 86
   Boilers, 87
   Burners, 87
   Combustion, 84
   Combustion Products, 90, 101
Coalescence, 208
Collection Efficiency (see also
     specific devices), 221
Combustion, 75
   Boilers, 87
   Coal, 84
   Emission Minimization, 154
   Flashback, 158
   Fluidized Bed, 91
   Gas, 104
   Heating Value, 145
   Motor Fuel, 109
   Oil, 99
   Temperature, 148
   Waste, 114
Condensation Force, 385
Conditioning Gases, 403
Conservation, 73
Control Devices, 271
Control Systems, 399
   Sales, 400
Conversion Factors, 472
Corona, 190
Correction Factors, 152, 168
Cost Index, 459
Costs, 455
   Absorbers, 443
   Acid Gas Control, 432
   Adsorbers, 445

Capital Cost Factors, 453, 455
Chemicals, 350, 459
Chimneys, 406
Cyclones, 402
Direct, 455
Dry Injection, 441
Escalation, 459
ESPs, 413
Extrapolation, 456
FGD, 432
Filters, 421
Generalized, 455
Incinerators, 448
Index, 459
Indirect, 455
Material of Construction, 458
Mist Eliminator, 444
Operating, 451
Quencher, 426
Scrubber, 425
Sizing Exponents, 457
Spray Dryer, 441
Stack, 407
Testing, 460
Utilities, 459
Waste Treatment, 459
Wet scrubbers, 425
Cunningham Correction, 168, 211
Cut Diameter, 3, 209, 321
Cyclones, 272
   Costs, 402
   Dimensions, 273
   Efficiency, 277
   Pressure Drop, 276
   Separation Factor, 179

Daltons Law, 15
Deutsch-Anderson Equation, 195
Diameter
   Aerodynamic, 2, 175
   Cut, 3, 209
   Equivalent, 3
   Sedimentation, 3
   Stokes, 3
Diffusional Mobility, 177
Diffusiophoresis, 195
Diffusive Deposition, 177, 184
Diffusivity
   Gas, 220, 263
   In Liquids, 223
   Particle, 177

Dimensionless Numbers, 167
Dispersion, 44

Electrical
    Charge, 191
    Migrational Velocity, 192
    Wind, 192
Electrostatic Filters, 376
Electrostatic Force, 189
Electrostatic Precipitation, 206
    Efficiency, 288
Electrostatic Precipitators, 281, 408
    Costs, 413
    Design, 292, 409
    Electrodes, 284
    Installation, 410
    Maintenance, 413
    Models, 408
    Operation, 410
    Power Density, 411
    Sizing, 409
    Spark Over, 284
    Specifications, 282
    Specific Collecting Area, 291, 412
    Voltage, 293
ESPs (see *Electrostatic Precipitators*)
Emission Correction Factors, 152, 168
Emission Minimization, 154
Emission Standards, 35
Energy, 73, 82
Entrainment Separators (see *Mist Elimination*)
Environmental Auditing, 61
Equivalent Diameter, 3
Escalation, Cost, 459
Euler Number, 391
Evacuation, 60
Excess Air, 78

Fan
    Power, 209, 325
Fanning Equation, 47
Filters (see also *Baghouses*), 294, 418
    Air-to-Cloth Velocities, 304, 418
    Costs, 421
    Fabric, 306, 422
    Gas Removal, 423
    Installation, 420
    Maintenance, 421
    Operation, 302, 420
    Permeability, 293

Porosity, 298
Pressure Drop, 298
Properties, 307
Relative Costs, 308
Sizing, 302, 418
Wet, 378
Flashback, 158
Flue Gas
    Composition, 76
    Energy Loss, 82
    Quantity, 81
Flue Gas Desulfurization, 432
    Dual Alkali, 438
    Lime/Limestone, 437
    Material of Construction, 440
    Sludge, 439
    Spray Dryer, 441
    Systems, 439
    Wellman/Lord, 438
Fluidized Bed Combustion, 87, 131, 136, 162
Froude Number, 391
Fugitive Emissions, 34

Gas, 104
    Absorption, 225
    Adsorption, 255
    Burners, 105
    Chemical Removal, 265
    Combustion, 106
    Combustion Products, 107
    Conditioning, 380, 403
    Control, 219
    Diffusivity, 220, 266
    Gases, 13, 19
    Mean Free Path, 18
    Moisture Content, 248
    Viscosity, 18
Gas Laws, 13
    Constant, Universal, 15
Gas Saturation, Adiabatic, 382
Gasification, 145
Gravitational Settling, 199

Hazardous Air Pollutants, 25
Heat of Vaporization, 13, 234
Heating Value, 145
Henry's Law, 228, 231
    Constants, 229
Hesketh Equation, 149, 291, 326, 332
Holland Equation, 48
Humidity, 249

Ideal Gas Law, 15
Impaction Parameter, 169
Impingement, 188
Incineration, 128
    Combustion Products, 138
    Controlled Air, 131
    Heating Value, 130, 145
    Terminology, 129, 154, 155
    Waste-to-Energy, 132
Incinerators, 448
    Controlled Air, 131
    Cost, 449
    Design, 138
    Fluidized Bed, 136
    Maintenance, 451
    Operation, 157, 451
    Rotary Kiln, 134
    Volumetric Heat Release, 139
Indoor Air Quality, 41
Inertial Force, 185
Inertial Impaction, 169, 187
Insolation, 51
Interception, 201
Interfacial Area, 232

Knudsen Number, 168
Kozeny Carmen Equation, 299

Liquefaction, 145
Log Mean Pressure Difference, 233

Mach Number, 16
Maintenance, 451
Marshall & Swift Index, 459, 460
Mass Transfer, 224
    Coefficients, 230, 245
    Stage Efficiency, 245
Material of Construction, Costs, 458
Mean Free Path, 18
Mechanical Collectors, 272, 402
Mechanical Mobility, 178
Migration Velocity, 282
Mist Elimination, 333
    Cost, 444
    Efficiency, 340
    Entrainment Separators, 335
    Flow Factors, 336
    Gas Velocities, 339
    Pressure Drop, 339
Modeling, 58
Modifications, Process, 401

Moisture Content of Gases, 248
Molecular Sieves, 258
Molecular Weight, 17
Motor Fuels, 109
    Combustion, 111
    Combustion Products, 112
Municipal Solid Waste (MSW), 114

Nozzles, Spray (see also *Atomization*), 310

Oil, 99
    Burners, 99
    Combustion, 100
    Combustion Products, 101
Operation, 451

Packed Towers (see *Absorbers*)
Packing, 351, 357
    Pressure Drop, 364
Particle
    Collection, 199, 202, 203, 204, 205
    Control, 386
    Diffusivity, 177
    Drag Coefficient, 171
    Friction Factor, 171
    Non-Spherical, 3
    Non-Steady State, 178
    Size Distribution, 6
    Steady State, 179
    Thermal Mobility, 178
Particulate Control, 165, 271, 386
Particulates, 2, 19
Peclet Number, 204
Phoretic Boundary Layers, 197
Phoretic Force, 195
Photophoresis, 196
Plume Rise, 48
Pneumatic Atomization, 319
Polychlorinated Biphenyls (PCBs), 28
Prandtl Number, 197
Pressure Drop
    Flow, 47
    Venturi, 326
Pressure Effects, 210, 266
Primary Combustion Chamber, 155
Primary Pollutants, 22
Process Modifications, 401
Psychometric Chart, 250
Puff Release, 56
Pump, Power, 317

Quencher, 426, 441
Quenching, Gas, 380, 403

Raoult's Law, 226, 233
Receptor Models, 60
Refuse Derived Fuel, 114, 120
Reheat, 406
Relaxation Time, 179
Reporting, 24
Resistivity, 287
Reynold's Number, 47, 167
Risk Assessment, 38

Saturation, Adiabatic Gas, 382
Scrubbers, 308, 425
  Cost, 425
  Installation, 428
  Maintenance, 429
  Operation, 426
Scrubbers, Dry, 347
  Chemicals, 350
  Dry/Dry, 348
  Wet/Dry, 348
Scrubbers, Wet, 309, 425
  Catenary Grid, 200, 341
  Charged, 341
  Chemicals, 341
  Collision Scrubber, 341
  Contacting Power, 322
  Ejectors, 379
  Ionizing Scrubber, 341
  Mist Elimination, 333, 444
  Packed Tower, 351, 357
  Preformed Spray, 343
  Sieve Tower, 355
  Spray Towers, 386
  Tower Scrubbers, 346
  Venturi, 318
  Waterloo Scrubber, 341
Secondary Combustion Chamber, 155
Secondary Particles, 165
Sedimentation Diameter, 3
Shape Factor, 3
Size Distribution, 4, 6
Sizing, Device, 391
Sizing Exponents, Cost, 457
Solutions, 226
Sources, 26, 73
Specific Heat, 16
Spray Systems, 204
Spray Towers, 386
Stability Categories, 49

Stack Gas Reheat, 406
Stacks, 44, 406
Standard Geometric Deviation, 9
Standard Industrial Classification, 26
Standard Temperature and Pressure, 13
Standards, 20
Stoichiometric Combustion, 75
Stokes Number, 169
Stopping Distance, 169
Surface Tension, 13

Temperature, Combustion, 148
Temperature Effects, 210, 266
Theoretical Air, 75
Thermal Conductivity, 267
Thermal Mobility, 177
Thermophoresis, 196
Toxic Air Pollutants, 24
Toxic Chemicals, 479
Toxic Waste Sites, 33
Transportation, 107

Vapor Liquid Equilibrium, 233
Vapor Pressure, 234
  Water, 252
Venturi Scrubber, 318
  Absorption, 329
  Contacting Power, 322
  Cut Diameter, 321
  Design, 331
  Droplet Size, 310
  Efficiency, 325, 327
  Liquid-to-Gas Ratio, 328
  Penetration, 323
  Pressure Drop, 326
  Throat Length, 332
Viscosity, 18
Volumetric Heat Release, 139, 449

Waste
  Bioinfectious, 123
  Classification, 119
  Combustion, 114
  Combustion Products, 124
  Composition, 121
  Fuel, 117
  MSW, 114, 117, 120, 134, 144
  RDF, 114, 120
  -to-Energy, 114
Water Content of Gases, 248
Wet Bulb Temperature, 249

T - #0154 - 101024 - C0 - 229/152/27 [29] - CB - 9781566764131 - Gloss Lamination